The Evolution of Agriculture: Social and Agroecological Perspectives

The Evolution of Agriculture: Social and Agroecological Perspectives

Clint Giffen

Callisto Reference
www.callistoreference.com

Callisto Reference,
118-35 Queens Blvd., Suite 400,
Forest Hills, NY 11375, USA

Visit us on the World Wide Web at:
www.callistoreference.com

ISBN: 978-1-64116-753-6 (Hardback)

Cataloging-in-Publication Data

The evolution of agriculture : social and agroecological perspectives / Clint Giffen.
p. cm.
Includes bibliographical references and index.
ISBN 978-1-64116-753-6
1. Agriculture--History. 2. Agriculture--Social aspects. 3. Agricultural ecology. I. Giffen, Clint.
S419 .E96 2023
630.9--dc23

Table of Contents

Preface

This book has been an outcome of determined endeavour from a group of educationists in the field. The primary objective was to involve a broad spectrum of professionals from diverse cultural background involved in the field for developing new researches. The book not only targets students but also scholars pursuing higher research for further enhancement of the theoretical and practical applications of the subject.

Agriculture refers to the practice of raising livestock, cultivating soil and growing crops. It involves preparing plant and animal products for human consumption and distributing them to marketplaces. Fibers, raw materials, foods and fuels are the four major agricultural products. Agroecology is the application of ecological principles to agricultural processes and methods as well as the application of social justice principles to entire food systems. This strategy has the potential to significantly improve farming system resilience against the effects of climate change and reversing land degradation trends. Agroecological techniques such as agroforestry have beneficial effects on nitrogen levels and fertility, water regulation, soil carbon, and biodiversity. These techniques also reduce the severity of temperature extremes. This book is compiled in such a manner, that it will provide in-depth knowledge about the social and agroecological perspectives on the evolution of agriculture. It will provide comprehensive knowledge to the readers.

It was an honour to edit such a profound book and also a challenging task to compile and examine all the relevant data for accuracy and originality. I wish to acknowledge the efforts of the contributors for submitting such brilliant and diverse chapters in the field and for endlessly working for the completion of the book. Last, but not the least; I thank my family for being a constant source of support in all my research endeavours.

Clint Giffen

1

Adapting a Metabolic Approach to Agriculture

In this chapter, we describe the theoretical and methodological background of this work. Our aim was to offer a new perspective on the evolution of Spain's agricultural sector over the last century, moving the focus away from growth capacity towards sustainability. We wished to discover whether the sector had been able to grow over time without deteriorating its social and ecological resources. To tackle this question, we chose the framework of *Social Metabolism*. Within so-called *Sustainability Science*, this framework currently provides the broadest capacity of analysis. The present research was thus conducted using the methodological tools included within Material and Energy Flow Accounting (MEFA). MEFA tools have been largely developed by the Sustainable Europe Research Institute (SERI), the Wuperthal Institute and the IFF-Social Ecology of Vienna (*Economy-Wide. Material Flow Accounting, EW-MEFA*). The methodologies are designed to obtain data that support the analysis of the biophysical trajectories of economies and societies, both today and throughout history. They also measure the biophysical relationships between territories, describe and characterize resource consumption and the ways in which resources are appropriated. Sufficiently consistent data can be obtained allowing to evaluate the degree of sustainability of the relations between a given society and its environment. MEFA thus constitutes an appropriate instrument to study the material aspects of socio-ecological transitions.

Few studies, however, have adopted this methodology to analyze biomass production and its role in the economy at large (Schandl and Schultz 2002; Krausmann et al. 2008a, b, 2011; Kovanda and Hak 2011; Gierlinger and Krausmann 2012; Singh et al. 2012; Infante et al. 2015). Most studies focus on the present and their time frame of analysis is fairly limited (Risku Norka 1999; Risku-Norja and Mäenpää 2007; for a review of the state of the art, see Infante et al. 2015). Estimates of food system metabolisms (Wirsenius 2003) or agri-food (Heller and Keoleian 2003; Infante et al. 2014) and analyses of global and continental biomass flows (Krausmann et al. 2008a, b) have also been carried out. Nevertheless, none of these works neither applied the methodology to the agrarian sector nor have they gone into sufficient historical depth to study the transition to the industrial metabolic regime. One work analyzes the changes in land uses and the energy transition in agriculture between 1830 and the year 2000 in Czechoslovakia (Kusova et al. 2008). Recently, a complete study entitled *Social Metabolism of Czech Agriculture in the period 1830–2010* (Greslova et al. 2015) was published for the same country.

Moreover, the limited number of studies which have addressed the evolution of land biomass flows throughout the 20th century have been conducted mostly on a local scale (Krausmann 2004; González de Molina and Guzmán 2006; Cunfer and Krausmann 2009; Garrabou and González de Molina 2010; Infante 2011; García Ruiz et al. 2012; Tello et al. 2012; Infante et al. 2014c). Virtually no study has addressed the issue in a wider context, i.e., the nation-state. Therefore, very few works have attempted to adapt Social Metabolism methodologies to agriculture. The metabolic approach has, in fact, an enormous heuristic potential. Nonetheless, the methodology is not yet capable of providing biophysical data and indicators that account for the specificity and complexity of agricultural activity. In this book, we present a preliminary calculation method adapted to the distinctive features of agriculture. This method combines different metabolic traditions (EW-MEFA, MuSIASEM, etc.) with the arsenal of knowledge, theories, and concepts proper to Agroecology. We call this completely original adaptation and synthesis approach *Agrarian Social Metabolism* or *Agrarian Metabolism.*

1.1 Agriculture and Social Metabolism: The Metabolism of Agroecosystems

Social Metabolism (hereon SM) refers to the set of theories and methodological tools that allow analyzing a society's biophysical behavior (Adriaanse et al. 1997; Matthews et al. 2000; Haberl 2001; Weisz et al. 2006). It provides valuable information to assess a society's environmental sustainability and has even turned into a new perspective on human beings' relationships with their physical environment, that is, with flows of energy, materials and information (Fischer-Kowalsky and Haberl 1997, 2007; Sieferle 2011; González de Molina and Toledo 2011, 2014). The term emerged from an analogy with the biological concept of metabolism, given that relationships that humans establish with nature are always twofold: individual or biological and collective or social. On an individual scale, humans extract sufficient amounts of oxygen, water, and biomass from nature per unit of time to survive as organisms, and excrete heat, water, carbon dioxide, mineralized, and organic substances. At the social level, groups of individuals connected in different ways, through relationships or links, are organized in such a way as to guarantee their subsistence and reproduction. They also group together to extract matter and energy from nature through meta-individual structures or artefacts and excrete a whole range of waste or residues. Social Metabolism can be defined, therefore, as the way in which human societies organize their exchanges of energy and materials with their natural environment (Fisher-Kowalski and Haberl 1994; Fisher-Kowalsky 1998, 2002; Giampietro and

Mayumi 2000; Giampietro et al. 2011) with the purpose of reversing the entropic process they are subject to, like all living beings (González de Molina and Toledo, 2014). In this sense, human societies carry out two basic material tasks: on the one hand, they produce goods and services and distribute them among society's individuals; on the other, they reproduce the conditions that make production possible, thus gaining stability over time. A substantial share of social relations is therefore oriented towards the organization and maintenance of exchanges of energy, materials, and information. Its mission is to configure and feed the "funds" (Georgescu-Roegen 1971) that societies build to generate goods and services, i.e. to counteract the law of entropy and thus generate order. Such funds are fed, that is, they are maintained and reproduced through the exchange of energy and materials with the environment. This enables us to understand the relationship between society and nature as a metabolic relationship. Social Metabolism is, in fact, a metaphor that borrows a concept taken from biology and applies it to the world of relations between society and nature.

SM is thus an analytical tool applied to socio-ecological relationships, whatever their scale or territorial scope. It can, therefore, be applied to agriculture. Consequently, Agrarian Social Metabolism (ASM) or *Agrarian Metabolism* (AM) can be described as the exchange of energy, materials, and information that agroecosystems perform with their socio-ecological environment. The purpose of metabolic activity is that of appropriating biomass to satisfy human species' endosomatic consumption directly or indirectly through livestock while providing basic ecosystem services. AM has also tried to satisfy the exosomatic demand (raw materials and energy) of societies with an organic metabolism and continues to do so, to a lesser extent, in industrial societies. To accomplish this, society colonizes or seizes a part of the available land. Within this territory, it establishes varying degrees of intervention or interference in the ecosystems' structure, functioning and dynamics, giving rise to different types of agroecosystems. In other words, AM refers to the appropriation of biomass by members of society by managing the agroecosystems present on the land (Guzmán Casado and González de Molina 2017).

Why should we consider, however, "agroecosystems" to be the subjects of AM? Because they constitute the basic unit of metabolic activity: these ecosystems are manipulated and artificialized by human beings in order to capture and convert solar energy into some particular form of biomass that can be used as food, medicine, fiber or fuel (Altieri 1989). From a thermodynamic point of view, they can also be considered as complex adaptive systems that dissipate energy to counteract the law of entropy (Prigogine 1978; Jørgensen and Fath 2004). To do this, they exchange energy, materials, and information with their environment (Fath et al. 2004; Jørgensen et al. 2007; Swannack and Grant 2008; Ulanowicz 2004). Compared to ecosystems, that still retain their capacity to self-sustain, self-repair and self-reproduce, agroecosystems are unstable, requiring external energy, materials, and information (Toledo 1993; Gliessman 1998).

The flows are exchanged through work or manipulations that aim at ensuring the production of biomass and its reiteration over successive cycles of cultivation or breeding, interfering in the carbon, nutrients and hydrological cycles and in the mechanisms of biotic regulation. In traditionally managed agroecosystems, this input of additional energy and materials comes from biological sources: human work and animal labor. Dependence on the land is maintained in a strict sense. In industrially managed agroecosystems, additional energy and materials also come from the direct and indirect use of fossil fuels as well as metallic and non-metallic minerals. In such systems, most of the energy generated as biomass is directed out of the system in the form of food or fiber as well as crop residues. The latter are not allowed to remain within the system and this contributes to important internal processes. Therefore, it is also necessary to import a large amount of biomass from other agroecosystems to ensure reproduction. These agroecosystems are mere "energy transporters" and can hardly be considered sustainable (Gliessman et al. 2007, 17). In short, they are part of society's general metabolism, specifically dedicated to the appropriation of photosynthesis products. From a metabolic point of view, the reproductive dynamic of agroecosystems is peculiar. Their sustainability, as artificialized ecosystems, also depends on their level of biodiversity, the maintenance of fertile soil, etc. This means that part of the generated biomass must be recirculated to meet both productive and basic reproductive functions of the agroecosystem itself: seeds, animal labor, organic matter in the soil, functional biodiversity, etc. The thermodynamic rationale underlying this characteristic was developed by Ho and Ulanowicz (2005) and, later by Ho (2013) when they related sustainability to dissipative low entropy structures. Ecosystems, as dissipative structures that can consume large amounts of energy or the reverse, can be structured in such a way that their entropy is low. This characteristic of ecosystems also works at different scales for agroecosystems and even for AM as a whole. Like ecosystems, agroecosystems constitute an arrangement of biotic and abiotic components in which living systems predominate and respond to what has been called "thermodynamics of organized complexity" (Ho and Ulanowicz 2005, 41, 45). This means that, going beyond the point raised by Prigogine (1962), an agroecosystem can be "far from thermodynamic equilibrium on account of the enormous amount of stored, coherent energy mobilized within the system, but also that this macroscopically non-equilibrium regime is made up of a nested dynamic structure that allows both equilibrium and non-equilibrium approximations to be simultaneously satisfied at different levels". In this sense, the really decisive aspect of ecosystems is not only the flows of energy and materials that keep them away from thermodynamic equilibrium but also their capacity to capture and store the energy that circulates inside them and transfer it to its different components (Ho and Ulanowicz 2005, 41, 45). This depends on the quality and quantity of circuits or internal loops through which the energy flows circulate as well as whether they are able to compensate for the entropy generated somewhere in the ecosystem by the negative entropy generated in another system within a given period of time. As Bulatkin (2012, 332) argues, "the agroecosystem as a natural-anthropogenic system has its own biogeocenotic and biogeochemical mechanisms and self-regulation structures, which should be used to reduce anthropogenic energy costs". That is, it contains cycles that, according to Ulanowicz (1983), have a "thermodynamic sense": "Cycles enable the activities to be coupled, or linked together, so that those yielding energy can transfer the energy directly to those requiring energy, and the direction can be reversed when the need arises. These symmetrical, reciprocal relationships are most important for sustaining the system" (Ho and Ulanowicz 2005, 43).

For example, in organic or agrarian metabolic regimes (González de Molina and Toledo 2011, 2014), agroecosystems used to function in an integrated manner. Biogeochemical cycles clearly went beyond the cultivated lands and extended over large parts of the territory. The increase in entropy that occurred in the most intensively cultivated areas (irrigation or hedges, in the case of the Mediterranean) was usually compensated by the import of nitrogen through livestock (manure) from other areas of low entropy such as forest areas. The result was a metabolic regime that was also of low entropy. Spatial heterogeneity and agrosilvopastoral integration were key for articulating the different circuits that captured, stored and transferred energy.[1]

This explains why, when different agroecosystem components are adequately articulated, it is possible to substantially reduce incurred land costs whenever biomass is produced, and thus generate the largest amount of biomass at a minimal land cost (Guzmán and González de Molina 2009; Garzón et al. 2011, Guzmán et al. 2011). In this sense, net primary production is found to correlate positively with the functional integration of different land uses in terms of territorial efficiency. The bigger the amount of energy is captured and stored in the internal cycles of agroecosystems, the smaller the amount of energy that will have to be imported from outside (Guzmán Casado and González de Molina 2017). For this reason, it is often commented (Gliessman 1998) that the more an agroecosystem resembles natural ecosystems in its organization and functioning, the greater its sustainability.

Each of the AM's forms of organization leaves their *particular mark* on the territory, configuring specific landscapes and specific agroecosystem arrangements. The landscape is the visible mark left on the territory, although hidden marks may materialize in a different, sometimes distant, territory from which natural resources (*land embodied, virtual land*) are imported (Guzmán and González de Molina 2009; Garzón et al. 2011, Guzmán et al. 2011; Infante et al. 2018). In pre-industrial agriculture, agroecosystems needed to appropriate large amounts of land to produce useful biomass and function correctly. Industrialized agriculture does not incur this cost, as it is fed by energy sources and materials that come from the subsoil. In this sense, the extent to which industrialized agriculture landscapes are simplified depends on the extent to which these internal circuits are reduced within the agroecosystems.

In short, agroecosystems can be understood as dissipative structures built and maintained by humans (Prigogine 1947, 1978; Jørgensen and Fath 2004) in order to provide energy and useful materials for society, generating order or negentropy. This way, agroecosystems can be "improved" in order to increase their net primary productivity (NPP): for example, by providing them with productive infrastructures that maximize water (irrigation channels) or available land (terraces) or biodiversity (hedges, landscape mosaics), or recreating outdoor environmental conditions

[1]As pointed out by Sieferle (2001, 20), different land uses were linked to different types of energy. Cultivated lands were associated with the production of metabolic energy to provide human food; the pasture land that fed farm animals was associated with mechanical energy and forests with the thermal energy that provided the fuel needed for cooking, heating and manufacturing.

(greenhouses). Toledo and Barrera-Bassols (2008) gathered a large number of cases where peasants improved the productive capacity of agroecosystems across the globe for thousands of years. This investment in "built capital" has also been called *Landesque Capital* (Widgren 2007). The concept should include the infrastructure of roads, houses and warehouses or buildings with productive uses that have been built over time and are essential for agricultural activity itself as well as for current soil productivity levels. However, due to the difficulty in achieving suitable accounting, we chose to leave this section aside, recognizing nonetheless that an increasingly significant share of energy and materials are dedicated to the maintenance and improvement of these infrastructures.

1.2 Funds and Flows in Agrarian Metabolism

We have defined AM as the exchange of energy, materials, and information between agroecosystems and their socio-ecological environment. This exchange is composed of flows that go in and out, as described in the EW-MEFA methodology. However, our proposal not only quantifies these flows but also carefully measures whether or not these flows maintain the dissipative structures or fund elements they are endowed with. The distinction between flows and funds was borrowed from Georgeuscu-Roegen (1971) and Giampietro et al. (2014) who incorporated it into the MuSIASEM methodology. According to Georgescu-Roegen, the economy's ultimate goal is not the production and consumption of goods and services, but the reproduction and improvement of the processes necessary for their production and consumption. This different understanding of economic activity's main objective implies that from a biophysical point of view, we need to shift our attention away from energy and material flows and instead focus on fund elements: we must center our analysis on whether fund elements are improved or at least reproduced during each productive cycle. In other words, our focus switches from the production and consumption of goods and services to sustainability, and whether both production and consumption can be maintained indefinitely. Within this framework, it is essential to distinguish between flows and funds. Flows include energy and materials that are consumed or dissipated during the metabolic process, such as raw materials or fossil fuels. The rhythm of these flows is controlled by external factors—relating to the accessibility of the environment's resources in which the metabolic activity unfolds—and by internal factors—related to the processing capacity of energy and materials, relying in turn on the technology used and the knowledge to manage it. Fund elements are dissipative structures that use inputs to transform them into goods, services, and waste, i.e., into outputs, within a given time scale; they remain constant during the dissipative process (Scheidel and Sorman 2012). They process energy, materials and information at a rate determined by their own structure and function. To do so, they need to be periodically renewed or reproduced. This means that part of the inputs must be used in the construction, maintenance and reproduction of the fund elements, limiting, of course, their own processing rhythm (Giampietro et al. 2008). The quantities of energy and materials invested in the maintenance and reproduction of the fund elements cannot be employed for end uses. These types of elements can even be improved over time, when energy and materials are allocated for this purpose.

The MEFA methodology, the best known social metabolism methodology, has been criticized (Giampietro et al. 2014, 29) and with good reason: it does not take into account dissipative structures proper to an agroecosystem as in our case, nor the different types of flows that feed them. The MEFA methodology reduces all flows to units of energy (MJ) or weight (t) to be able to add them and does not differentiate between their varying qualities and purpose. What this approach actually does is transfer a prevailing idea from conventional economics, whereby only the existence of inputs and outputs are taken into account in metabolic accounting. This preferential consideration of the transfer of energy and materials has no thermodynamic basis and is therefore not useful when it comes to describing the biophysical functioning of a society's economic activity and its degree of sustainability.

Agroecosystems are, as we have said, dissipative structures that can be decomposed, in turn, into other structures, be they social or ecological, which compose them. To consider each one of them individually would render our proposition so complex, it would lose its heuristic nature. For this reason, only essential dissipative structures or fund elements for the reproduction of agroecosystems themselves and the provision of their services in AM are included. However, such funds may be decomposed into other fund elements in order to refine, if deemed necessary, the analysis. We will come back to this point when examining the environmental impacts of Spanish agriculture's industrialization. For the purposes of this research, four fund elements were taken into account: land, livestock, agrarian population and technical means of production (or technical capital today). However, it is relevant to differentiate between fund elements of a biophysical nature and fund elements of a social nature since they are not reproduced in the same way. The four funds are closely connected and represent the fullest manifestation of the socioecological relationships at the heart of each agroecosystem and at the center of the metabolic exchange. The articulation between the four fund elements is fundamental, as we shall see later, to explain metabolic dynamics.

Each fund element has a different either biophysical or social nature, and, therefore, each fund element works with different quality flows and different metrics. As pointed out by Giampietro et al. (2014, 29), the flows' characteristics are closely related to the fund they come from. A territory is colonized or land is appropriated by society to generate useful biomass flows; it is usually measured in hectares and subdivided into different uses that produce vegetal biomass, expressed in tons of vegetal biomass per hectare (t ha^{-1})—or its equivalent in energy, MJ ha^{-1}—or net primary productivity (NPP). The livestock fund element is the source of flows directed to society as well as to the agroecosystem itself, providing animal biomass for raw materials, food and, to a much lesser degree, energy or services such as traction or manure. It is usually measured in standard livestock units of 500 kg (LU_{500}) and the flows it generates are expressed in kg or t of animal biomass ha^{-1} or LU or MJ ha^{-1} if the flows are expressed in energy units. The agrarian population is the fund element which represent the human work flows. They are usually measured in hours or days of work (hours or days/year^{-1}). Finally, the "means of production" fund brings together the supply of production tools that generate mechanical work flows or vegetal health services and other services: its size and composition vary significantly according to whether the agrarian system is industrial or pre-industrial. In our case, we will examine a period of transition from traditional organic agriculture to industrial production, during which mechanical instruments stand out from the rest of the work tools. It is usually measured, for example, in terms of installed capacity, expressed in Kw of power or cv, and its flows in Kw/h^{-1} or MJ ha^{-1}, etc.

Whatever they may be, fund elements require a quantity of energy in terms of biomass and human work that must be taken into account for each production process. The process of industrialization of agriculture has consisted in substituting the agroecosystems' biogeochemical circuits with working capital that depends on resources outside the agrarian sector, usually via markets. This explains a fundamental difference in the metabolic functioning of traditional and industrialized agroecosystems: the reproduction of fund elements was possible through biomass flows in organic metabolic regimes; but under the industrial metabolic regime, external fossil energy flows are widely reproduced by social funds and can cause environmental deterioration when attempting to reproduce biophysical funds, especially agroecosystem services. For example, trophic chains that support both edaphic life and the agroecosystem's biodiversity can generally only be fed with biomass. Deterioration of colonized or appropriated land cannot be compensated using energy and external materials or any other resource than vegetal biomass. In this way, the industrialization of agriculture can be interpreted as the process of replacing dissipative structures of a biophysical nature, that belong to agroecosystems and have been maintained by peasants through integrated management, with man-made dissipative structures or, to put it in economic terms, with means of technical production obtained through markets and, to a lesser degree, from State intervention.

In AM not only are biophysical flows exchanged, but also flows of information. They are usually excluded from metabolic methodologies, perhaps due to their complexity and difficulty in measuring them. However, these flows have the capacity to order and organize components of physical, biological and social systems. They are therefore essential to understand not only the specific configuration of metabolic regimes but also their dynamics. We have thus attempted to integrate them into our proposition. As they cannot all be taken into account given their inherent complexity, we selected the information flows with the biggest explanatory capacity regarding farmers' decisions. Agricultural work is also considered as a workflow containing decisive information that organizes agroecosystem structure and functioning. Consequently, we assumed that these decisions were directly influenced by the monetary remuneration that farmers receive in exchange for the sale of their products. Therefore, they constitute a suitable proxy for synthesized information flows. In other words, for the purposes of this research, information flows are defined as follow: flows originating in the agrarian population fund element, in the form of work and incorporated management decisions; and monetary flows stemming from the agroecosystem's social environment and ending up in this population fund in the form of money obtained in exchange for production.

Monetary flows represent a specific type of flow which, from a metabolic standpoint, have multiple purposes: they allow to closely articulate the biophysical and social components of social metabolism itself. As suggested by Swanson et al. (1997), money is a commodity but it is also information. Even if it is understood as a measure of the value of entropy, it can be considered according to its capacity to reduce the levels of a social system's future entropy (González de Molina and Toledo 2014). Money, expressed in relative prices, has transmitted information that has enabled to largely explain—especially in societies with monetized exchanges—the behavior of social agents, in our case that of farmers. This does not mean that markets, as they are organized today, have determined farmers' behaviors based on relative prices. Markets were not always the main or only way to exchange goods and services. Therefore, their dynamics only explain productive decisions in contexts of commodified economies. In organic metabolism societies, seigneurial rights, for example, imposed monetary levies on farmers and were based on feudal law, not on markets; but ultimately, they constituted monetary exactions that forced peasants to sell part of the harvest to be able to pay them. The price of agricultural products, even in "imperfect" markets, forced peasants to take production decisions. The situation is less ambiguous in market societies, where relative prices are the most relevant indicator or source of information. Nevertheless, relative prices determine farmers' behaviors, according, for example, to cost-benefit calculations; the decisions they make are usually based on multiple and weighted criteria, as in the case of other economic agents. We are also aware of other non-monetary information flows that also have a bearing on farmers' decisions, such as public policies, the institutional framework, etc., and even the rural worlds' successive cultural values along the twentieth century. Monetary flows also reflect the large array of cultural and institutional factors but the extreme complexity of the subject would make it impossible to consider them all. Consequently, we will use the agrarian sector account, elaborated from national accounts, to quantify and analyze AM monetary flows. The prices received by farmers, the prices paid for inputs and agricultural income will constitute the major macromagnitudes used as information flows in the sector, as we will see in Chap. 4.

To summarize, we have defined four fund elements belonging to agroecosystems. They receive flows of energy, materials and originate, in turn, outflows in the form of biomass that society remunerates through monetary flows. These monetary flows, which are also information flows, stop at farmers, who through their agricultural labor, that is through their tasks, maintain and reproduce the funds. Monetary flows are input information flows and agroecosytems' management decisions are output flows. Consequently, energy and material flows, both inputs and outputs, can be broken down into productive, reproductive or maintenance flows of fund elements; meanwhile, information flows are composed of monetary flows (input) and human work flows (output) and normally have a reproductive role. The main hypothesis in this research is that the reproductive capacity of these latter flows determines the dynamics of agroecosystems and the dynamics of the agrarian sector as a whole.

1.3 The Appropriation of Biomass and Colonization of the Territory. Biophysical Funds (Land and Livestock)

The territory is the main element in any agroecosystem, since biomass production requires land for the photosynthetic process to take place. Therefore, the first task of AM consists of colonizing the ecosystems (Cook 1973) and appropriating a part of their net primary productivity (NPP). Through colonization and appropriation, farmers conduct three basic types of intervention on the territory, thus affecting different ecosystems (González de Molina and Toledo 2011). The first intervention consists in hunting, fishing, gathering, as well as certain forms of extraction and livestock by foraging on the original vegetation: this type of intervention does not cause substantial changes in the structure, architecture or dynamics of ecosystems. The second type of appropriation consists in disarticulating or disorganizing ecosystems to introduce groups of domesticated species or species in the process of being domesticated, as in the case with all forms of agriculture, livestock, forestry, and aquaculture. While in the case of the first type of appropriation, ecosystems' intrinsic or natural capacity to self-maintain, self-repair, and self-reproduce is not affected, in the second case, ecosystems lose these latter abilities and require external energy (whether human, animal or fossil), as well as materials and information to sustain themselves. Over the last decades, a third form of appropriation has emerged responding to conservationist actions by public administrations or non-governmental organizations that seek to preserve natural or regenerating areas. They also aim at providing ecosystem services.[2] The distinction between the first two types of intervention is of special interest in our work because as we will observe later, they have coexisted and been combined in different ways throughout the study period.

Consequently, an agroecosystem may also contain appropriated areas where the ecosystem is minimally subjected to manipulation or intervention while remaining inseparable from its territorial arrangement. Different units of biomass appropriation can coexist there, some obtained by means of hunting and gathering and others by plant manipulation. This situation is more visible when beyond the scale of plots, we move up to observe the territorial arrangements an agroecosystem must necessarily be made of in terms of different land uses. We need to specify this because agroecosystems are commonly confused with cultivated areas. Agroecosystems, however, make up coherent and articulated units of analysis: biogeochemical flows circulate within them and, therefore, human appropriation gives rise to different degrees of intervention (Guzmán and González de Molina 2000; González de Molina and Toledo 2011, 2014; Guzmán Casado and González de Molina 2017). This also applies to plants that inhabit agroecosystems, since cultivated plants are often the only ones to be taken into consideration. Moreover, only the aerial parts of these plants are taken into account, while the root biomass, often the crop residues or the adventitious flora

[2] Although it is possible to find examples in the past of protection of natural spaces through religious-cultural practices such as sacred forests.

are ignored. An agroecological approach to AM must be based on all the biomass produced within its limits, that is, all the net primary productivity. As we shall see, the reason is that when defined as a fund element, territory or land reproduction depends directly or indirectly on the total amount of produced biomass, not only on the harvest. Consequently, the colonized territory a society disposes of to develop its AM tends to be fragmented into different categories or land uses, such as croplands, pasture lands, and forest lands depending on the degree and type of its dedicated management. Each one of them can be subdivided in turn into different categories depending on their specific or multiple uses and labor intensity.

The other biophysical fund element under consideration is livestock, whose basic function is to provide useful animal biomass to society in the form of food (meat, milk, fat, etc.) or raw materials (wool, skins, etc.) and certain services, among which some essential ones such as manure to renew soil fertility. As mentioned earlier, livestock volume is measured in international livestock units (LU of 500 kg), as it allows to reduce different livestock species to a common denominator. In societies with organic metabolism, livestock was adapted to the soil and climate conditions of each territory and to the interests of the society on which it depended. Livestock, thanks to its functional biodiversity, mediated the agroecosystem's different ecological processes. As a result, its composition had to be diversified, to take advantage of the different available food resources (herbaceous, arboreal pastures, crop residues, feed, etc.), in very different environments (wetlands, steep slopes, etc.) and generate goods and services of a varying nature (food, work, fiber, etc.). Therefore, the herds had to present a certain degree of diversification and balance based on land availability and society's needs and resources, without being excessively simplified or specialized. This explains the diversity of livestock breeds, species and multiple uses made of animals (traction, reproduction, production of food and raw materials). This was the situation when no biomass flows were imported from third countries. Today, growing transport capacity and differences in production costs have facilitated and generalized the import of large quantities of vegetal biomass to feed livestock that in certain cases can no longer be maintained, due to its volume and nature (monogastric or granivorous), by native agroecosystems. A huge amount of livestock is maintained based on these imported biomass flows. This livestock is highly specialized in meat or milk production, it is contained and composed of a few breeds, designed to meet the growing demands of meat and dairy products. This growing consumption has characterized western diets, including the Spanish diet, over the last decades. Therefore, livestock, in its role of biophysical fund, can only be reproduced, maintained or increased through a constant flow of biomass—whether domestic or imported—that is mostly vegetal, but not only vegetal.

The biophysical fund elements of an agroecosystem require therefore a certain amount of energy and materials in the form of biomass for reproduction and maintenance. In addition, trophic chains that support both edaphic life and biodiversity can generally only be fed with biomass. In this sense, it is worth recalling the idea, expressed by ecological economists, that natural capital cannot be replaced by manufactured capital (Ayres 2007; Häyhäa and Franzese 2014, 125), just as not all types of energy are interchangeable or have the same use (Giampietro et al. 2010). That is, the biophysical elements of an agroecosystem cannot be sustained using oil or coal or the fuels that derive from them.

On the other hand, ecosystems, including those appropriated by society (agroecosystems), generate ecosystem service flows, some of which are directed towards its renewal (De Groot et al. 2003; Ekins et al. 2003; Millennium Ecosystem Assessment 2005; Folke et al. 2011). According to Schröter et al. (2014), each agroecosystem has a specific capacity to provide these services, in accordance with their edapho-climatic conditions. Since agroecosystems are human-being-dependent ecosystems, the quantity and quality of the services that they provide depend on the way they are managed. An adequate provision of services will depend on the agroecosystem's health, that is, on the sustainable state and size of its fund elements (Cornell 2010; Costanza 2012). Conversely, the degradation of an agroecosystem's fund elements can lead to reductions in the amount of ecosystem services it provides (Burkhard et al. 2011). Services are usually grouped into four categories: provisioning, regulating, supporting and cultural services. Provisioning includes the extraction of goods (e.g., wood, firewood, food, and fiber); regulatory services help to modulate ecosystem processes (e.g., carbon sequestration, climate regulation, pest and disease control and recycling of waste or residues); support services sustain the provision of all other categories (e.g., photosynthesis, soil formation, nutrient recycling); while cultural services contribute to spiritual well-being (e.g., recreational, religious, spiritual and aesthetic) (de Groot et al. 2010).

1.4 Social Fund Elements (Human Work and Technical Means of Production)

Agroecosystems process energy and materials to produce biomass thanks to human labor. As we have seen, human work has a characteristic that distinguishes it from other funds: it incorporates information flows. The origin of these flows is not farmers alone but also the household they are part of. Consequently, the fund element considered in this work—unlike that established by other metabolic methodologies—is the "agrarian population", composed of domestic groups or households that are dedicated to this activity. There are three reasons for this, based on the distinction between flows and funds. Firstly, because the continuation of the human work flow depends on the time investment in other tasks carried out by the entire household. For example, time devoted to care, which are reproductive tasks from the physiological point of view (overheads), or to social and educational activities, which from a social perspective, would correspond to reproductive activities. Second, because maintaining agroecosystems in good productive conditions requires performing maintenance tasks that are not usually considered to be part of working hours directly related to agricultural production or are effectively paid. And lastly, because agricultural labor has usually been performed by farmers with the help of the family, so agrarian work is essentially family work. As a result, in our research, we not only considered the number of individuals engaged in agricultural work but also their families who are responsible for "producing" the agricultural workers and who can engage in other paid and unpaid activities to achieve it. In fact, for small and medium agricultural producers, the family is above all the basis of their economy and the objective of their productive strategies.

To quantify human work flows, and their reproduction cost, the Time Budget methodology should be used. It calculates the population's Time Use at any moment in time. The aggregate nature of this methodology, however, based on the scale of a state, makes it difficult to apply this technique. The aggregate information we dispose to quantify workflows is the standard information provided by Spanish statistics: the population that is active, employed or unemployed in the agricultural sector. We have assumed that such flows (whether paid or not, and whether directly agrarian or not) originate in farmer households or field workers whose reproductive costs must be covered by the monetary income received in exchange for production. Consequently, we have assumed that the reproduction of a farmer household is viable when agricultural income allows reaching the national average of household expenditure.

Human labor logically requires energy, basically endosomatic energy, to maintain and reproduce itself. In fact, this is the amount of energy that we used to calculate the energy efficiency of each of the successive metabolic arrangements over the study period, i.e. a century and a few years. Nevertheless, as human societies have been gaining in complexity, cost of reproduction has also increased to include all exosomatic energy incorporated in that process (or its equivalent in monetary terms). As the metabolic profile of contemporary societies has increased, the cultural consumption of energy and materials has been gaining importance—thus so has its monetary cost.

The fourth and last fund element considered is the technical means of production. Today it could be called "Technical Capital" as referred by Mario Giampietro and it includes instruments or substances that aim at replacing certain ecosystem functions and services such as pest control, fertility replacement, etc., which are manufactured outside the agricultural sector and through the use of fossil fuels or mineral sources. It also includes the set of artifacts created to perform all kinds of agricultural tasks. For example, they may consist in devices capable of converting fuel flows into tractor power flows in a localized manner to perform tasks that human work would not usually perform, either because they require a lot of power or because they save work. The capacity of this fund element is measured as we have seen in kW or cv (or hp) and has worked and still works with fossil fuels both for its manufacturing and functioning. The maintenance of this fund requires investment in energy and materials and, unlike the other funds, its replacement occurs thanks to metabolic processes that take place outside the agricultural sector itself. In this fund, and for purely formal purposes, we include cattle when used for fieldwork, measurable in kW or horse power, which depends instead on biomass metabolized in the form of grains, straw or grass for maintenance. As a biophysical fund, its size depends on the availability of food produced by agroecosystems and the nutrient needs and stocks of manure and other fertilizers; as a social fund, its size depends on traction needs (Graph 1.1).

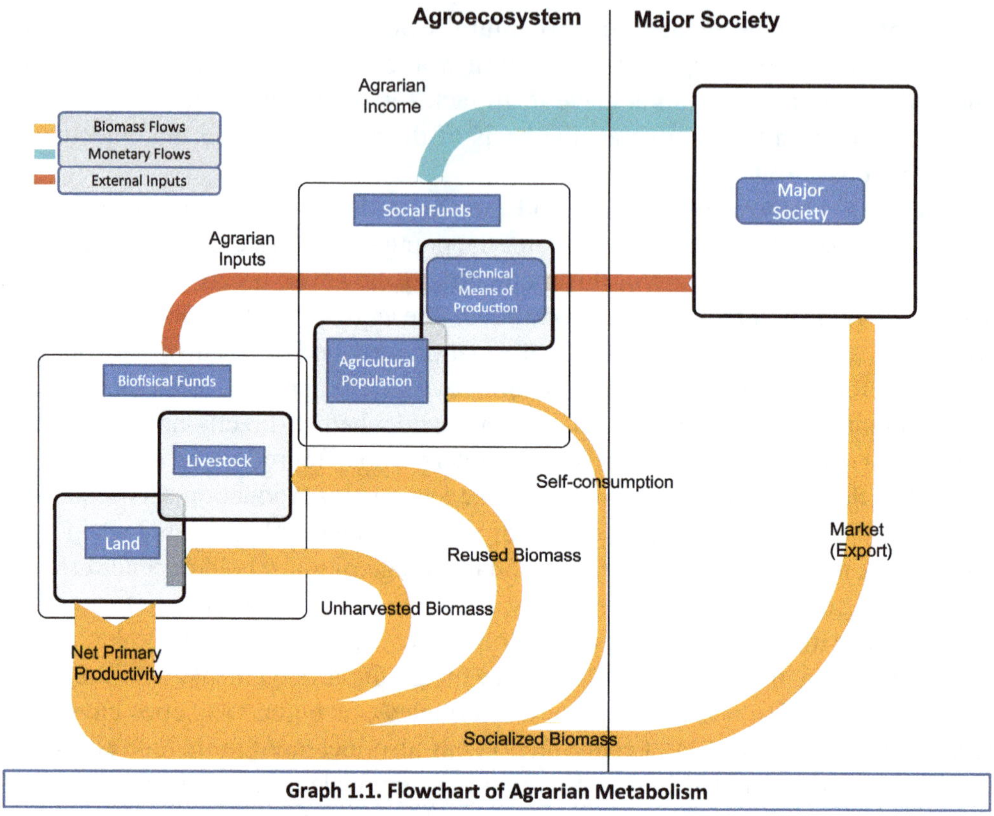

Graph 1.1 Flowchart of Agrarian Metabolism

1.5 The Organization and Dynamics of Agrarian Metabolism

The four fund elements that operate in AM mutually affect each other. This interrelation is expressed in each metabolic arrangement in a particular fashion, giving rise to a sort of unstable equilibrium that makes the AM function as a whole in a specific way. The ability to process energy and materials that each fund element has determines not only the magnitude of the flows it generates but also generates restrictions for others. Land availability and its capacity to produce biomass has a direct impact on the magnitude of the flows originating in human work, livestock or means of production. Low availability of human work can, for example, limit the capacity of the land to produce useful biomass, favoring, for example, livestock use of the land and vice versa; very strong traction power can lead to a more intensive agricultural use of the land that would not correspond to the size of the population and its capacity to work; while an excessive volume of the technical means of production or livestock may require importing energy from the outside, in the form of biomass or fossil fuels. This is what occurred, as we will see, with Spanish agriculture over the last half-century.

The fund elements' mutual dependence not only explains the structure, functioning and dynamics of AM, but also constitutes the key to its degree of sustainability i.e., whether if each fund element is capable of providing the services required by AM itself to function (Giampietro et al. 2014, 19) and to do so in a balanced manner over time. The imbalance between each fund's capacity and the generated flows can make it necessary to seek a new equilibrium, thus causing changes that compromise the medium and long-term viability of the established metabolic arrangement. This consistency between flows and funds and among funds themselves is not found in the EW-MEFA methodology but in the MuSIASEM methodology, made operational through the so-called "sudoku effect" (Giampietro et al. 2014; Giampietro and Bukkens 2015). We incorporated this idea into our AM proposition to evaluate the sustainability or at least get an idea of it: a variable part of the energy and material flows must necessarily be invested in the maintenance and reproduction of the funds and these, in turn, must preserve the necessary relation of congruity in such a way that the flows originated in one fund can help the others to function and vice versa. In our view, the imbalances or lack of correspondence between the funds and their flows explain the metabolic dynamics and, therefore, tendencies towards unsustainability and towards metabolic change.

The study of metabolic dynamics and the processes of change indeed lies at the heart of this work. The objective is to understand why certain decisions were made leading AM to be in the situation it is in now. In social sciences, theories that explain long-term changes in human societies using the concept of transition have become increasingly relevant (Bergh and Bruinsma 2008; Lachman 2013). A academic trend that analyzes transition towards sustainability from a metabolic perspective has also been developed. According to this approach, which is linked to the IFF-Social Ecology of Vienna, socio-ecological transition processes are processes of structural change affecting the configuration of energy, material and information flows that societies exchange with their environment (Fischer-Kowalski and Rotmans 2009; Fischer-Kowalski 2011). However, this approach has been criticized by Lachman (2013, 274), and rightly so, because it interprets socio-metabolic transition at a system level that is overly abstract (social metabolism) and leaves no place for social actors. Factors such as beliefs, political or economic interests or culture—we call them information flows—are not taken into account. This inordinately general and abstract framework can hardly generate advice to its users on concrete policies to advance the transition. Our proposition, in line with this criticism, considers that social agents undeniably play a fundamental role. If we keep this in mind, information flows become central to the analysis.

Either way, Socioecological Transition is a conceptual tool that aims at making social and environmental change comprehensible by reducing its complexity. In this sense, the notion of socioecological transition allows us to understand the mechanisms underlying AM's shift from its organic configuration to its current industrial configuration, adopting in the process hybrid forms of variable duration, where AM is not totally organic nor totally industrial. With the industrialization of agriculture, AM specialized in the production of biomass to satisfy the endosomatic consumption of individuals as well as demands for industrial raw materials and services. AM has gone from being at the heart of the sociometabolic process, and its main source of energy, to constituting an apparently marginal segment of the process due to fossil fuel exploitation (González de Molina and Toledo 2011, 2014). The metabolic functionality of agrarian activities has therefore changed. They constitute just another input of the metabolism of materials. Though the market does not reward them for it, they offer essential environmental services (carbon sink, climate regulation, water purification, maintenance of certain levels of biodiversity, etc.) for the stability of industrial metabolism. Perhaps because of this, they have tended to deteriorate with the process of industrialization and commodification of agriculture (De Groot et al. 2002; Pagiola and Platais 2002).

According to the literature available, AM's socioecological transition that culminated with the industrialization of agriculture followed three major "waves" (González de Molina 2010): the first wave was driven by the institutional change that accompanied the implementation of liberalism, occurring within the limits of organic AM. It entailed "*optimizing*" its capacity to increase biomass production. During the second wave, AM underwent its first great metamorphosis related to partially unstructured biogeochemical cycles. The vector of this process was the appearance and diffusion of artificial fertilizers in the final decades of the nineteenth century and the beginning of the twentieth century. Its introduction meant "overcoming" the most common limiting factor of production until then, i.e., the lack of nutrients, and breaking the link of dependency between fertility replacement and land. A long transition process began in which agricultural production would *shift from being dependent on soil to being dependent on the subsoil*, in other words on fossil fuels and minerals. The third milestone of metabolic change opened the path towards the definitive metamorphosis of traditional organic agriculture. The energy transition was completed: fossil fuels replaced a large part of the workforce and all animal traction. AM was industrialized and the distinctive limitations of organic metabolism disappeared. It was associated with changes in energy patterns where coal was replaced with oil and natural gas, providing higher energy densities. Two basic industrialization innovations linked to these changes allowed to massively subsidize agriculture based on external energy: electricity and the internal combustion engine. Developments during the second and third of these three waves draw our special interest as they correspond to our study period.

All these changes eventually took place at different scales. At the crop scale, highly significant changes took place that mainly affected the genetic material, that is, seeds. Crop varieties or livestock breeds offering higher yields or certain productive characteristics were sought. At the farm level, connections between crops and polycultures were drastically reduced, rotations were simplified and they were subsequently replaced with crop alternations governed by market demands. Heterogeneous crops and plants as well as their combined distribution over fields gave way to monoculture, significantly reducing genetic, structural and functional diversity (Gliessman 1998). At the landscape scale, land uses were segregated and the productive and functional synergies generated by agrosilvopastoral integration were lost. The trend towards productive specialization grew ever stronger, imposing specialized land uses according to market demands rather than according to land capabilities or the presence of natural resources. Geodiversity and spatial heterogeneity were consequently lost. Thus, energy and material flows, which tended to be local and contained (renewable) ended up being global and originating from fossil sources. We will come back to this issue in the following section. Finally, on a more aggregate scale, agrarian systems have been integrated into a nationwide market, and this has fostered productive specialization based on comparative advantages and opportunity costs; this process has reached its peak today with the constitution of a global agrarian market and a global agri-food regime, in which agroecosystems are vertically integrated (McMichael 2009).

1.6 The Forces of Change

This research is dedicated to analyzing socio-metabolic changes in Spanish agriculture since 1900 and to uncover its driving forces. As we will see in the following chapters, the history of Spanish agriculture in the last century could be understood as a continuous effort to increase and specialize the volume of biomass production to face both the endosomatic and exosomatic demands of society. As a result, the main forces that drove the change, that is, the socioecological transition, were intensification and productive specialization. These occurrences were all similar to those in industrialized countries and to some extent to those in neighboring agricultures (IPES-Food 2016). They seem to have responded to a general trend.

In fact, Spanish agricultural developments since 1900 should be studied in the light of a more universal evolution. The notion of intensification refers to more intensive agricultural cultivation and growing use of inputs that generally lead to greater agricultural productivity. This seems to have been a general trend that has followed similar patterns across space and time (Mustard et al. 2004). To understand this process, Foley et al. (2005, 571) proposed a land-use sequence called *Land Use Transitions* mainly powered by productive intensity (Rudel et al. 2009; Lambin and Meyfroidt 2010; Jepsen et al. 2015). In a similar line, Ellis et al. (2013, 7980) recently advanced a *Land Intensification Theory*. Most of these processes include sustained increases in land productivity associated with more intensive management (Currie et al. 2015; Federico 2009). But the process does not seem to have been completely linear. Net primary productivity has not clearly increased in all cases (Smil 2011; Krausmann et al. 2013); and productivity doesn't seem to have grown continuously but has rather followed phases of involution and crisis (Ellis et al. 2013). Intensive land use has also been linked to drops in labor productivity in pre-industrial contexts, while in industrial contexts the opposite has occurred, labor productivity has increased at the same time (Fischer-Kowalski et al. 2014).

The most interesting debate, however, regards the causes of intensification. The most widely circulated hypothesis is based on the theories of Boserup (1965), according to which soil use intensity increases have been a response to population growth and a fall in available land. Many hypotheses are based on "demographic pressure" as the explanatory variable (Currie et al. 2015, 26). Nevertheless, in recent years the theses of Boserup and his followers have attracted growing criticism. Lambin et al. (2001) have fought against the idea that population and poverty are responsible for deforestation or that population growth drives unsustainable intensification, especially among small farmers. Other authors have limited the relationship between deforestation and population density to pre-industrial societies (Kaplan et al. 2009, 3031) or the viability of the Boserupian theory generally (Fischer-Kowalski et al. 2014). Ellis et al. (2013, 7980) suggest understanding intensification of land uses as an adaptive response, not only to demographic pressures but also to social and economic pressures. They do however warn that the intensification process is not linear, nor continuous or uniform. It can undergo steps forward and reversals. This explains, together with social, demographic and economic conditions, the existence of different land-use regimes. On the other hand, Kay and Kaplan (2015) maintain that the reasons for intensification are more complex than either the population or the social and economic variables. They suggest that a specific combination of a cluster of factors would explain land-use patterns. They described, based on archaeological records and other sources, the characteristic patterns of sub-Saharan Africa for 2500 years. According to the authors, these patterns reflect specific combinations of diet, technology, culture, subsistence or urbanization, which together lead to differentiated soil intensity patterns. Finally, some authors argue that land uses are a reflection of raw material flows, energy, population, information, etc., and call for new theoretical and methodological approaches to address their analysis (Friisa et al. 2015, 4).

In line with these latest contributions and our theoretical and methodological approach, particular configurations of land uses are more easily analyzed when they are understood as a specific metabolic design; productive intensification and specialization should be understood as a more general socioecological transition process (Fischer-Kowalski and Haberl 2007; Fischer-Kowalski et al. 2014; González de Molina and Toledo 2014). We must, therefore, address the factors underlying Spanish agriculture's intensification and productive specialization since 1900. These processes were the main drivers of the shift towards an industrial metabolic regime. The hypothesis we intend to verify is as follows: the agrarian population, a part of society and a center of metabolic activity, is at the intersection between the social and biophysical dimensions of AM; as such, the agrarian population forms its bond with the rest of society, providing certain quantities of animal and vegetal biomass and obtaining certain amounts of money in return in order to reproduce and, where appropriate, improve the fund elements. If the amount of monetary flow allows the agrarian population to easily reproduce itself, the surplus will be invested in improvements to the other funds or in increasing domestic consumption. But if the amount of monetary flow is insufficient for the agrarian population to reproduce itself based on the society's overall average exosomatic consumption levels, the most likely response is to intensify and/or specialize agricultural or livestock production or use more inputs, or all three at the same time. This can cause the reproduction cost of other funds to increase, bringing down household incomes, causing a vicious circle with increasingly intensive and specialized use of agroecosystems.

In this work, we also contend that unequal access and distribution of agricultural resources and, therefore, of the satisfiers to meet the historically changing needs of farmers has had a decisive influence on intensification and productive specialization. Unequal allocation of AM fund elements among the agrarian population (internal inequality) is not the only cause of inequality, as is usually assumed. It is also the result of unequal exchanges between the agrarian sector and other urban-industrial sectors (external inequality) over the last century. In fact, the market has been a historical means of transfer of income from the agricultural sector to other sectors of activity (Bernstein 2010; Hornborg 2011, 5). Inequality, both internal and external, deteriorated farmers' incomes, who compensated by developing various strategies that stimulated intensification and/or productive specialization, pushing towards a socio-ecological transition. In short, the hypothesis we wish to establish in this book is that intensification and specialization efforts were linked to attempts to compensate for varying drops in agricultural income and employment during the 20th century. There has been a negative impact on the four funds: destruction of employment, strong livestock imbalances, disproportionate increase in technical means of production and deterioration of the land's environmental quality and its associated resources. The congruent links between the funds were broken and the reproduction conditions of AM seriously compromised.

1.7 Sources and Methods

To make the biophysical analysis of Spanish agriculture operational, it was necessary to adapt and combine existing metabolic methodologies (EW-MEFA, MuSIASEM) and add further instruments. Our proposal integrates several aspects not included in these latter methodologies and that result from hybridization with Agroecology.

1.7.1 The Specificities of Agrarian Matabolism

First, in accordance with the importance we have attributed to the fund element of reproduction, special attention is paid to what happens inside agroecosystems. We thus developed a specific methodology for calculating net primary productivity (NPP). It shares some of the assumptions of the HANPP (Human Appropriation of Net Primary Productivity) but distinguishes itself in some pivotal aspects. The HANPP only takes into account a part of net primary production, mainly Domestic Extraction (DE), and not the whole. This leaves out a large part of the impacts generated by different metabolic arrangements on the agrarian environment. Our methodological proposal takes into account all the NPP, both aerial and root parts. From the AM perspective, the NPP is of interest as a whole since it supports trophic chains. NPP establishes the limits of the maintenance capacity of heterotrophic populations: all members of the animal kingdom (human population, domesticated animals, and wild fauna), the fungi, a large part of the bacteria and the archaea. Therefore, appropriation by human societies of NPP affects the maintenance of the rest of heterotrophic organism populations that depend on the same resources.

To perform accurate calculations, we had to consider the productivity of both the cultivation areas and those dedicated to pastures and forestry. When all the biomass produced in agroecosystems is considered, a somewhat distinct perspective from that usually adopted in Social Metabolism (SM) studies can be taken. SM studies tend to center on society and the resources it appropriates. The problem is that they leave aside the structure and functioning of ecosystems, whose healthy ecological status depends on the supply of services that society receives. In other words, from an agroecological perspective, levels and sustainability of DE also depend on the biomass *that is not extracted* thus remaining available in the ecosystems for its other heterotrophic components. A detailed guide on how the calculation was performed for Spain and how it can be calculated for other territories can be found in Annex I. A detailed description of the different sections into which it has been broken down is also provided. Finally, we took into account *all* the year's biomass produced in the territory under study, that is, the totality of actual NPP. This approach distinguishes itself from that of usual social metabolism studies (Schandl et al. 2002; Imhoff et al. 2004; Haberl et al. 2007), which only take into account the amount of harvested, reused or useful biomass.

A final difference should be highlighted referring to the way in which biomass should be accounted for. The standard methodology (E-W MEFA) is based on distinguishing fresh matter products from dry matter products (mainly pastures and forage plants) (Eurostat 2013), thus adding different kinds of weights. The most rigorous way to consider all types of biomass is to reduce them all to dry matter, as commonly done in specific studies on agriculture from a biophysical perspective (Krausmann et al. 2008a, b; Smil 2013). This avoids distortions produced by the different water contents among types of biomass, especially between pastures and crops (between 15 and 95%). Furthermore, this consideration is necessary to study, as in our case, the evolution of agricultural production. Crops with greater water content have become highly relevant (in horticultural production for example), to the detriment of other crops with lesser water content (cereals and legumes).

Secondly, in line with the Agroecology definition of agroecosystem, our AM proposition includes economic aspects that have been left out until now. This integration is performed via the study of the monetary flows that enter and leave the system. Society remunerates the agrarian sector for its products, whether agricultural, livestock or forestry products, and this remuneration constitutes the main inflow. Expenditures outside the sector, including expenses related to current means of production acquired on markets (commercial seeds and seedlings, that are not farm-reproduced nor exchanged between farmers, fuels and lubricants, fertilizers and amendments, feed, phytosanitary ware, material maintenance expenses, etc.) constitute the outflows. To quantify both flows, we used the macromagnitudes provided by Spain's national accounting and, more specifically, the agricultural sector accounts. The difference between both flows constitutes the gross value added (GVA) produced within agroecosystem boundaries. Agrarian income is obtained once depreciations, land taxes, fixed assets, or employed labor have been deducted, and the subsidies added together. This constitutes the flow of money with which to reproduce and improve, where appropriate, agrarian metabolism fund elements and, especially, the agrarian population. In this way, our AM proposal also integrates information flows that traditional methodologies do not contemplate.

Third, special attention is given to inequality in access and distribution of these monetary flows. Land tenure and means of production play a crucial role here. We already mentioned internal and external inequality: internal inequality was measured according to farm structures and the unequal distribution of agrarian income it introduces. To further refine this distribution, we distinguished between paid wages and entrepreneurial income. External inequality was measured through terms of trade with other sectors of activity and the evolution of agricultural income in constant values.

Fourth and last, our proposal makes it possible, unlike other methodologies, to measure both the social and environmental impacts generated by a specific metabolic arrangement on agroecosystems. The perspective we adopted here is as follows: an agroecosystem will be more sustainable if its fund elements are reproduced adequately through energy flows, materials, and information. This means that the agroecosystem's capacity to maintain biomass production in the long term, without relying on external inputs, is the chief manifestation of sustainable management. So far, methods for assessing agricultural sustainability have mainly relied on a battery of indicators of potential natural resource degradation problems, e.g., the "Framework to Assess Natural Resources Management Systems incorporating Sustainability Indicators" (*Marco para la Evaluación de Sistemas de Gestión de Recursos Naturales que Incorporan Indicadores de Sostenibilidad*, López-Ridaura et al. 2002) or the Sustainability Assessment of Farming and the Environment (SAFE) (van Cauwenbergh et al. 2007). However, their usefulness is limited because they do not reflect the functioning of the agroecosystem or describe the funds' reproduction mechanisms and, therefore, they do not reveal much about the processes that cause degradation or the interrelationships between them. Specific modes of calculation are detailed in the corresponding chapters.

1.7.2 Scale and Bounderies of the Study

Most metabolic studies have considered biomass as one of society's most essential materials. These studies have been conducted on a local or nationwide scale, covering all materials, biotic and abiotic, without specifically considering the agricultural sector as a unit of analysis. This work focuses on agroecosystems and its bounderies are those of any society's agrarian sector. It is characterized by the "production" of living organisms (biomass), a specificity shared by no other productive sector (except fishing). Given that the sector has required non-biotic inputs since the beginning of the twentieth century, our proposal of agrarian metabolism also considers both biotic and abiotic materials. This somewhat confounds the analysis as it is not possible to add apples to oranges, i.e., to mix biotic with abiotic factors without taking into account their different nature and indirect costs. But if we do not embrace this twofold analysis, it will be impossible to capture the significant change undergone by agrarian metabolism during its industrialization: from functioning almost exclusively on biotic materials at the start, it has become increasingly dependent on abiotic materials (metallic and non-metallic minerals and, above all, fossil fuels) for the manufacturing and operation of inputs. Future sustainable agriculture is faced with the challenge of minimizing the use of abiotic inputs and relying on biotic materials. Knowing the reasons for its progressive adoption in contemporary agriculture and its degree of dependence is decisive in planning the transition towards sustainable agrarian systems.

To summarize, AM includes the energy, materials and information exchanges contained within all agricultural activities in a given territory in order to produce vegetal or animal biomass for society, whether in the form of human or animal food, raw materials or fuels, and to provide basic ecosystem services, as for other ecosystems. Thus, any inputs not included in the list above, even if they originate—as in our case—in the country itself, are considered imports in the same way that all plant or animal products that go out of the territory are considered exports, even if they end up in local society. What is the reason for this rule? In Agroecology, the closure of agroecosystem cycles and their autonomy regarding the market or other external agroecosystems are known to be an essential attribute of agrarian sustainability (González de Molina and Guzmán Casado 2017).

1.7.3 Sources of Information

The analysis of the evolution of Spanish agriculture has required the collection and processing of a huge amount of data. Until the end of the 19th century, no statistical information had been collected on surface areas, yields, and production of various crops. We dispose of annual production series for the most important crops—cereals, legumes, grapes, and olives—dating from the end of the 19th century until the 1930s (GEHR 1991). The missing information was reconstructed from complementary sources, namely annual reports published on various topics by the Agronomic Advisory Board. In this way, we have been able to make estimations of Spanish production for three periods: 1900, 1910, and 1922. As of 1929, annual series of agricultural production were published: as from 1929 in the Statistical Yearbooks of Agricultural Productions and as from 1972 in the Yearbooks of Agricultural Statistics. Based on these yearly sources, we calculated five-year averages around the years 1933, 1940, 1950, 1960, 1970, 1980, 1990, 2000, and 2008. We also used five-year averages to calculate biomass exports and imports, based on foreign trade sources.

To draw up magnitudes in biophysical terms we also considered all agricultural residues that were utilized in some way (mainly livestock), such as straw and stubble, foliage and branches or tuber and horticultural residues. We used historical sources for straw, foliage, and shoots. For the remaining residue, we used the converters provided by the literature, that we compiled (Guzmán et al. 2014, available at www.seha.info). Land uses were reconstructed based on the same sources, from which we calculated the production of pastures and fallows. For firewood and wood, we used our own estimate (Infante-Amate et al. 2014b) and that of Iriarte and Iriarte-Goñi and Ayuda (2008).

We reconstructed the subsector of livestock productions based on livestock censuses. To do this, we used data on meat and milk productions available in the 1930s Spanish Statistical Yearbooks of Agricultural Production applying coefficients corresponding to previous censuses. For fertilizer production, livestock food needs, and average weight of different species, we applied converters elaborated from the 1891 and 1917 livestock records. The biophysical approach presents a methodological advantage in that it enables detecting and correcting possible problems in the sources. By comparing livestock food needs, animal labor needs for agriculture, and the availability of livestock feed, we were able to correct or validate the amount of livestock provided by the censuses and livestock counts. A detailed description of the sources and our methodological decisions can be found in Annex I.

References

Adriaanse A, Bringezu S, Hammond A, Moriguchi Y, Rodenburg E, Rogich D, Schütz H (1997). Resource flows: the material basis of industrial economies. World Resour Inst, Washington

Altieri MA (1989) Agroecology: the science of sustainable agriculture. Westview Press, Boulder, Colorado

Ayres RU (2007) On the practical limits to substitution. Ecol Econ 61:115–128

Bernstein H (2010) Class dynamics of Agrarian change. Fernwood Publishing, Halifax, N. S

Boserup E (1965) The conditions of agricultural growth: the economics of agrarian change under population pressure. Aldine, Chicago, Ill

Bulatkin GA (2012) Analysis of energy flows in agroecosystems. Herald Russian Acad Sci 82(4):326–334

Burkhard B, Fath BD, Müller F (2011) Adapting the adaptive cycle: hypotheses on the development of ecosystem properties and services. Ecol Model 222:2878–2890

Cook S (1973) Zapotec stoneworkers: the dynamics of rural simple commodity production in modern mexican capitalism. Univesity Press of América, Mexico

Cornell S (2010) Valuing ecosystem benefits in a dynamic world. Climate Res 45:264–272

Costanza R (2012) Ecosystem health and ecological engineering. Ecol Eng 45:24–29

Cunfer G, Krausmann F (2009) Sustaining soil fertility: agricultural practice in the old and new worlds, Glob Envir 4:8–47. http://www.environmentandsociety.org/node/4791

Currie TE, Bogaard A, Cesaretti R, Edwards NR, Francois P, Holden PB, Hoyer D, Korotayev A, Manning J, Moreno Garcia JC, Oyebamiji OK, Cameron P, Turchin P, Whitehouse H, Williams A (2015) Agricultural productivity in past societies toward an empirically informed model for testing cultural evolutionary hypotheses. Cliodynamics 6(1):24–56

de Groot RS, Wilson MA, Boumans RMJ (2002) A typology for the classification, description and valuation of ecosystem functions, goods and services. Ecol Econ 41:393–408

de Groot R, van der Perk J, Chiesura A, van Vliet A (2003) Importance and threats determining factors for criticality of natural capital. Ecol Econ 44:187–204 de Groot R, Alkemade R, Braat L, Hein L, Willemen L (2010) Challenges in integrating the concept of ecosystem services and values in landscape planning, management and decision making. Ecol Complex 7:260–272

Ekins P, Simon S, Deutsch L, Folke C, de Groot R (2003) A framework for the practical application of the concepts of critical natural capital and strong sustainability. Ecol Econ 44:165–185

Ellis EC, Kaplan JO, Fuller DQ, Vavrus S, Goldewijk KK, Verburg PH (2013) Used planet: a global history. PNAS 110(20):7978–7985

Eurostat (2013) Economy wide material flow accounts (EW-MFA): Compilation guide 2013. Luxemburgo: European Statistical Office

Fath BD, Jørgensen SE, Patten BC, Straškraba M (2004) Ecosystem growth and development. Biosystems 77:213–228

Federico G (2009) Feeding the world. Eco Hist Agric 1800–2000. Princeton. Princeton University Press

Fischer-Kowalski M (1998) Society's metabolism: the intellectual history of materials flow analysis, part I, 1860–1970. J Ind Ecol 2:61–77

Fischer-Kowalski M (2002) Exploring the history of industrial Metabolism. In: Ayres RU, Ayres LW (eds) A handbook of industrial ecology. Edward Elgar Publishing, Cheltenham, UK; Northampton MA, USA, pp 16–26

Fischer-Kowalski M (2011) Analyzing sustainability transitions as a shift between sociometabolic regimes. Environ Innov Soc Transit 1(1):152–159

Fischer-Kowalski M, Haberl H (2007) Socioecological transitions and global change: trajectories of social metabolism and land use. Institute of Social Ecology, Edward Elgar Publishing, Vienna, Austria

Fischer-Kowalski M, Haberl H (1997) Tons, joules, and money: Modes of production and their sustainability problems. Soc Nat Resour 10 (1):61–85

Fischer-Kowalski M, Rotmans J (2009) Conceptualizing, observing, and influencing social-ecological transitions. Ecol Soc 14(2), art. 3

Fischer-Kowalski M, Krausmann F, Pallua I (2014) A sociometabolic reading of the anthropocene: modes of subsistence, population size and human impact on earth. Anthropocene Rev 1(1):8–33

Foley JA, Defries R, Asner GP, Barford C, Bonan G, Carpenter SR, Chapin FS, Coe MT, Daily GC, Gibbs HK, Helkowski JH, Holloway T, Howard EA, Kucharik CJ, Monfreda C, Patz JA, Prentice IC, Ramankutty N, Snyder PK (2005) Global consequences of land use. Science 309(5734):570–574

Folke C, Jansson Å, Rockström J, Olsson P, Carpenter SR, Chapin FS, Crepín AS, Daily G, Danell K, Ebbesson J, Elmqvist T, Galaz V, Moberg F, Nilsson M, Österblom H, Ostrom E, Persson Å, Peterson G, Polasky S, Steffen W, Walker B, Westley F (2011) Reconnecting to the biosphere. Ambio 40:719–738

Friisa C, Østergaard Nielsen J, Otero I, Haberl H, Niewöhner J, Hostert P (2015) From teleconnection to telecoupling: taking stock of an emerging framework in land system science. J Land Use Sci 11(2):131–153. https://doi.org/10.1080/1747423X.2015.1096423

Garcia-Ruiz R, González de Molina M, Guzmán G, Soto D, Infante-Amate J (2012) Guidelines for Constructing Nitrogen, Phosphorus, and Potassium Balances in Historical Agricultural Systems. J Sustain Agric 36(6):650–682

Garrabou R, González de Molina M (2010) La reposición de la fertilidad en los sistemas agrarios tradicionales. Icaria, Barcelona

Garzón Casado B, Iniesta-Arandia I, Martín-López B, García-Llorente M, Montes C (2011) Entendiendo las relaciones naturaleza y sociedad de dos cuencas hidrográficas del sureste semiárido andaluz desde la historia socio-ecológica. VII Congreso Ibérico sobre Gestión y Planificación del Agua 'Ríos Ibéricos +10': mirando al futuro tras 10 años de DMA. Talavera de la Reina, España, pp 16–19 de febrero de 2011

GEHR (Grupo de Estudios de Historia Rural) (1991) Estadísticas Históricas de la producción agraria española, 1859–1935. Ministerio de Agricultura Pesca y Alimentación, Madrid

Georgescu-Roegen N (1971) The entropy law and the economicprocess. Harvard University Press, Cambridge

Giampietro M, Bukkens S (2015) Analogy between Sudoku and the multi-scale integrated analysis of societal metabolism. Ecol Inf 26(1):18–28

Giampietro M Mayumi K, Sorman AH (2010) Assessing the quality of alternative energy sources: energy return on the investment (EROI), the metabolic pattern of societies and energy statistics. In: Working papers on environmental sciences, ICTA. Barcelona, Spain

Giampietro M, Mayumi K (2000) Multiple-scale integrated assessment of societal metabolism: introducing the approach. Popul Environ 22:109–154

Giampietro M, Mayumi K, Ramos-Martin J (2008) Multi-scale integrated analysis of societal and ecosystem metabolism (MUSIASEM). An outline of rationale and theory. In: Document de treball, Departament d'Economia Aplicada de la Universitat Autònoma de Barcelona. Barcelona

Giampietro M, Mayumi K, Sorman AH (2011) The metabolic pattern of societies: where economists fall short. Routledge, London

Giampietro M, Aspinallis RJ, Ramos-Martin J, Bukken SGF (eds) (2014) Resource accounting for sustainability assessment: the nexus between energy, food, water and land use. Routledge, London

Gierlinger S, Krausmann F (2012) The physical economy of united states of America. J Ind Ecol 16(3):365–377

Gliessman SR (1998) Agroecology. Ecological processes in sustainable agriculture. Lewis Publishers (CRC Press), Boca Ratón

Gliessman SR (2007) Agroecología: promoviendo una transición hacia la sostenbilidad. Ecosistemas 16(1):13–23

González de Molina M (2010) A guide to studying the socio-ecological transition in European agriculture. Documentos de Trabajo de la Sociedad de Estudios de Historia Agraria, nº 10–06

González de Molina M, Guzmán Casado GI (2006) Tras los pasos de la insustentabilidad. Agricultura y Medio ambiente en perspectiva histórica (siglos XVIII-XX). Editorial Icaria, Barcelona

González de Molina M, Guzmán Casado G (2017) Agroecology and ecological intensification. A discussion from a metabolic point of view. Sustain 9(1):1–19

González de Molina M, Toledo V (2011) Metabolismos, naturaleza e historia. Una teoría de las transformaciones socio-ecológicas. Icaria, Barcelona

González de Molina M, Toledo V (2014) Social metabolism: a theory on socio-ecological transformations. Springer, New York

Grešlová P, Gingrich S, Krausmann F, Chromý P, Jančák V (2015) Social metabolism of Czech agriculture in the period 1830–2010. AUC Geogr 50(1):23–35

Guzmán Casado GI, González de Molina M (2000) Introducción a la Agroecología como desarrollo rural sostenible. Mundi-Prensa, Madrid

Guzmán Casado GI, González de Molina M (2009) Preindustrial agriculture versus organic agriculture: the land cost of sustainability. Land Use Policy 26:502–510

Guzmán Casado GI, González de Molina M, Alonso AM (2011) The land cost of agrarian sustainability: an assessment. Land Use Policy 28:825–835

Guzmán GI, González de Molina M (2017) Energy in agroecosystems: a tool for assessing sustainability. CRC Press, Boca Raton, FL

Guzmán GI, Aguilera E, Soto D, Cid A, Infante-Amate J, García-Ruiz R, Herrera A, Villa I, González de Molina M (2014) Methodology and conversión factors to estimate the net primary productivity of 112 historical and contemporary agro-ecosystems (I). Documento de Trabajo de la Sociedad Española de Historia Agraria, nº 14–06. Disponible en: www.seha.info

Haberl H (2001) The energetic metabolism of societies. I: accounting concepts. J Ind Ecol 5:11–33

Haberl H, Erb KH, Krausmann F, Gaube V, Bondeau A, Plutzar C, Gingrich S, Lucht W, Fischer-Kowalski M (2007) Quantifying and mapping the human appropriation of net primary production in earth's terrestrial ecosystems. Proc Nat Acad Sci 104(31):12942–12947

Häyhäa T, Franzese PP (2014) Ecosystem services assessment: a review under an ecological-economic and systems perspective. Ecol Model 289:124–132

Heller MC, Keoleian GA (2003) Assessing the sustainability of the US food system: a life cycle perspective. Agric Syst 76:1007–1041

Ho MW (2013) Circular thermodynamics of organisms and sustainable systems. Systems 1:30–49

Ho MW, Ulanowicz R (2005) Sustainable systems as organisms? BioSystems 82:39–51

Hornborg A (2011) Global ecology and unequal exchange: fetishism in a zero-sum world. Routledge, London

Iriarte-Goñi I, Ayuda MI (2008) Wood and industrialization. Ecol Econ 65(1):177–186

Infante-Amate J (2011) Los temporeros del olivar: una aproximación al estudio de las migraciones estacionales en el sur de España (siglos XVIII-XX). Revista de Demografía Histórica 29(2):87–118

Infante-Amate J, Aguilera E, González de Molina M (2014a) La gran transformación del sector agroalimentario español. Un análisis desde la perspectiva energética (1960–2010). Documento de Trabajo de la Sociedad de Estudios de Historia Agraria 14–03. https://ideas.repec.org/p/seh/wpaper/1403.html

Infante-Amate J, Soto D, Iriarte Goñi I, Aguilera E, Cid A, Guzmán G, García Ruiz R, González de Molina M (2014b) La producción de leña en España y sus implicaciones en la transición energética. Una serie a escala provincial (1900–2000). Sociedad Española de Historia Agraria, Documento de Trabajo nº 14-16. http://econpapers.repec.org/paper/ahedtaehe/1416.htm

Infante-Amate J, González de Molina M, Vanwalleghem, T, Soto Fernández D, Gómez Calero JA (2014c) Reconciling Boserup with Malthus. Agrarian change and soil degradation in olive orchards in Spain (1750–2000). In: Fischer-Kowalski M, Reenberg A, Schaffartzik A, Mayer A (eds) Ester Boserup's legacy on sustainability: orientations for contemporary research. Springer, New York, pp 99–116. https://doi.org/10.1007/978-94-017-8678-2_7

Infante-Amate J, Aguilera E, Palmeri F, Guzmán G, Soto D, García-Ruiz R, de Molina M González (2018) Land embodied in Spain's biomass trade and consumption (1900–2008): historical changes, drivers and impacts. Land Use Policy 78:493–502

Infante-Amate J, Soto, D, Aguilera, E, García Ruiz R, Guzmán, G, Cid A, González de Molina M (2015: The Spanish transition to industrial metabolism long-term material flow analysis (1860–2010). J Ind Ecol 19(5):866–876. https://doi.org/10.1111/jiec.12261

IPES-FOOD (International Panel of Experts on Sustainable Food Systems) (2016) From uniformity to diversity: a paradigm shift from industrial agriculture to diversified agroecological systems www.ipes-food.org

Jepsen MR, Kuemmerle T, Müller D, Erb K, Verburg PH, Haberl H, Vesterager JP, Andrič M, Antrop M, Austrheim G, Björn I, Bondeau A, Bürgi M, Bryson J, Caspar G, Cassar LF, Conrad E, Chromý P, Daugirdas V, van Eetvelde V, Elena-Rosselló R, Gimmi U, Izakovicova Z, JANˇCÁK V, Jansson U, Kladnik D, Kozak J, Konkoly-Gyuró E, Krausmann F, Mander U, McDonagh J, Pärn J, Niedertscheider M, Nikodemus O, Ostapowicz K, Pérez-Soba M, Pinto-Correia T, Ribokas G, Rounsevell M, Schistou D, Schmit C, Terkenli TS, Tretvik AM, Trzepacz P, Vadineanu A, Walz A, Zhllima E, Reenberg A (2015) Transitions in European land-management regimes between 1800 and 2010. Land Use Policy 49:53–64

Jørgensen SE, Fath BD (2004) Application of thermodynamic principles in ecology. Ecol Complex 1:267–280

Jørgensen SE, Fath BD, Bastianoni S, Marques JC, Müller F, Nielsen SN, Tiezzi E, Ulanowicz RE (2007) A new ecology: systems perspective. Elsevier Press, Amsterdam

Kaplan JO, Krumhardt KM, Zimmermann N (2009) The prehistoric and preindustrial deforestation of Europe. Quatern Sci Rev 28(27-28):3016–3034

Kay AU, Kaplan JO (2015) Human subsistence and land use in sub-Saharan Africa, 1000 BC to AD 1500: a review, quantification, and classification. Anthropocene 9:14–32

Kovanda J, Hak T (2011) Historical perspectives of material use in Czechoslovakia in 1855–2007. Ecol Ind 11(5):1375–1384

Krausmann F (2004) Milk, manure, and muscle power. Livestock and the transformation of preindustrial agriculture in Central Europe. Hum Ecol 32:735–772

Krausmann F, Erb KE, Gringrich S, Lauk C, Haberl H (2008a) Global patterns of socioeconomic biomass flows in the year 2000: a comprehensive assessment of supply, consumption and constraints. Ecol Econ 65:471–487

Krausmann F, Fischer-Kowalski M, Schandl H, Eisenmenger N (2008b) The global sociometabolic transition: past and present metabolic profiles and their future trajectories. J Ind Ecol 12(5–6):637–656

Krausmann F, Gingrich S, Nourbakhch-Sabet R (2011) The metabolic transition in Japan. J Ind Ecol 15(6):877–892

Krausmann F, Erb KH, Gingrich S, Haberl H, Bondeau A, Gaube V, Lauk C, Plutzar C, Searchinger TD (2013) Global human appropriation of net primary production doubled in the 20th century. PNAS 110(25):10324–10329. http://dx.doi.org/10.1073/pnas.1211349110

Kusova, P, Gringrich, S, Krausmann F (2008 Long term changes in social metabolism and land use in Czechoslovakia, 1830–2000: An energy transition under changing political regimes. Ecol Econ 68(1–2):394–407. https://doi.org/10.1016/j.ecolecon.2008.04.006

Lachman DA (2013) A survey and review of approaches to study transitions. Energy Policy 58:269–276

Lambin EL, Turner BL, Geist HJ, Agbola SB, Angelsen A, Bruce JW, Coomes OT, Dirzo R, Fischer G, Folke C, George PS, Homewood K, Imbernon J, Leemans R, Li X, Moran EF, Mortimore M, Ramakrishnan PS, Richards JF, Skanes H, Steffen W, Stone GD, Svedin U, Veldkamp TA, Vogel C, Xu J (2001) The causes of land-use and land-cover change: moving beyond the myths. Glob Environ Change 11(4):261–269

Lambin EF, Meyfroidt P (2010) Land use transitions: Socio-ecological feedback versus socio-economic change. Land Use Policy 27(2):108–118

López-Ridaura M, Masera O, Astier M (2002) Evaluating the sustainability of complex socio-environmental systems. The MESMIS framework. Ecol Ind 2(1-2):135–148

Imhoff ML, Bounoua L, Ricketts T, Loucks C, Harriss R, Lawrence WT (2004) Global patterns in human consumption of net primary production. Nat 429 (6994):870–873

Matthews E (2000) The weight of nations: material outflows from industrial economies. World Resources Institute, Washington DC

Millennium Ecosystem Assessment (2005) Ecosystems and human well-being: synthesis. Island Press, Washington, DC

Mustard JF, Defries RS, Fisher T, Moran EF (2004) Land-use and land-cover change pathways and impacts. In: Gutman G, Janetos AC, Justice CO, Moran EF, Mustard JF, Rindfuss RR, Skole D, Turner BL II, Cochrane MA (eds) Land change science. observing, monitoring and understanding trajectories of change on the earth's surface. Kluwer Academic Publishers, Boston, pp 411–430

Pagiola S, Platais G (2002) Payments for environmental services, environment strategies, No. 3. The World Bank, Washinton, DC

Prigogine I (1947) Etude thermodynamique des phénomènes irreversibles. Dunod, París

Prigogine I (1962) Non-equilibrium statistical mechanics. Interscience, New York

Prigogine I (1978) Time structure and fluctuations. Science 201:777–785

Risku-Norja H (1999) The total material requirement-concept applied to agriculture: a case study from Finland. Agric Food Sci Finland 8:393–410

Risku-Norja H, Mäenpää I (2007) MFA model to assess economic and environmental consequences of food production and consumption. Ecol Econ 60(4):700–711. https://doi.org/10.1016/j.ecolecon.2006.05.001

Rudel TK, Schneider L, Uriarte M, Turner BL II, Defries R, Lawrence D, Geoghegan J, Hecht S, Ickowitz A, Lambin EF, Birkenholtz T, Baptista S, Grau R (2009) Agricultural intensification and changes in cultivated areas, 1970–2005. PNAS 106(49):20675–20680

Schandl H, Schulz N (2002) Changes in the United Kingdom's natural relations in terms of society's metabolism and land-use from 1850 to the present day. Ecol Econ 41(2):203–221

Schandl H, Grünbühel C, Haberl H, Weisz H (2002) Handbook of physical accounting. Measuring bio-physical dimensions of socio-economic activities MFA-EFA-HANPP. In: Social ecology, Working Paper 73. Vienna. IFF

Scheidel A, Sorman AH (2012) Energy transitions and the global land rush: ultimate drivers and persistent consequences. Glob Environ Change 22(3):559–794

Schröter M, Barton DN, Remme RP, Hein L (2014) Accounting for capacity and flow of ecosystem services: a conceptual model and a case study for Telemark, Norway. Ecol Ind 36:539–551

Sieferle RP (2001) The subterranean forest: energy systems and the industrial revolution. White Horse Press, Cambridge, UK

Sieferle RP (2011) Cultural evolution and social metabolism. Geogr Ann Ser B Hum Geogr 93(4):315–324

Singh, SJ, Krausmann F, Gingrich S, Haberl H, Erb KH, Lanz P (2012) India's biophysical economy, 1961–2008. Sustainability in a national and global context. In: Ecological economics, No 76, pp 60–69. https://doi.org/10.1016/j.ecolecon.2012.01.022

Smil V (2011) Harvesting the biosphere: the human impact. Popul Dev Rev 37(4):613–636. https://doi.org/10.1111/j.1728-4457.2011.00450.x

Smil V (2013) Harvesting the biosphere: what we have taken from nature. The MIT Press, London

Swannack TM, Grant WE (2008) Systems Ecology. In: Jørgensen SE, Fath BD (eds) Encyclopedia of ecology. Elsevier, Oxford, pp 3477–3481

Swanson GA, Bailey KD, Miller JG (1997) Entropy, social entropy and money: a living systems theory perspective. Syst Res Behav Sci 14(1):45–65

Tello E, Garrabou, R, Cussó X, Olarieta JR, Galán E (2012) Fertilizing methods and nutrient balance at the end of traditional organic agriculture in the Mediterranean bioregion: Catalonia (Spain) in the 1860s. Hum Ecol 40(3):369–383. http://dx.doi.org/10.1007/s10745-012-9485-4

Toledo V (1993) La racionalidad ecológica de la producción campesina. In: Sevilla E, González de Molina M (eds) Ecología, campesinado e Historia. La Piqueta, Madrid, pp 197–218

Toledo V, Barrera-Bassols N (2008) La memoria biocultural. La importancia agroecológica de las sabidurías tradicionales. Icaria, Barcelona

Ulanowicz RE (1983) Identifying the structure of cycling in ecosystems. Math Biosci 65:210–237

Ulanowicz RE (2004) On the nature of ecodynamics. Ecological Complexity 1:341–354

van Cauwenbergh N, Biala K, Bielders C (2007) SAFE—a hierarchical framework for assessing the sustainability of agricultural systems. Agr Ecosyst Environ 120(2–4):229–242

van der Bergh J, Bruinsma F (2008) Managing de transition to renewable energy. Theory and practice from local, regional and macro perspectives. MPG Books Ltd., Bodmin

Weisz H, Krausmann F, Amann C, Eisenmenger N, Erb KH, Hubacek K, Fischer-Kowalski M (2006) The physical economy of the European Union: crosscountry comparison and determinants of material consumption. Ecol Econ 58(4):676–698

Widgren M (2007) Precolonial landesque capital: a global perspective. In: Hornborg A, Mcneill JR, Martínez-Alier J (eds) Rethinking environmental history: world-system history and global environmental change. Altamira Press, Lanhan, pp 61–78

Wirsenius S (2003) The biomass metabolism of the food system. A model-based survey of the global and regional turnover of food biomass. J Ind Ecol 7(1):47–80

2

Animal and Vegetal Biomass Flows of Spanish Agroecosystems

In this chapter, we review the animal and vegetal biomass flows of Spanish agroecosystems based on the evolution of net primary productivity, its different components and livestock production. This enables us to examine the performance of biophysical, land and livestock fund elements as well as the changes that took place throughout the twentieth century to meet society's biomass requirements. As explained in the introduction, our analysis moves away from the standard narrative on Spain's agricultural sector and its contribution to economic growth. A historiographic review of the main transformations in Spanish agriculture since 1900s was, however, presented in the first section of this chapter. Based on both traditional discourse and new historiographical contributions, we determine the milestones of change and then submit them to the discussion. The following sections aim at laying the foundations of an alternative narrative based on the study of Spanish agroecosystem productivity since 1900.

The second section is dedicated to changes in types of land use since 1900 until now. We discuss whether Spain's evolution really did follow the so-called forest transition and its associated theory of productive intensification. We later analyze the NPP of agroecosystems, breaking it down into categories to verify whether production efforts affected their health. We then center on the domestic extraction entity over time and its progressive concentration in cultivated lands. Subsequently, we focus on these lands' specialization in specific Mediterranean crops oriented towards both national and international markets. Next, we turn our attention to livestock, a critical activity in recent decades, focusing first on its evolution before examining livestock production. To finish, we propose a holistic understanding of the biophysical implications of the sector's industrialization.

2.1 Traditional Historiographical Narrative of Agricultural Transformations During the Twentieth Century

When using agricultural production increase as the main indicator and applying conventional measurements, i.e. in tons of fresh products or in money, the outlook is frankly positive. Spanish agriculture strongly intensified and multiplied its production by 3.3 between 1900 and 2008, reaching a peak at the start of this century turning out over 104.2 million tons, almost four times more than in 1900. These figures emerge from the analysis of agricultural production measured in fresh matter, without compensating the distortion relating to the water content of agricultural products. According to our own data collected in Table 2.1, all crops increased their yields because cultivated areas decreased in size, as we will see later. Cereals, fruits and vegetables, forage plants and olive groves grew particularly sharply, reflecting Spain's progressive specialization.

Livestock production grew even further, multiplying by a factor of 8.2 over the same period (Table 2.2). It also progressively reached a heavier weight, from 7% of

Table 2.1 Evolution of agricultural production in million tonnes (Mt) of fresh matter

	1900	1960	1990	2000	2008	2015*
Cereals	6.24	8.62	18.83	21.24	23.00	20.14
Legumes	0.50	0.90	0.24	0.33	0.23	0.50
Vegetables and tubers	5.37	9.77	16.77	15.95	14.99	17.09
Fruit trees	1.85	3.66	8.59	9.68	10.20	10.81
Vineyard	3.78	2.63	5.33	6.14	5.40	5.80
Olive groves	1.09	1.95	2.93	5.23	6.26	7.35
Industrial and others	1.27	4.43	9.39	9.42	6.73	4.92
Forage crops	8.40	25.52	35.61	36.34	28.33	25.41
Total	28.50	57.48	97.70	104.32	95.14	92.04
	1900	1960	1990	2000	2008	2015
Cereals	100	138	302	341	369	323
Legumes	100	179	49	66	46	100
Vegetables and tubers	100	182	312	297	279	318
Fruit trees	100	197	463	522	550	584
Vineyard	100	70	141	162	143	153
Olive groves	100	179	270	482	577	735
Industrial and others	100	349	739	741	529	387
Forage crops	100	304	424	433	337	302
Total	100	202	343	366	334	323

Source Yearbooks of agricultural statistics and own calculations
*The 2015 data correspond to provisional estimates based on the Yearbook of Agricultural Productions (MAPAMA 2016)

agricultural production to almost 17%. Meat and eggs were the major players in this unusual growth of livestock production. Milk production increased fivefold, despite already being the main livestock product in 1900. On the other hand, traditional products associated with extensive breeding such as wool declined throughout the century and became residual (Table 2.2).

Growth in agricultural production seems even more remarkable when measured in monetary terms. We do not dispose of a Final Agricultural Production (FAP) series covering our entire period under study. We do dispose of sector accounts drawn up by the Ministry of Agriculture since the early fifties, but they use three different methodologies, as we will see in Chap. 5. By connecting the annual values of the three series (see Chap. 4), we can conclude that the value of FAP multiplied by 5.1 (Graph 2.1a) at constant prices between 1953 and 2003. Furthermore, agricultural production in Spain continued to grow until the turn of the century (Graph 2.1b), unlike in the rest of Western Europe where growth came to a halt at the end of the eighties (Martín-Retortillo and Pinilla 2015). In Spain, growth was interrupted as of 2003. Between that year and 2014, the Agricultural Industry's production dropped by 6.4% in real terms (Graph 2.1b). At the beginning of the fifties, Spain was still subject to Franco's dictatorship's autocratic economic policy anyway. Consequently, agricultural production was low in relation to that achieved during the Second Republic. Despite this, the figures reveal spectacular growth that must be attributed to the economic effects of agriculture's industrialization. According to estimations by Prados De La Escosura (2003), Spanish agriculture GDP multiplied by 3.6 between 1900 and 1990.

Table 2.2 Evolution of livestock production, in thousands of tons of fresh matter

	1900	1960	1990	2000	2008	2015*
Meat	517	971	5.200	7.427	8.005	8.701
Milk	1.355	3.252	6.737	7.196	7.308	7.865
Eggs	54	197	558	565	656	639
Wool	26	31	29	31	28	23
Honey and wax	6	8	24	34	34	35
Total	1.959	4.460	12.548	15.253	16.031	17.264
	1900	1960	1990	2000	2008	2015
Meat	100	188	1.007	1.438	1.550	1.682
Milk	100	240	497	531	539	580
Eggs	100	366	1.034	1.048	1.217	1.183
Wool	100	117	109	119	106	88
Honey and wax	100	130	377	528	517	583
Total	100	228	641	779	818	881

Source Yearbooks of agricultural statistics and own calculations
*The 2015 data correspond to provisional estimates based on the Yearbook of Agricultural Productions (MAPAMA 2016)

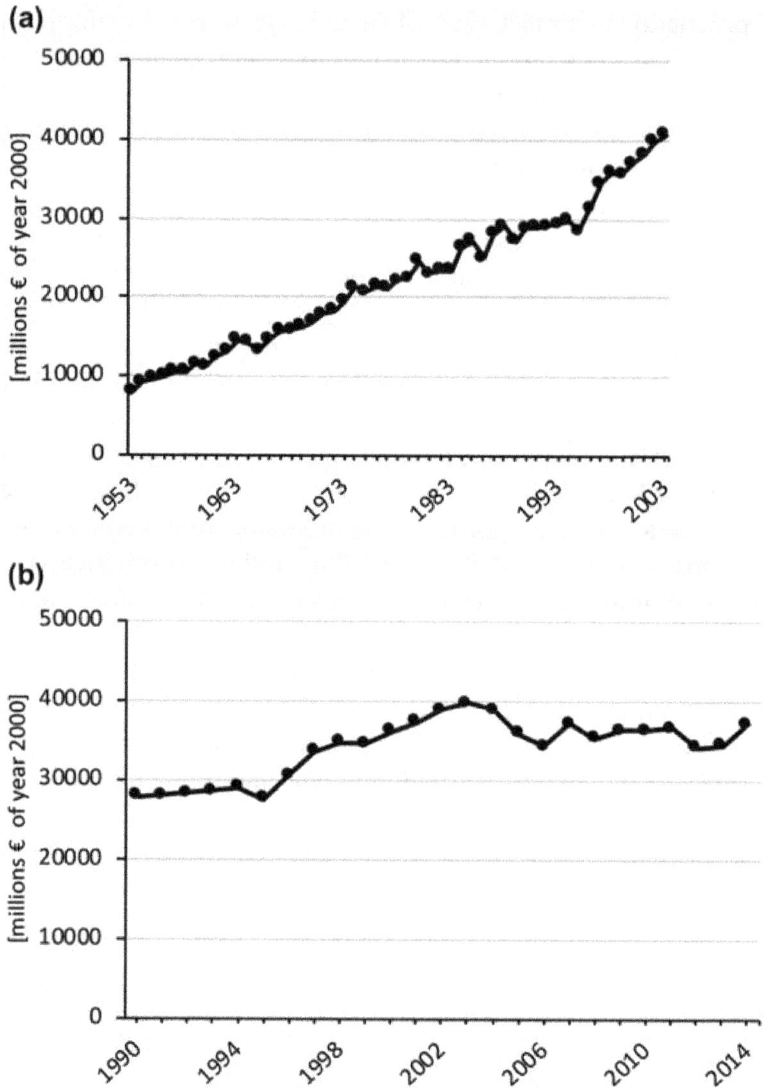

Graph 2.1 Value of agricultural production (**a**) and production of the agricultural branch (new accounting system) (**b**), in millions of constant year 2000 euros

The upward trend can be broken down into four different periods. The first period corresponds to the years between 1900 and 1936, when production grew by 52%, confirming findings from previous studies (GEHR 1991; Simpson 1997) according to which this period was one of the agricultural growth and sector "modernization". Between 1900 and 1931 land productivity grew by 1.2% per year and labor productivity by 1.9% (Gallego Martínez 1993, modified by Soto 2006). Both productivities could have increased more, but according to traditional historiographical narrative, there were insufficient incentives to spread chemical fertilizers or mechanization. However, cereal production, for which rainfed lands offered low yields and little competitiveness, maintained itself and even increased thanks to the domestic demand of a growing population. Spain became the biggest wine exporting country and the second biggest oil-exporting country in the 1890s after Italy (Infante and Parcerisas 2013; Infante-Amate 2014). But this avenue would not lead to continued expansion in export markets, because of stagnant demand and powerful competitors in the sector (Pinilla and Serrano 2008; Infante-Amate 2014). Spain began to specialize in fruits and vegetables. This activity was to play a critical role decades later, but it was burdened at the time by scarce public investment in irrigation as well as fresh export food transport difficulties. In comparative terms, the factor productivity evolved in parallel, similar in size to that of other European countries (Bringas 2000; Clar et al. 2016, 183).

The traditional narrative generally describes this period as one of slow or insufficient growth and "modernization". The Spanish economy's sluggish industrialization meant that urban demand for labor and agricultural products was not sufficiently stimulated, which in turn failed to drive an in-depth modernization of agriculture. Moreover, high capital costs and the abundance of labor in the countryside failed to trigger speedier transformations (Gallego 2001). Nevertheless, it is important to remember that with a few exceptions, demand for agricultural products during this period came from the domestic market. Undoubtedly, population growth and the satisfaction of food and energy demands were decisive. The overcoming of the depression at the turn of the century and agricultural growth itself must have nevertheless increased the purchasing power of a large share of the rural population, substantially improving its diet during the first third of the twentieth century (Cussó 2005; González de Molina et al. 2014).

The second period covers the first two decades of Franco's dictatorship. According to an increasingly widespread historiographical consensus, this period can be qualified as simply tragic. It began with the end of the Civil War, the regime's international isolation and its autarchic economic policy. Spanish agriculture's "modernization" experienced a sharp reversal, widening the gap with other European countries. The lack of chemical fertilizers, due to the shortage of foreign currency and obstacles to international trade, partly explain the decline in yields. Livestock activity was also affected by the decreasing availability of animal feed, which, in turn, aggravated the shortage of fertilizers—organic fertilizers in this case (Fernández Prieto 2007). Powerful State intervention, through the National Wheat Service (*Servicio Nacional de Trigo*), guaranteed low prices to avoid wage increase sand ended up subjecting agricultural policy to industrial policy, providing poorly remunerative prices that failed to stimulate agricultural production growth (Barciela 1986; Baricela and López 2003). Falling agricultural and livestock production caused shortages in the domestic market and a sharp drop in food availability led to spells of hunger and malnutrition affecting different segments of the population, an unprecedented chapter in the country's recent history (González de Molina et al. 2013). The data we provide and present later indisputably confirms this.

Things would begin to change in the fifties, once autarchic policy had proved to be unviable. At the end of that decade, coinciding with the start of the so-called "Stabilization Plan" (1959), our third period of study began, i.e. Spanish agriculture's last industrialization phase. Liberalization led to the sector's rapid transformation. It was financed by foreign investment, remittances from emigrants and income from tourism. Yields per unit area multiplied thanks to the use of the green revolution's complete package: improved seeds, synthetic chemical fertilizers, phytosanitary treatments, and irrigation. Production quickly recovered, doubling between 1900 and 1960, tripling by 1980 and almost quadrupling by the end of the century (Table 2.1). Cereal production, so far on the decline, forewarning the abandonment of traditional agriculture, promptly rebounded thanks to the growing demand of feed grains.

The expansion of irrigation surfaces played a major part thanks to the construction of large hydraulic infrastructures. According to Cazcarro et al. (2014), two-thirds of Spanish agricultural production was obtained from irrigated lands. Changes in the agricultural policy encouraged the growth of livestock, leading to an increase in its share of final agricultural production, eventually reaching one-third of total production (Clar et al. 2016, 190). The support given to the imports of feed and to intensive livestock farming based on little or no land allowed to maintain incentives for traditional production, especially wheat. Agrarian trade balances became increasingly negative, although this fact was compensated by the loss of the relative weight of the agricultural sector in the overall trade balance and Spain's economy generally.

Final agricultural production also grew at unprecedented rates, while its share of total employment and Spanish GDP declined. This behavior is viewed positively, as it is characteristic of agricultural growth and industrialization, and common to most economically advanced countries. During the first decades of industrialization, the sector was driven by the green revolution's new technologies and in the eighties, the impetus was given by resulting factor productivity gains. Despite this, agricultural incomes are still today on the decline, compensated by CAP aid. Major productivity improvements facilitated the transfer of labor from rural areas to urban-industrial activities. Between 1960 and 2008, the active agricultural population fell from almost five million to just over one million individuals. This population transfer from the countryside to cities has also been judged as a highly positive contribution of the agricultural sector to Spain's economic growth.

Overall, the role of agriculture was redefined during its industrialization process. Agriculture shifted from providing food and labor—its basic functions during the first two periods under study—to becoming mainly a supplier of raw materials to an agri-food complex, a business made up of the agro-industry, food distribution, and agricultural input manufacturers. This shift is considered to be another notable contribution to Spain's economic growth. In fact, the agri-food trade has been generating a surplus in monetary terms since the beginning of the 1980s, based on quality differentiation and greater external competitiveness. The sector's transformation also sparked other concerns relating to environmental management and rural development, that is no longer directly linked to agricultural activity. Public administrations have become worried about nature conservation and rural depopulation, though these issues are only vaguely related to the model of agricultural growth that has been followed since the late fifties.

2.2 The Evolution of Land Uses

A biophysical study of this evolution, however, projects a less comfortable picture and gives us more reason to be critical. We can begin with changes in net primary productivity that originated, in turn, in modified land uses and land productivity differences.

In Graph 2.2 we summarize the changes in land uses based on six differentiated categories: unproductive lands, pastures, meadows, coppice, timberlands, and cultivated areas. A detailed account of the methodology used to estimate forestland mass can be found in Infante-Amate et al. (2014). In this work, we propose an unprecedented long-term estimate of nationwide forestland uses. The task was challenging. As already pointed out by other historians who have worked extensively with these types of indicators: "It is difficult to determine the current state of Spanish woodlands because sources are based on different methodologies and figures on woodlands differ significantly. In addition, as recognized in agricultural statistics, changes in the criteria used to establish forest statistics make it impossible to connect data series" (Barciela et al. 2005: 256–257).

Our strategy was to draw up estimates of unproductive surface areas and cultivated areas in each province using Agrarian Statistics data. These data are more consistent than those relating to woodlands. We then assumed that remaining areas would be wooded or forested. Subsequently, we extracted all the sources of information on forest areas at provincial levels. We soon observed the inconsistencies described by Barciela et al. (2005) as we encountered changing definitions of forest mass or unlikely shifts over time of the same forest mass. Our job was to adjust the least plausible changes within each province and create a series of woodland data falling into the four forest categories mentioned above.

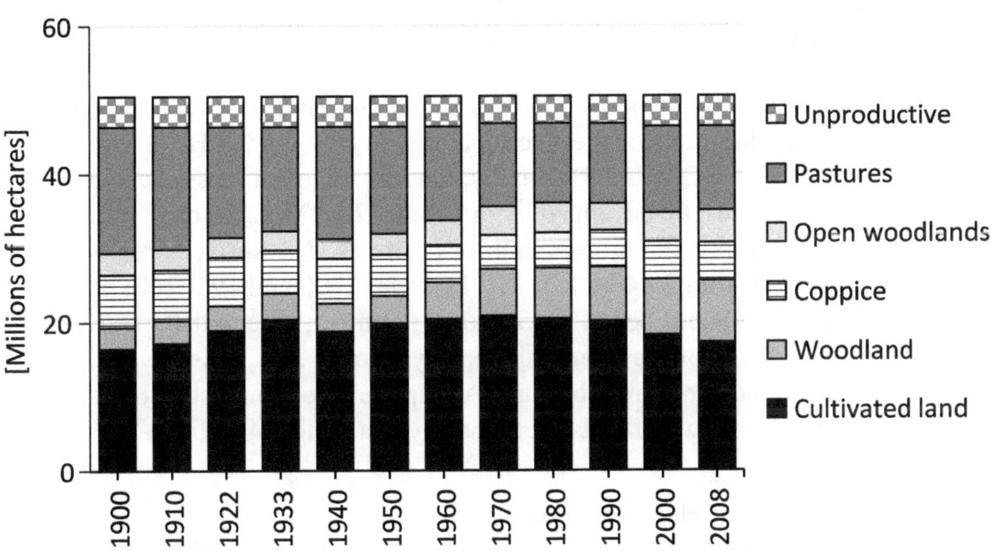

Graph 2.2 Evolution of land uses, in million hectares

Generally, the main process of change emerged from opposing trends between cultivated areas and woodlands. Cultivated areas grew continuously from 1900 to the 1970s, reaching almost 21 million hectares (Table 2.3). They have steadily declined ever since reaching a little more than 17 million hectares by 2008. The fall in cultivated areas is attributable to the drop in cereal system surface area. As we shall see later, this does not imply a fall in cereal production. Rather, rainfed lands in the interior of the country, with low yields, have decreased, often because agricultural activity has been abandoned. We will come back to this issue in Chap. 4.

As mentioned, woodlands followed the opposite trend. In the mid-nineteenth century, Spain's total forested area was about 32 million hectares (MH), occupying almost two-thirds of the land.[1] By 1900, the first year for which we can make a reasonable estimate, it had fallen to 29.85 MH. In 1960, the figure was 25.92 MH (Table 2.3). More than 6 MH, one-seventh of the country's territory, had been deforested. Deforestation processes have taken place all over the world. It is estimated that since the invention of agriculture, 15–45% of woodlands have disappeared, at a different pace and according to distinct regional patterns throughout history (McNeil 2001: 229). In Europe, by the mid-seventeenth century, more than a third of the entire continent had been deforested (Williams 1990), thus accentuating a process initiated in ancient times (Kaplan et al. 2009). Since 1500, the disappearance of woodlands and the consequent shortage of forest resources seems to have become clearly problematic for most European countries, as it kept worsening well into the nineteenth century (Allen 2003; Warde 2006: 41–42; Radkau 2008: 139, 2011; Parrotta and Trosper 2012: 216). Spain's deforestation, therefore, is part of a process common to the whole continent in which increasing pressure on resources and the expansion of agricultural borders ended up substantially reducing woodland areas (Boserup 1965, 1981). Spain, however, presented some peculiarities. Loss of forest mass lasted until the mid-twentieth century (Table 2.3), while in much of Europe it had already slowed down throughout the nineteenth century and forests were beginning to be managed intensively, concentrating on timber production once fossil fuels began to replace firewood (Table 2.3).[2]

In all events, according to the reconstructed data used in this work, the total forest area fell during the first half of the twentieth century by 3.38 MH, of which only 0.86 MH were forested (Graph 2.3). That is, the loss of forest masses concentrated in bush and pasture areas, that dropped by more than 2.52 MH. Deforestation during these years did not imply, therefore, a big loss of trees. The surface area of timber woodlands increased from 2.84 to 3.70 MH between 1900 and 1950 (Graph 2.4).

Thus, Spanish woodlands began to specialize in timber, a trend that greatly intensified in the second half of the twentieth century (Graph 2.4). This development is much better accounted for mainly because "complete" woodland statistics began to be published, and timber woodlands grew sharply, reaching 7.46 MH in 2000. Thus,

---

[1] Though we do not know which parts were forested.

[2] According to Warde (2006: 37) we can get a glimpse of this process in Central Europe in the 14th century, though it really boomed at the end of the 19th century. See: Williams (2003: 164), Radkau (2008: 214), Agnoletti et al. (2011) or Parrotta and Trosper (2012: 219–20).

Table 2.3 Land use in Spain in thousands of hectares

	Cultivated lands	Timber woodlands	Coppice	Meadows	Wooded	Bush and pasture	Forest	UAA	Unproductive	Total
	1	2	3	4	5	6	7	8	9	10
1900	16.479	2.836	7.126	2.989	12.951	16.901	29.852	46.331	4.169	50.500
1910	17.228	2.980	6.872	2.814	12.666	16.437	29.103	46.331	4.169	50.500
1920	18.799	3.336	6.549	2.702	12.587	14.845	27.432	46.231	4.169	50.500
1930	20.368	3.558	5.843	2.586	11.987	13.976	25.963	46.331	4.169	50.500
1940	18.782	3.759	6.115	2.626	12.501	15.049	27.550	46.331	4.169	50.500
1950	19.856	3.697	5.619	2.776	12.092	14.383	26.476	46.331	4.169	50.500
1960	20.413	4.929	5.076	3.320	13.325	12.594	25.919	46.331	4.169	50.500
1970	20.885	6.240	4.640	3.835	14.715	11.190	25.905	46.789	3.680	50.470
1980	20.499	6.741	4.824	4.033	15.598	10.691	26.289	46.788	3.684	50.472
1990	20.172	7.189	4.979	3.636	15.805	10.746	26.550	46.723	3.746	50.469
2000	18.304	7.460	5.055	3.893	16.408	11.645	28.053	46.357	4.143	50.500

1 **Cultivated**: Arable and crop lands. *2* **Timber woodlands**: Wooded areas where benefit methods are applied, that is, cutting trees so they regenerate. Not generally used as pastures and aimed at timber production. Strong presence of conifers. *3* **Coppice**: areas where coppicing methods are applied. Aimed at firewood production. The landscape's morphology depends on the part of the country. Some scrublands are intensive while others are more similar to pastures. Mainly oak trees. *4* **Meadows**: usually meadows, though this category also includes another type of farm characterized by low tree density combined with pastures. Strong presence of oak trees. *5* **Wooded**: All three previous categories together. Spanish statistics, since 1973, refer to this heading as forest, including only forested areas. *6* **Bush and pasture**: land of agrarian use, not tilled and not wooded. It includes natural meadows, pastures, wasteland pastures, esparto fields and areas with a variety of scrubs. *7* **Forest**: sum of wooded and bush and pasture areas. Corresponds to non-cultivated agricultural areas in our categorization. *8* **UAA**: Useful agricultural area. Includes forests, pastures and bush lands. *9* **Unproductive**: Agricultural areas that are unproductive and areas with other uses such as rivers, lakes or urbanized areas. 10. **Total**: Spain's geographical area. Sum of unproductive areas and the UAA

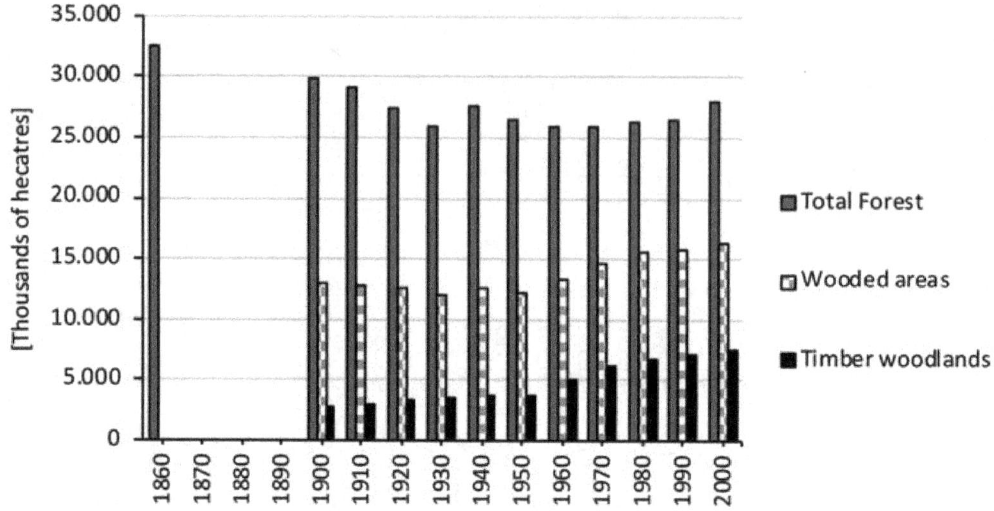

Graph 2.3 Total forest, wooded and timber woodlands in Spain in thousands of hectares. *Source* see text

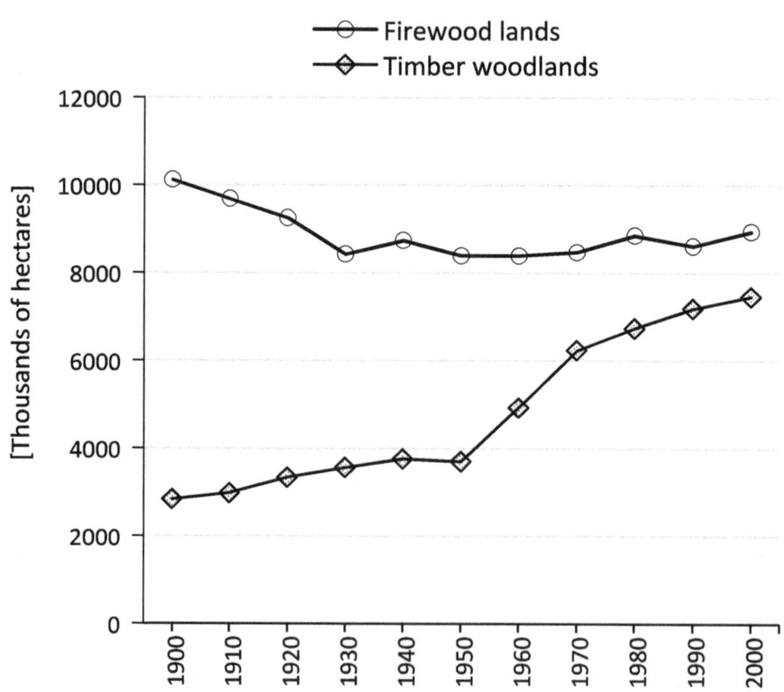

Graph 2.4 Surface area of timber woodlands and wooded areas in thousands of hectares. *Source* see methodology section

during the second half of the twentieth century, total forested areas in Spanish woodlands reversed their downward trend, increasing from 12.09 to 16.41 MH (Graph 2.3). Reforestation plans that started to be implemented at the beginning of the twentieth century were implemented from the 1940s onwards, so forest policies focused on the upsurge of conifers and fast-growing timber species. The Spanish economy demanded increasing amounts of timber as a raw material while firewood scrublands lost relevance (Graph 2.4).

This growth of woodlands is a textbook case of the process we describe today as forest transition (see Lambin and Meyfroidt 2011), that is, a shift from decreasing to increasing forest areas. The case of Spain epitomizes what happened in many other industrialized countries where the advance of agricultural areas eventually halted, giving way to the recovery of forest areas. However, as we will see later, the process can often be explained by the fact that cultivated areas have been displaced to other countries. Strictly speaking, countries that increased their forest areas actually displaced deforestation to other parts of the world (Meyfroidt et al. 2010). This seems to be the case in Spain. As estimated in another study (Infante-Amate et al. 2018), at the beggining of the 20th century Spanish imports represented a little over 700 thousand hectares of embodied land (due to biomass imports). In 2008, the figure reached over 11 million hectares, a much higher figure than that of woodland growth, and that has not exceeded 3 million hectares since that date.

2.3 Evolution of Actual Net primary Productivity

Contrary to interpretations in Sect. 2.1, agricultural production as a whole, that is, actual net primary productivity (NPP_{act}), did not grow as spectacularly as fresh matter or monetary statistics seem to imply. NPP_{act} grew by only 28.5% between 1900 and 2008, though not in a uniform way over the period (Graph 2.5). During the first half of the century, productivity increased by only 5%, while in the second half of the century, growth was much more significant, reaching 22%, coinciding with major changes in the sector (Graph 2.5). At the turn of the century, growth seems to have slowed down: actual NPP only grew by 1% between 2000 and 2008. In any case, industrialization did not actually lead to greater yields per unit area if we consider the agroecosystems' overall capacity to producc biomass. This observation should challenge the view that industrialization is associated with the massive use of external inputs and yield increases. What actually occurred, as we will see at the end of this chapter, was that production efforts were concentrated in a handful of plants, and biomass was translocated focusing on the socially useful parts of plants; this phenomenon had greater repercussions than the increase in yields per unit area and crop.

Table 2.4 shows comparative data according to the evolution of land use. Worthy of note, woodlands include forested or woodlands directed mostly at timber use, while pasture lands include not only meadows and grasslands but also forests for livestock use and pasture plots; therefore, the data on uncultivated surfaces in Table 2.5 and the data in Table 2.4 are not the same. Given that woodland and livestock use are both included in woodlands, when calculating the NPP/ha, we assumed that woodland

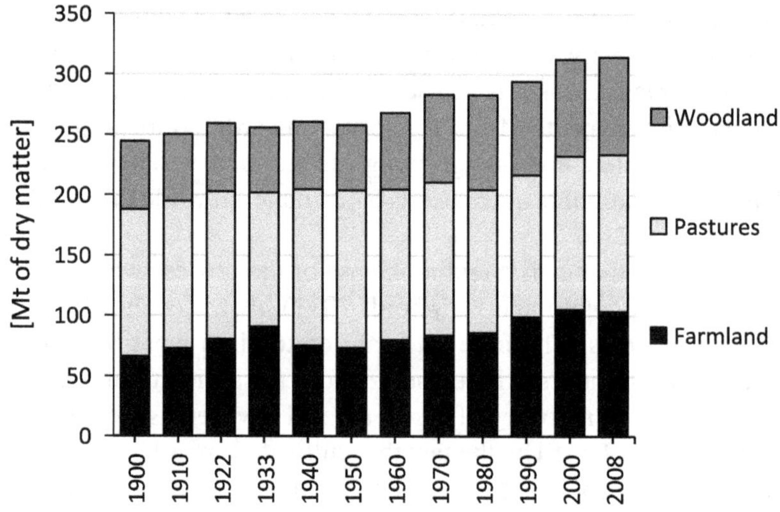

Graph 2.5 Net primary productivity according to production origin, in Mt of dry matter

Table 2.4 Net primary productivity according to land use, in Mt of dry matter

	1900	1933	1950	1970	1990	2008
Farmlands	66	91	73	84	99	104
Pastures	122	111	131	127	117	130
Woodlands	57	54	54	73	76	81
Total	245	256	258	283	293	314
Interannual variation	1900	1933	1950	1970	1990	2008
Farmlands	100	138	111	127	150	157
Pastures	100	91	107	104	96	107
Woodlands	100	95	96	129	134	142
Total	100	105	105	116	120	128
Real NPP/ha	1900	1933	1950	1970	1990	2008
Farmlands	4.0	4.5	3.7	4.0	4.9	6.0
Pastures	4.5	5.0	5.7	6.4	6.1	6.3
Woodlands	1.9	2.1	2.0	2.8	2.9	2.8
Total	4.8	5.1	5.1	5.6	5.8	6.2
Farmlands	100	112	92	100	123	150
Pastures	100	111	127	142	136	139
Woodlands	100	109	108	148	151	146
Total	100	106	106	117	121	128

Source Agrarian statistics

Table 2.5 Net primary productivity according to origin, in Mt of dry matter

	1900	1933	1950	1970	1990	2008
Woody crops (aerial)	1.1	1.4	1.7	2.0	2.0	2.1
Woody crops (roots)	0.5	0.5	0.7	0.7	0.7	0.6
Woodlands (aerial)	0.5	−0.1	0.7	9.3	10.4	9.8
Woodlands (roots)	9.6	9.1	9.0	9.7	10.4	11.1
Total accumulated biomass	11.7	10.8	12.0	21.7	23.4	23.7
% of real NPP	4.8	4.2	4.6	7.7	8.0	7.5
Crops (aerial)	19.0	26.9	19.9	15.5	20.6	27.0
Crops (roots)	21.1	29.6	22.5	21.2	22.1	22.1
Pastures (aerial)	53.6	44.9	49.9	64.3	60.3	64.6
Pastures (roots)	54.2	49.4	58.0	56.3	52.2	57.8
Woodlands (aerial)	16.3	15.4	15.4	19.3	20.0	21.6
Woodlands (roots)	19.2	18.1	18.5	26.9	27.6	29.0
Total non-harvested biomass	183.4	184.3	184.2	203.5	202.9	222.1
% of real NPP	75.0	72.0	71.4	71.8	69.3	70.7
Domestic extraction	50	61	62	58	66	68
% of real NPP	20.2	23.7	23.9	20.5	22.6	21.8
Total aerial biomass	140	149	149	168	180	194
Total root biomass	105	107	109	115	113	121
Total real NPP	245	256	258	283	293	314

Source Agrarian Statistics and author's own compilation

corresponded to all land classified as such, including pasture woodlands. Therefore, we subtracted forest from these woodlands to calculate potential grazing areas. Obviously, the actual NPP per hectare grew in the same proportion as in absolute terms as the stock of useful land is fix, except for marginal fluctuations of unproductive surfaces. However, annual amounts of biomass production differed according to the type of land use (Table 2.4). The actual NPP increased more in cultivated areas (57%) and in woodlands (42%) than in the pastures which grew by only 8% (Table 2.4). This means that human pressure to increase production concentrated mainly in cultivated areas, where biomass utility is greater. It measured essentially the same area size in 2008 than in 1900. In fact, by 2008, at the end of the series, productivity per hectare had grown by 50% compared to 1900. This growth took place despite relative extensification pointed out when analyzing agricultural production in fresh matter in recent years (Table 2.1).

The increase in woodland biomass, however, was due, as we have seen, to the increase in woodland area, the increase in forests, improved management and, paradoxically, the abandonment of many traditional types of use, especially firewood collection. When comparing woodland productivity per hectare between 1900 and 2008, an increase of 46% can be observed compared to the beginning of the century (Table 2.4), resulting from forested area growth from 2.8 million hectares to 8.3 million hectares. The small increase in pasture NPP can be explained both by the decrease in surface area and, paradoxically, by Spain's abandonment and underusage of lands dedicated to this use. This becomes more evident when looking at productivity per hectare, which increased by almost 40% since 1900 (Table 2.4).

Interesting aspects of actual NPP's evolution come to light when broken down into different categories. As we have already observed, the net primary productivity of woodlands grew the most, after farmlands. This was largely due to a notable growth of forested areas and accumulated biomass, which doubled over the period, revealing a growing percentage of total net primary productivity, from 4.8 to 7.5%. Accumulation of biomass in the aerial parts of woodlands, which multiplied almost 20-fold, was the main factor (Table 2.5). Woody crops, which hardly doubled, made a much less significant contribution. Biomass accumulated in the roots of woodlands and woody crops grew even less (Table 2.5). This accumulation can be explained by a dual phenomenon: on the one hand, areas with trees multiplied threefold, on the other, the use of firewood from Spanish forests was disappearing as the energy transition progressed in households and firewood was replaced with butane gas as well as electricity (though to a lesser extent). Biomass accumulated in trees and in general in forests because it was no longer harvested for firewood. This accumulation greatly intensified from the sixties onwards and explains the proliferation of recorded forest fires in public woodlands ever since (González de Molina and Casero 1992). Although forested areas grew throughout the century, this progression accelerated over the last decades for two reasons: a peak in household energy transition and the implementation of public policies of conservation and declaration of protected natural spaces.

Unharvested biomass, i.e., biomass that is not appropriated for human use, supports trophic chains. It provides food for the herbivorous species inhabiting the agroecosystems and therefore it also helps to sustain their diversity. As we will see later, unharvested biomass is a good indicator of agroecosystem health. Its evolution throughout the period somewhat reflects the deterioration of the land fund

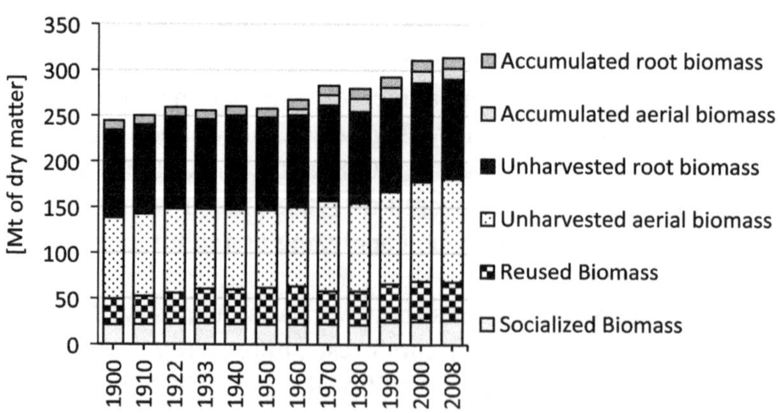

Graph 2.6 Evolution of Net primary productivity according to the use, in Mt of dry matter

element. Overall, this biomass type grew by 21% in absolute terms, but in relative terms, it contributed the least to the growth of total NPP, clearly below the average (Graph 2.6). This means that its importance dropped in relative terms from 75 to 70.7%. This drop was worst in pastures and farmlands and we will see the consequences of this later. Until the 1960s, Spain's annual biomass production remained essentially stable, growing by only 2% during those decades (Graph 2.6). It actually began to grow from that period onwards and would accelerate in the 1990s (Graph 2.6). The growth was due, as in the case of accumulated biomass, to the expansion of woodlands, i.e., of forested mountains, especially aerial parts and the falling use of harvest residues in croplands.

Only approximately one-fifth of total NPP was appropriated throughout the period by humans (Graph 2.6), that is, it was extracted from agroecosystems. This amount of biomass is called Domestic Extraction, which we analyze below.

2.4 Evolution of Domestic Extraction

Graph 2.7 shows the evolution of vegetal biomass extracted from Spanish agroecosystems between 1900 and 2008, in percentage. As pointed out earlier, dry matter is the most suitable way of measuring the real significance of the changes. This is because current varieties have higher water contents and a shift towards crops that require more water affects total production weight. In addition, irrigation allowed multiplying yields per unit area. Crops that are impossible to produce in rainfed conditions have turned into Spanish agriculture's main specialization, as is the case of

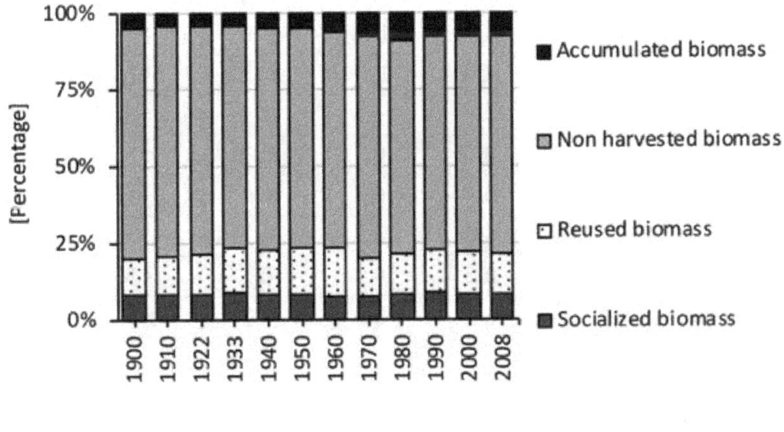

Graph 2.7 Evolution of net primary productivity according to the use, in percentage

fruits and vegetables.[3] Domestic Extraction (DE) of biomass, that is, biomass appropriated directly or indirectly by Spanish society, grew by 38% throughout the period, i.e., more than NPP (28%) (Graph 2.7). The increase in agricultural production based on our measurements differs from the increase obtained using conventional calculations: while in monetary terms "agricultural production" grew almost sixfold, it grew only by 38% in physical terms. Agriculture's great leap forward following its industrialization can be put into perspective if we consider not only commercial biomass, but all the biomass appropriated by Spanish society: both that with end uses in society and biomass reused in agroecosystems to feed livestock, used as seeds, etc. The difference is even more notable when considering the total biomass necessary to reproduce the fund elements that maintain agroecosystems production capacity, i.e., the NPP. These "costs" are usually not included in national accounts nor, therefore, in agricultural sector accounts (Table 2.6).

We can, therefore, argue that Domestic Extraction growth was, in fact, modest, which explains that its share in Spanish agroecosystem NPP hardly increased, going from 20.2% in 1900 to 21.8% in 2008. The period of maximum levels of relative

Table 2.6 Biomass domestic extraction in Mt of dry matter

	1900	1933	1950	1970	1990	2008
Crops	11	17	14	25	36	38
Harvest residues	13	16	15	19	18	14
Pastures	14	17	23	6	5	8
Woodlands	11	11	11	8	8	9
Total	50	61	62	58	66	68
Crops	100	149	124	222	318	336
Harvest residues	100	123	111	148	138	108
Pastures	100	123	160	43	35	57
Woodlands	100	102	96	70	68	83
Interannual variation	100	123	125	117	134	138
Productivity per ha						
Crops	0.7	0.8	0.7	1.2	1.8	2.2
Harvest residues	0.5	0.7	0.6	1.0	0.9	0.7
Pastures	5.0	4.8	6.1	1.0	0.7	0.9
Woodlands	0.2	0.2	0.2	0.2	0.2	0.2
Domestic Extraction per ha	6.4	6.5	7.7	3.3	3.5	4.0

Source Agrarian statistics

[3]Water in semi-arid climates like that of Spain has been one of the key factors of agricultural industrialization. Irrigation is of fundamental importance. However, the complexity of the subject requires a separate study and we do not address the issue in this book. We are currently fine-tuning our approach to water and its role in Spanish agriculture from the metabolic point of view. A future work is thus under way on this question.

extraction took place in the 1950s, coinciding with the end of traditional agriculture (23.9%), when production difficulties caused by Francoism led to the highest level of appropriation of biomass, in a context of falling yields. However, in absolute terms, the largest amounts of biomass were extracted in the year 2000, reaching almost 70 million tons of dry matter, mostly from crops (77%) (Graph 2.8).

A disaggregated analysis shows that Domestic Extraction growth concentrated on the marketable portion of the plants, increasing by 236% compared to 1900, due to a limited 8% growth in residues. In contrast, biomass appropriated by pastures and forests decreased, respectively by 46 and 17%. The different behaviors of biomass types become clear when observing the evolution of productivity per hectare. Primary crop productivity increased threefold, while residue productivity only increased by 40%. Pasture abandonment and underuse explain an extraction drop of up to 81% for these lands, below one ton per hectare, despite having reached more than six tons in the 1940s and 1950s. Forest conservation policies and household energy transition, as mentioned above, explain the decrease in woodland extraction per hectare (17%). Either way, the evolution of Domestic Extraction uncovers significant changes where primary crops were favored over other kinds of biomass. In this sense, the industrialization of agriculture led to a significant increase in produced biomass, but this increase concentrated in the cultivated lands and more specifically in the most commercially oriented crops. In fact, extraction in these lands went from 22.5 to 54.8% of total Domestic Extraction in 2008, reflecting Spanish agriculture's production specialization, as we will see in the following section.

Important changes also took place in the end uses of Domestic Extraction (Graph 2.9). Biomass aimed at human food accounted for 9% in 1900 rising to 14% in 2008. Similarly, use of biomass as a raw material went from 1% in 1900 to 4% today. The major change, however, concerned biomass aimed at animal feed. As early as 1900, it represented 56% of extracted biomass, which is logical since main

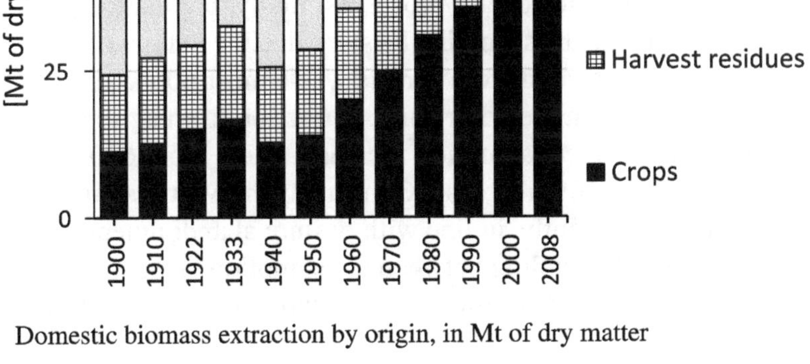

Graph 2.8 Domestic biomass extraction by origin, in Mt of dry matter

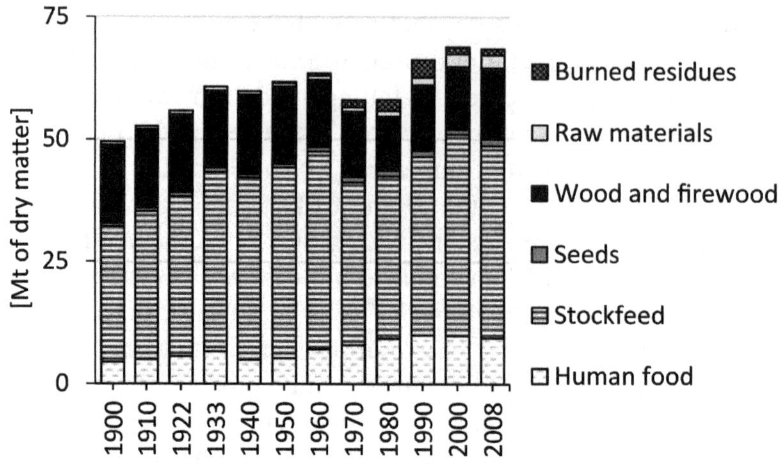

Graph 2.9 Destination of biomass domestic extraction in Mt of dry matter

agricultural tasks were carried out using working animals. In 2008, that percentage rose to 57.5%, reaching its highest percentage in 1960 with a share of almost two thirds (Graph 2.9). Since that decade, around 40 million tons of dry matter have been used to feed livestock, despite animal traction no longer being used. As we will see below, growth of livestock has reached unprecedented levels in recent decades, linked to notable diet changes, specializing in meat and dairy products.

On the other hand, Domestic Extraction aimed at timber and firewood decreased considerably, from 32% in 1900 to 21% in 2008 (Graph 2.9), mainly due to the lesser weight of biomass energy use, thanks to the arrival of gas to Spanish households. However, the drop in firewood use, especially from woodlands, was partially offset by the use of wood as a raw material. This fact is consistent with the growth of forested woodlands mentioned previously. Spanish forest Domestic Extraction thus dropped by only 17%, less than firewood consumption based on this source. Strikingly, as from the 1950s, burnt harvest residues eventually reached 3.6 megatons in the 1990s, i.e. 5.5% of total Domestic Extraction, a real wastage. The prohibition of stubble-burning, a measure taken to prevent forest fires, led burning to fall by half (Graph 2.9).

As shown in Graph 2.10 and Table 2.7, Domestic Extraction decreased per capita by 26%. This drop was due to several processes: firstly, population growth, despite total consumption increase, and secondly the steep decline in firewood use. Per capita extraction of wood and firewood from forests was almost 600 kg in 1900 dropping to almost 200 kg in 2008. On the other hand, extraction from agricultural land went up in accordance with food consumption increase, going from 600 kg per capita in 1900 to 815 kg in 2008, a 36% increase. Conversely, harvest residue extraction and pasture use dropped significantly, in line with pasture abandonment and the burning of residues. Progressive decoupling between consumption and extraction, especially notable since 1960 also had a major impact, as we will see, on the decline of Domestic

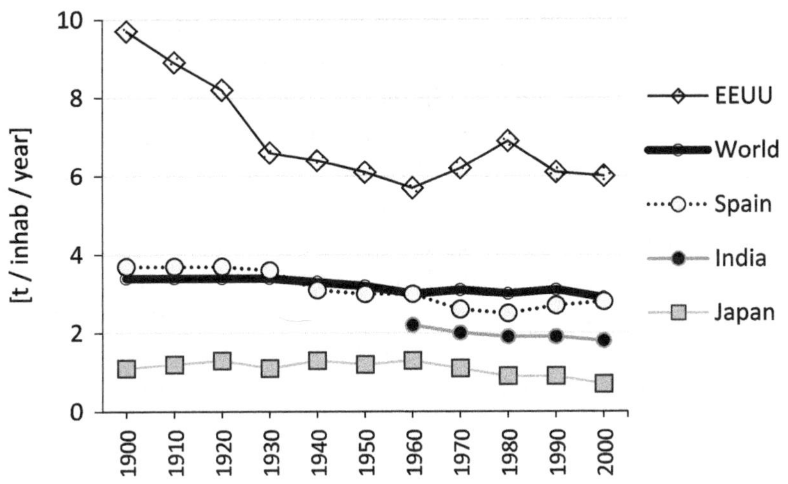

Graph 2.10 Domestic biomass extraction per inhabitant in several countries and the world average, in tons per inhabitant per year

Table 2.7 Destination of domestic extraction in Mt of dry matter

	1900	1933	1950	1970	1990	2008
Human food	4.4	6.6	5.3	8.0	10.0	9.4
Animal feed	27.8	36.6	39.0	33.2	36.6	39.4
Seeds	0.6	0.9	0.7	1.1	1.2	1.3
Wood and firewood	16.0	15.7	15.9	13.4	13.4	14.6
Raw materials	0.7	0.8	0.6	0.8	1.5	2.6
Burned residues	0.0	0.0	0.2	1.5	3.6	1.3
Total	49.5	60.7	62	58.1	66.3	68.5
Human food	100	149	119	179	224	212
Animal feed	100	132	140	120	132	142
Seeds	100	155	115	180	191	209
Wood and firewood	100	98	99	84	84	91
Raw materials	100	114	98	126	229	392
Burned residues	0	0	100	756	229	616
Interannual variation	100	123	125	117	134	138

Source Agrarian statistics and author's own compilation

Extraction. While Domestic Extraction per capita fell by 6.5% between 1960 and 2000, Domestic Consumption grew by 5.7%.[4]

In comparative terms, the drop in Domestic Extraction per capita throughout the twentieth century was less significant than in other countries such as Japan or the USA, but higher than in the rest of the world (Graph 2.10). One reason is the partial

[4]Consumption and trade dynamics are analyzed in detail in Chap. 6.

replacement of biotic materials with abiotic materials in the Spanish economy's metabolism flows. Vegetal biomass lost part of its energy functions (especially as a domestic fuel source or stockfeed), but its use as a raw material for industry increased, especially in the case of wood, which in many industrial processes, is very difficult to replace (Infante-Amate et al. 2014b). In other words, Spanish agroecosystems that once provided most of society's needs in energy and raw materials now specialized in human and animal food as well as raw materials for industry. This is the reason why Domestic Extraction and production efforts concentrated in crops and, to a lesser extent, in wood production. In this sense, the most significant transformations undergone by agroecosystems are related to the growing importance of animal feed in Spanish agriculture, as we will see in the following sections, and in industrial wood demand.

2.5 The Specialization of Spain's Agricultural Production

The analysis of the evolution of net primary productivity and Domestic Extraction in Spanish agriculture has shown how growth has concentrated mainly in crops, and especially in grains rather than residue. In this section, we analyze the evolution of different crop groups and their specialization patterns. The analysis of dry matter, unlike conventional approaches based on fresh matter described in the first section of this chapter, facilitates the appreciation of specialization patterns. By combining dry and fresh matter analyses, we can uncover some of the problems of Spain's agricultural specialization during its industrialization process.

According to the data in Table 2.8, virtually all crops increased their production throughout the twentieth century. The crop that grew the most was olive trees, in line with its expansion throughout the twentieth century (Infante-Amate 2014). The only exceptions were legumes, which went from 4% of total production at the beginning of the century to becoming marginal, as well as potatoes. Potato production increased until mid-century, undoubtedly due to its key role in peasant diet, and its weight decreased thereafter. Production levels in 2008 were similar to those of 1900. This was a general European phenomenon for two essential reasons: a reduction in per capita consumption, and the increase of imports. Beyond this rapid growth in overall crop production, especially visible in the second half of the twentieth century, dry matter figures draw a quite different picture of specialization compared to that of fresh matter. Indeed, the most commercial and export-oriented crops (horticultural crops above all) incorporate water quantities, as well as water demands, that are far higher than other crops such as cereals, with a much higher content of dry matter. In 2008, these accounted for 53.8% of dry matter, compared to 21.9% of fresh matter. This different way of presenting the data shows that Spanish agriculture specialization has been based on water export and the increasing use of water resources in an eminently semi-arid country, as pointed out in the literature (Cazcarro et al. 2015; Duarte et al. 2014).

Table 2.8 Crop domestic extraction (without residues) in Spain in Mt of dy matter, 1900 = 100 and in percentage

	1900	1910	1922	1933	1940	1950	1960	1970	1980	1990	2000	2008
Cereals	5.5	6.3	6.8	7.9	5.0	5.9	7.6	10.7	13.1	16.5	18.6	20.2
Legumes	0.5	0.6	0.7	0.7	0.5	0.5	0.7	0.6	0.3	0.2	0.3	0.2
Grapes	1.1	0.8	1.1	1.0	0.9	0.7	0.8	1.2	1.8	1.6	1.8	1.6
Olives	0.6	0.6	0.9	1.0	0.8	0.9	1.1	1.1	1.4	1.6	2.8	3.4
Potatoes	0.5	0.7	0.7	1.1	0.8	0.8	1.0	1.1	1.3	1.2	0.7	0.5
Fruit and vegetables	0.7	0.9	1.1	1.5	1.5	1.2	1.5	1.7	2.0	2.5	2.6	2.7
Industrial and other	0.5	0.5	0.6	0.7	0.5	0.7	1.3	1.9	2.4	3.5	3.2	2.4
Artificial meadows and fodder	1.8	2.3	3.0	2.9	2.7	3.1	6.1	6.5	8.5	8.5	8.5	6.7
Total	11.2	12.6	14.9	16.7	12.7	13.8	19.9	24.8	30.8	35.6	38.7	37.5
	1900	1910	1922	1933	1940	1950	1960	1970	1980	1990	2000	2008
Cereals	100	115	124	144	92	107	138	195	239	302	340	369
Legumes	100	121	141	143	102	116	143	129	73	48	63	44
Grapes	100	72	102	87	79	67	70	108	161	141	162	143
Olives	100	100	149	168	143	147	179	191	239	270	482	577
Potatoes	100	132	143	206	146	148	194	219	241	223	132	94
Fruit and vegetables	100	127	150	200	200	164	202	224	268	330	354	359
Industrial and other	100	94	125	146	108	144	268	392	502	725	669	486

(continued)

Table 2.8 (continued)

	1900	1910	1922	1933	1940	1950	1960	1970	1980	1990	2000	2008
Artificial meadows and fodder	100	128	168	163	151	174	339	362	471	475	475	371
Total	100	113	134	149	114	124	178	222	275	318	346	336
	1900	1910	1922	1933	1940	1950	1960	1970	1980	1990	2000	2008
Cereals	49.0	49.9	45.5	47.1	39.6	42.4	37.9	43.0	42.6	46.5	48.2	53.8
Legumes	4.2	4.5	4.4	4.0	3.8	3.9	3.4	2.4	1.1	0.6	0.8	0.6
Grapes	9.8	6.2	7.5	5.7	6.8	5.3	3.8	4.8	5.8	4.4	4.6	4.2
Olives	5.2	4.6	5.8	5.9	6.6	6.2	5.3	4.5	4.5	4.4	7.3	9.0
Potatoes	4.7	5.5	5.0	6.4	6.0	5.6	5.1	4.6	4.1	3.3	1.8	1.3
Fruit and vegetables	6.7	7.5	7.5	9.0	11.8	8.9	7.6	6.8	6.5	7.0	6.9	7.2
Industrial and other	4.3	3.6	4.1	4.3	4.1	5.1	6.5	7.7	7.9	9.9	8.4	6.3
Artificial meadows and fodder	16.1	18.2	20.2	17.6	21.3	22.5	30.5	26.2	27.5	24.0	22.1	17.7
Total	100	100	100	100	100	100	100	100	100	100	100	100

Source Agrarian statistics and author's own compilation

One of the most notable results is that not only has the weight of cereals in grain crop production actually not declined, but it also grew throughout the twentieth century (Table 2.8). The reason is not an increase in the share of cereals in the Spanish diet but rather a reorientation of the production of the cereal system. In fact, the percentage of cereals in all crops decreased until the 1960s because of diet changes, implying a greater varieties and the presence of other Mediterranean diet products (González de Molina et al. 2014, 2017). The percentage began to grow back from that date onwards. Among the crop groups, cereals grew the most between 1950 and 2008 (244%), coming only after olives. This growth must be linked to livestock specialization in the second half of the twentieth century, a factor that underpins most of the arguments put forward in this chapter. In fact, figures in Table 2.9 show that this livestock reorientation concentrated in farmlands. The production of crops for animal feed increased from 49% in 1900 to 65% in 2000 and 2008. In contrast, human food fell from 40% of production in 1900 to 25% in 2008. This reorientation mainly concerned fodder and cereals, although the evolution of fodder was different from that of cereals. Its share in crop production grew to 30% in 1960 and dropped to 17% in 2008 (although in absolute terms, this was one of the groups that grew the most throughout the period).

As we will see in the next section, this production reorientation was related to livestock transformations. It can also be observed in cereal group transformations (Table 2.10). The two cereals that grew the most were barley and corn, both of which are mostly used for stockfeed. In fact, barley replaced wheat as the main cereal since 1980, though over the last decades a growing share of wheat has also been used for animal feed. In the same way, cereals such as rye destined for human consumption and that used to be at the heart of Spanish organic agriculture have become marginal, as in the case of legumes.

2.6 Spanish Livestock in the Twentieth Century

In this section, we study the evolution of livestock and its relationship with the land. Livestock is the other biophysical fund element that we considered crucial to agroecosystems. It was a core element for the reproduction of traditional organic agriculture, providing services as essential as manure, traction for heavier work, transport or the provision of food and indispensable raw materials. In recent decades, livestock activity has played an unprecedented economic and nutritional role, which should have attracted greater attention from agricultural historians. There are numerous gaps, however, in this fundamental chapter of the sector's history.

We overview Spanish livestock transformations throughout the twentieth century, its changing role in the sector as a whole and its relationships with agriculture.[5] To perform the study, we compiled the data of all livestock censuses carried out in Spain between 1891 and 2012. In Annex I, we examine the reliability of these censuses and

[5]Starting with the pioneering works of GEHR (1978, 1979) and García Sanz (1991).

Table 2.9 Destination of domestic extraction of crops (without residues) in Spain in Mt of dry matter and in %

	1900	1910	1922	1933	1940	1950	1960	1970	1980	1990	2000	2008
Human food	4.4	5.0	5.7	6.6	4.9	5.3	7.1	8.0	9.3	10.0	9.9	9.4
Stockfeed	5.5	6.4	7.7	8.4	6.5	7.2	11.2	14.9	19.3	22.9	25.2	24.3
Seeds	0.6	0.7	0.8	0.9	0.6	0.7	0.9	1.1	1.1	1.2	1.0	1.3
Industrial use	0.7	0.5	0.8	0.8	0.6	0.6	0.8	0.8	1.1	1.5	2.6	2.6
Total	11.2	12.6	14.9	16.7	12.7	13.8	19.9	24.8	30.8	35.6	38.7	37.5
	1900	1910	1922	1933	1940	1950	1960	1970	1980	1990	2000	2008
Human food	40	39	38	40	39	38	36	32	30	28	26	25
Stockfeed	49	51	52	50	51	52	56	60	63	64	65	65
Seeds	5	6	5	6	5	5	4	4	4	3	3	3
Industrial use	6	4	5	5	5	5	4	3	4	4	7	7
Total	100	100	100	100	100	100	100	100	100	100	100	100

Source Agrarian statistics and author's compilation

Table 2.10 Domestic extraction of grain cereals in Spain in Mt of dry matter and in %

	1900	1910	1922	1933	1940	1950	1960	1970	1980	1990	2000	2008
Wheat	2.7	3.1	3.3	3.8	2.2	2.8	3.8	4.4	4.0	4.2	5.5	6.0
Barley	1.3	1.4	1.6	2.1	1.3	1.5	1.7	3.5	5.9	8.3	7.9	9.1
Oats	0.2	0.4	0.4	0.6	0.4	0.4	0.4	0.4	0.5	0.4	0.7	0.9
Rye	0.5	0.6	0.6	0.5	0.3	0.4	0.4	0.3	0.2	0.2	0.2	0.2
Corn	0.5	0.6	0.6	0.6	0.6	0.5	0.9	1.6	2.0	2.7	3.5	3.2
Rice	0.1	0.2	0.2	0.3	0.2	0.2	0.3	0.3	0.4	0.5	0.7	0.6
Others	0.1	0.0	0.1	0.1	0.0	0.0	0.0	0.2	0.2	0.2	0.2	0.2
Total	5.5	6.3	6.8	7.9	5.0	5.9	7.6	10.7	13.1	16.5	18.6	20.2
	1900	1910	1922	1933	1940	1950	1960	1970	1980	1990	2000	2008
Wheat	49.8	49.8	48.3	48.4	43.2	48.5	50.6	41.0	30.7	25.5	29.5	29.5
Barley	23.3	22.2	24.3	26.5	25.7	25.3	22.3	32.7	45.3	50.4	42.5	45.0
Oats	4.2	6.0	6.2	7.3	8.3	7.1	5.8	4.1	3.5	2.4	3.5	4.4
Rye	9.3	9.5	8.8	6.1	6.5	6.6	5.3	2.5	1.5	1.5	0.9	0.9
Corn	9.6	8.8	8.3	7.7	11.8	8.0	11.3	15.1	15.0	16.1	19.0	16.1
Rice	2.7	2.9	3.4	3.2	3.8	3.9	4.5	3.1	2.8	2.9	3.9	2.9
Others	1.0	0.8	0.8	0.7	0.8	0.5	0.2	1.5	1.3	1.1	0.8	1.1
Total	100	100	100	100	100	100	100	100	100	100	100	100

Source Agrarian statistics and author's compilation

livestock counts are corrected wherever reliability problems were found. The annex explains the method used to verify and correct the figures (Table 2.11).

Overall, Spain's livestock changed fundamentally throughout the twentieth century, shifting from organic-based livestock, closely linked to the territory, to livestock of an industrial nature, mostly stabled, landless and thus much more dependent on stockfeed, industrial inputs and international trade. This change brought about notable growth of total herds, from 54 million heads to 838 million heads, with significant transformations to their composition, destination of products and services, management and feeding. In terms of live weight, the livestock multiplied by 2.4, from 2.8 million tons at the beginning of the twentieth century (with peaks

Table 2.11 Evolution of livestock live weight in thousand of tons between 1900–2008 and 1900 = 100

Years	Bovine	Ovine	Goats	Porcine	Equine	Poultry	Rabbits	Total
1900	690	543	102	141	636	43	8	2163
1910	856	601	121	180	713	50	9	2530
1922	1173	550	116	251	756	58	11	2915
1933	1339	630	151	373	794	67	13	3366
1940	1454	662	225	430	723	54	15	3562
1950	1422	544	175	292	754	53	7	3248
1960	1198	648	100	309	655	90	78	3078
1970	1402	545	79	315	418	661	39	3458
1980	1681	485	70	634	203	1334	206	4612
1990	1876	858	124	1043	159	1258	45	5363
2000	2238	826	90	1370	101	2007	88	6720
2008	2215	705	91	1547	101	2078	89	6826
Inter annual variation 1900 = 100								
1900	100	100	100	100	100	100	100	100
1910	124	111	119	128	112	117	117	117
1922	170	101	114	178	119	135	135	135
1933	194	116	149	265	125	156	156	156
1940	211	122	221	305	114	126	186	165
1950	206	100	173	207	119	124	88	150
1960	174	119	99	219	103	209	966	142
1970	203	100	78	223	66	1533	479	160
1980	244	89	69	449	32	3093	2540	213
1990	272	158	122	739	25	2918	550	248
2000	324	152	88	971	16	4654	1092	311
2008	321	130	89	1097	16	4819	1101	316

Source Corrected livestock censuses (see Annex I)

of around 3.4 million in the 1930s), to almost 7 million in the middle of the last decade (Table 2.11). This huge increase would also have considerable environmental impacts, especially those generated by intensive livestock farming and the increase of GHG emissions (Lasaletta et al. 2014a, b).

The pace of change during the twentieth century, however, was irregular. Livestock increased by 55.6% between 1900 and 1940, stalled between 1950 and 1960 and rose again rapidly between 1970 and 2008 (97%) (Graph 2.11). The first third of the twentieth century was a period of livestock growth. Significantly, this increase was consistent with a notable rise in agricultural production. Throughout the first 30 years of the twentieth century, most of biomass Domestic Extraction growth concentrated in cultivated lands. These cultivated lands also continued to expand at the expense of pasture and coppice. However, this fact did not affect the growing numbers of livestock, since the biomass aimed at livestock feed also increased. The reason was the rising amounts of stockfeed harvested on farmlands (forage crops, artificial meadows, cereals and legumes for animal feed, harvested residues, etc.), leading to a rise in its share of total stockfeed from 49% in 1900 to 54% in 1930. This was due to a moderate growth inland productivity that took place during the first decades of the twentieth century and increasing domestic extraction in the pastures, despite reductions in their surface area. Imports of stockfeed were not significant during these years.

After 1936, the analysis of censuses and livestock production yields show opposite and seemingly incompatible results (Graph 2.11). Apparently, livestock must have declined considerably during the Civil War, a drop that is not reflected in the census after 1936. This census presents an even larger livestock than that of 1933. An examination of the censuses described in Annex I show that 1940 and 1950 data appear to be plausibly adjusted to livestock feeding capacity. How can we explain this ostensibly odd livestock evolution? Our data do not invalidate the sector's likely depression after livestock destruction during the war, but they show an abnormally high number of offspring in the 1940s' census. Our hypothesis, which would need to be confirmed by a subsequent study, is that considerable efforts must have been made

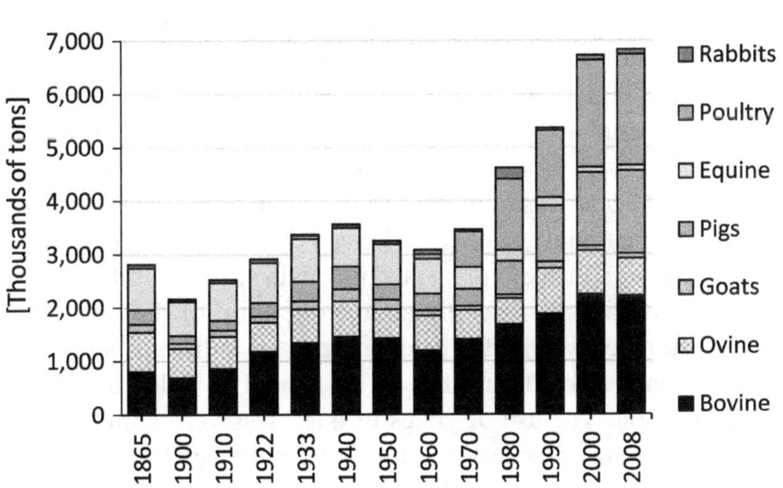

Graph 2.11 Evolution of live livestock weight in thousands of tons

during those years to recover the livestock, allowing many more offspring to reach adulthood. This would explain that livestock production was as low as shown in the statistics, while the censuses indicated such high values. On the other hand, during the second half of the twentieth century, the Western World adopted an industrial or intensive livestock model, closely linked to the considerable increase of products of animal origin in diets. The Civil War and the Autarchy delayed Spain's adoption of this model, in lockstep with the industrialization of agriculture and of the Spanish economy generally (Infante et al. 2015).

Circumstances under early Francoism help to understand the extent of the transformations undergone since the sixties. Contrary to previous reports (Domínguez 2001a, b), livestock industrialization, and that of agriculture as a whole, did not concern an agricultural sector such as that had existed before the war. The sector had gone back to being almost entirely organically based and its territorial balance had been shattered by two decades of autarchic agricultural policy. This point is relevant because peasants were much less able to adapt to market-induced changes and the State's food policies than in the first third of the twentieth century (Fernández Prieto 2007).

Livestock's internal composition and its end uses also underwent considerable transformations that we can categorize into four major groups. The first series of changes is related to labor and transport livestock, which was characteristic of agriculture prior to mechanization. Labor and transport livestock lost ground to livestock aimed at human food. The reason is that livestock was replaced by internal combustion engines, both for mechanization and transport (since trucks are used more than rail in Spain). In fact, equine livestock became almost marginal over time. Horses, mainly dedicated to recreation or sports activities, barely reached 240,000 heads nationwide in 2008, while mules did not exceed 28,000 heads and donkeys 55,000. In contrast, at the beginning of the twentieth century, equine livestock represented 29.4% of live livestock weight, to which we should add a large part of the cattle, which then represented 32% of the total.

The second major transformation was the loss of relative importance of traditional livestock breeding. Ovine and goat livestock, once tied to pasture and forest lands according to their feeding needs and breed and thus better adapted to land and climate conditions, shrunk and lost importance throughout the century. At the beginning of the twentieth century, both types of herds represented 29.1% of live weight. In 2008, sheep accounted for little more than 10% of herds, specialized in the production of lambs and to a lesser extent, of milk for cheese. Goat livestock, on the other hand, represented only 1.3% and was dedicated to the same purposes as sheep.

The third transformation was that experienced by cattle. In 1900 beef cattle represented 31.9% of live weight and represented the main species. Today, cattle's relative weight has increased slightly, reaching almost one-third of the total. However, big changes have taken place within this category, related to breeds, specialization, and feeding. A large share of cattle used to be dedicated to fieldwork and to a lesser extent to transportation. The use of yokes of oxen was very common for tillage or threshing at the beginning of the last century. Oxen-pulled carts were commonly used to transport goods. Cows were dedicated to breeding and punctual agricultural tasks.

Most of the milk was thus used to breastfeed calves, in turn rarely aimed at meat production. This livestock's life cycle exceeded 10/12 years. They were sacrificed once their useful life—whether in the field or for reproduction—had come to an end. Beef livestock predominated in the center and south of the peninsula, while mixed livestock, also dedicated to the production of meat and milk, was prevalent in the north. This diversified use was gradually substituted by livestock specialized in meat and milk. Breeds were improved to maximize production, they were stabled and fed with compound feed.

Porcine livestock followed a similar path. At the beginning of the century, pigs represented almost 6.5% of livestock in terms of live weight and the predominant breeds were those belonging to each agrarian area of the country. This type of livestock was fattened, slaughtered and preserved as cold meats and sausages to serve as a supply of quality proteins for the whole year. It was associated with the survival of the poorest strata of the peasantry. Pigs' diet was based on oak acorns in pasture land and forests and, to a lesser extent, on domestic organic waste. Today porcine herds represent 22.7% of livestock, coming after cattle and poultry in total livestock weight. Pig breeding has now become a highly industrialized activity that provides meat at affordable prices. It represents one of the pillars of the Spanish diet. They are bred in large intensive farms, are stabled on a permanent basis, given compound feed and undergo frequent veterinary treatments.

The fourth and final series of transformations concerned poultry. Breeding, especially that of poultry, is today a major livestock activity also involving intensive production and permanent stabling. In 2008, poultry represented 30.4% of total livestock live weight with almost 737 million heads dedicated to the production of meat and eggs, with increasingly shorter breeding cycles and fed with grains and concentrates. Poultry breeding, feeding, and use have radically changed. At the beginning of the last century, chickens were raised to produce eggs and when production declined they were slaughtered for meat or other dishes. Chickens' diet competed with human food, as in the case of pigs, hence their limited numbers. They were fed household organic waste or crops. Rabbits followed a similar trend but to a lesser extent. Once the most popular hunting product, they became farm animals.

To summarize, these transformations reflect the reorientation of Spanish livestock throughout the industrialization process from a multifunctional use of animals, typical of traditional organic agriculture (suppliers of food, labor, manure and raw materials such as wool, fats, skins, etc.) to livestock centered on the production of food mainly for human consumption. The species that grew the most in terms of live weight and number of heads were pigs and poultry, that is, monogastric livestock, to the detriment of herbivorous livestock, with the consequences that we will see below.

Livestock feeding thus depended mainly on pastures and harvest residues (45% and 25%, respectively) until the 1960s. From then on, it began to depend on quality feed from crops and industrial processing. Since the 1980s, an increasing share of this feed has come from foreign trade. In 2008, 48% of stockfeed came from primary crops and 15% from net imports (Graph 2.12). Meanwhile, a large part of the agricultural area used was abandoned or the pastures were underutilized, as shown by the drastic reduction in grazed biomass.

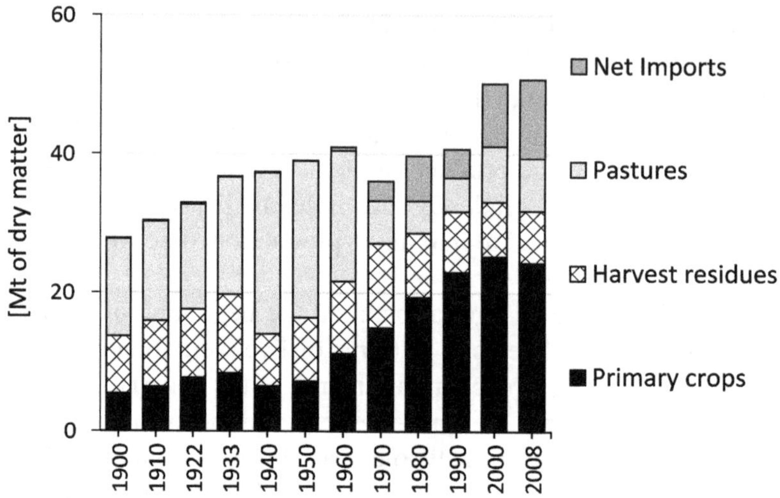

Graph 2.12 Origin of animal feed, in Mt of dry matter

Livestock, once closely linked to agriculture and pastures, lost part of its ties to the land and the food it produced. As is commonly known, in organic agrarian systems, livestock not only provided food products to society, but also played an essential part in sustaining agricultural activity, both in terms of work and replacement fertility. For this reason, simply comparing this type of livestock with later livestock, especially with intensive livestock farming, in terms of productivity and market links, is inappropriate. The industrial livestock model is compatible only with industrialized agriculture that relies on industrial inputs. As we have pointed out, the model's adoption is mostly linked to profound changes in the animal feeding model. The model favors breeds that are dependent on grain food thus increasing reliance on industrial feed. Over the last fifty years, this dependence has relied on growing biomass from crops to the detriment of pastures, and on the increase of feed imports.

Meanwhile, livestock has become increasingly central to Spanish agriculture as a whole. Its significance in monetary terms was highlighted in the first section of this chapter, but it can also be appreciated in biophysical terms. Domestic Extraction for animal feed went from 27.8 Mt of dry matter in 1900 to 39.4 Mt in 2008, from 56% to 57.5% of total extraction. Furthermore, livestock has grown at a much higher rate than the domestic availability of stockfeed (a 215% increase in live weight between 1900 and 2008 compared to an 82% increase of stockfeed). As a result, levels of Domestic Extraction have shown to be insufficient to maintain the livestock growth rate and its food requirements, explaining the need to resort to feed imports.

2.7 Livestock Production

The evolution of livestock production logically reflected the transformations reviewed above. We can decompose its analysis into three major periods: first, that of continuous production growth until the mid-thirties, marking the end of the agrarian crisis at the end of the century; second, the period of crisis including the Civil War and autarchy that lasted until the mid-sixties; and a third and final period of accelerated production growth, especially of meat and milk, from the 1960s until today. The first period was characterized, as we have seen, by sustained growth of livestock leading to increased livestock production. Nevertheless, the production downturn of the last third of the 19th century was not overcome until the 1920s (García Sanz 1991; Garrabou and González de Molina 2010). Meat production grew considerably, by 86% compared to 1900 (Table 2.12), which explains why apparent consumption could increase from 14.1 kg per capita per year in 1900 to 21.1 kg in 1933, that is an increase of 50% (González de Molina et al. 2014, 167). This level of consumption would not be reached until the 1960s (González de Molina et al. 2017, 2348). Pork and beef grew the most, by 157 and 86%, respectively (Table 2.12). Egg production also grew (55.6%) as well as milk production (42.1%). Within milk production, cow milk grew the most, by 70% (Table 2.12). During those years, milk breeds were introduced from Switzerland and Holland.[6] Farms specializing in these types of livestock emerged in or around main cities. Other livestock productions remained stable or grew slightly, for example in the case of wool or honey production.

As mentioned in the previous section, the Civil War had a major impact on livestock activity, in line with falls in production and agricultural productivity. Biophysical production data obtained from official post-war statistics show reductions in food available for livestock, due, above all, to the drop in grains and residues from culti-

Table 2.12 Livestock production in tons of dry matter, 1900–1950

	1900	1910	1922	1933	1940	1950
Beef	53.838	66.740	99.928	105.626	41.514	52.763
Lamb meat	44.487	49.257	50.942	52.892	26.429	39.289
Goat meat	9.468	11.238	11.956	14.121	5.228	6857
Pig meat	40.641	51.889	83.351	104.444	73.397	82.522
Horse meat	0	0	0	0	0	1.632
Poultry meat	4.115	4.813	5.545	6.403	6.255	4.673
Rabbit meat	491	574	661	764	3.342	4.453
Total meats	153.040	184.509	252.383	284.249	192.190	313.347
Eggs	12.728	14.886	17.150	19.805	19.347	31.435
Wool	21.933	24.285	22.190	25.442	32.605	24.762
Cow milk	104.417	127.514	134.290	177.739	226.804	259.956
Sheep milk	17.147	18.986	29.013	8.672	13.748	13.610
Goat milk	36.577	43.411	33.129	38.223	38.749	33.155
Honey	4.908	4.908	4.908	4.908	5.149	5.391

Source Agrarian statistics and author's compilation

[6] Ministerio de Fomento (1892, 1920).

vated areas. This led to more intense use of pastures than in the 1930s (Graph 2.12) and is consistent with the drop in total livestock production (12%) and especially meat (−45%) reflected in the statistics (Table 2.12) compared to 1933. A decline in agricultural production between 1940 and 1950 compared to 1936 and the impossibility to sustain raw material imports to manufacture chemical fertilizers made it unfeasible to maintain the livestock feeding model of the first third of the twentieth century as described above. Based on the data in Graph 2.12, maintaining the size of livestock as indicated in the 1940 and 1950 censuses implied a much better use of pastures than during the first third of the twentieth century.

The third and last period coincides with the industrialization of Spanish agriculture and livestock. As we will see in Chap. 6, this process was linked to big increases in demand form eat and milk (Domínguez 2001a) and to the rapid adoption of consumption patterns moving ever further away from the Mediterranean diet (González de Molina et al. 2013, 2017). The Franco regime also directly encouraged the model's adoption via its agricultural policies, which fostered intensive or industrial livestock and the production of cereals to manufacture feed, at the expense of cereals intended for human consumption (Clar 2005). Both livestock slaughtering (Graph 2.13) and milk production (Graph 2.14) grew considerably between the mid-1960s and the mid-1980s (Table 2.13).

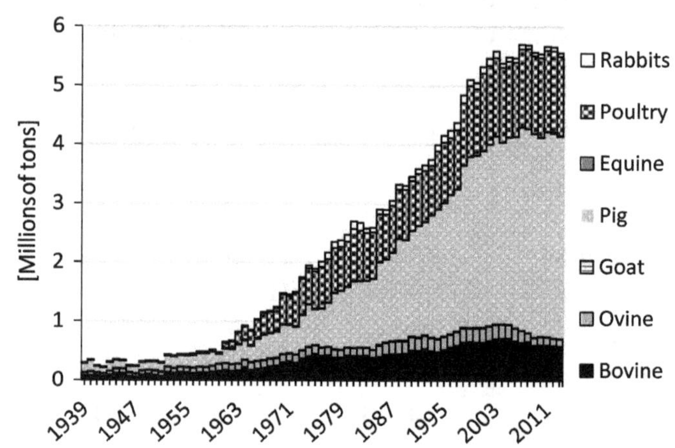

Graph 2.13 Slaughtered meat in tons of carcass weight. *Source* Agricultural Statistics Yearbooks

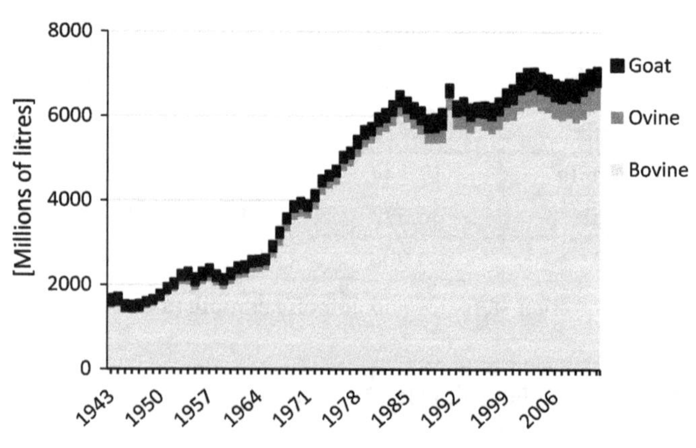

Graph 2.14 Milk production for direct consumption and transformation in Spain, in millions of liters. *Source* Agricultural Statistics Yearbooks

Table 2.13 Livestock production in tons of dry matter, 1960–2008

	1960	1970	1980	1990	2000	2008
Bovine meat	81.744	148.877	212.630	256.673	341.350	330.754
Sheep meat	58.492	74.284	77.361	128.693	132.289	97.780
Goat meat	5.844	6.401	5.731	8.819	7.890	5.203
Pig meat	135.268	256.874	547.244	1.013.979	1.611.507	1.888.429
Horse meat	6.298	5.458	3.853	2.325	2.475	2.216
Poultry meat	20.962	207.108	365.622	388.947	545.374	604.380
Rabbit meat	4.739	11.306	45.616	35.895	45.984	30.600
Total meat	313.347	710.308	1.258.058	1.835.332	2.686.869	2.959.362
Eggs	46.535	97.860	134.455	131.665	133.341	154.882
Wool	25.575	21.660	18.093	23.981	26.104	23.191
Cow milk	316.742	523.777	707.568	718.461	755.555	751.934
Sheep's milk	30.860	30.326	25.377	35.685	45.711	56.314
Goat's milk	36.208	33.155	33.291	43.694	50.486	56.372
Honey	6.295	6.816	9.883	18.255	25.071	25.081

Source Agrarian statistics and author's compilation

After this date, milk production stabilized because of Spain's entry into the European Economic Community and the establishment of the quota system. Meat production continued to grow, however, until the beginning of the century (Table 2.13). This growth is clearly correlated with the evolution of livestock herds (Graph 2.11). As we saw, sheep and goats lost relative weight in overall livestock. Although livestock grew by 75% between 1965 and 2008, poultry (354%) and pigs (452%) grew the most because of rising domestic demand for low-priced meat (Gonzâlez de Molina et al. 2017) and even for export. The industrial model undeniably permitted market-oriented livestock production to grow to an extent that would have been unthinkable in traditional organic agriculture. However, this growth came at the expense of a quasi-complete rupture of ties between livestock and the land, as well as an almost entire loss of traditional multifunctionality.

2.8 An Overview of Spanish Agriculture Industrialization

Changes caused by the industrialization of agriculture profoundly affected both the entity and the final use of biomass produced by agroecosystems. These changes were due to ongoing efforts to increase the size of commercially oriented biomass production to meet the growing demand for raw material, as well as human and animal food. Any increase in biomass production comes with its corresponding increase in land costs. Compensation was thus sought through changes in labor and technology. Agricultural production grew essentially in two different ways: on the one hand, the total amount of biomass per hectare increased and, on the other, an increasing share was appropriated. We can synthesize these changes taking into account the scale of the process, as described below.

At an aggregate scale, the changes consisted of the increase of NPP_{act}, which was more notable in croplands and lower in woodlands. This increase was at any rate far lower than the extent of growth based on a conventional analysis, measured in monetary terms or fresh matter, primary crops and some residues. Within NPP, croplands (47%) and woodlands (42%) grew as well as pastures, though to a much lesser extent (7%). Second, patterns of use of NPP_{act} changed. The growth of wooded areas and the loss of use of forest firewood, as well as the growth of woody crops, increased the accumulated biomass. On the other hand, Domestic Extraction increased moderately and unharvested biomass somewhat decreased, especially in croplands. Finally, reused biomass grew by 48%, i.e., more than total NPP, reflecting Spanish agroecosystems ever-greater orientation towards livestock. In turn, the increasing size of appropriated biomass was sustained because of extraction efforts, that took place more in croplands than in pastures and woodlands, despite the fact that woodland areas almost tripled since 1900. Croplands grew steadily until the 1970s and then regularly declined until 2008. Cultivated land surface areas rose by merely 5% when comparing surface areas in 2008 with those of 1900. Consequently, productivity per hectare of this type of land increased by almost 50%. Production pressure on Spanish agroecosystems thus centered on this type of land.

In terms of Domestic Extraction in croplands, production efforts centered in turn on primary crops, instead of "residues". Among all extracted biomass components, biomass from primary crops grew the most, more than tripling. In turn, the bulk of extracted biomass concentrated on crops of greater commercial value, resulting from production specialization and Spanish agriculture relations with markets. Added to this, agricultural production since the 1960s, and more intensely since the 1990s, centered ever more on stockfeed production, sustaining livestock specialization that, as we have seen, the Spanish agricultural sector has been experiencing in recent decades.

In summary, Spanish agriculture over the last century evolved towards increasing production commodification and significant changes in the patterns of biomass use. Both production and technological efforts have been directed towards maximizing the share of biomass of highest commercial value, entailing, as we will see, the reduction of crop multifunctionality. In other words, agricultural production growth has been much greater than the growth of agroecosystems' NPP. The process went through three distinct phases: the first period lasted until the 1960s and was characterized by the growth of agricultural production; over a second phase, livestock played a major role, and agricultural production was subjected to stockfeed; in the third and final phase, livestock production continued to increase, not only at the expense of agriculture, that stabilized production since the beginning of this century (as confirmed by the provisional data presented for 2015), but also at the expense of international trade, that is, of other countries, as commented earlier.

At landscape or at agroecosystem scale as a whole, Spanish agriculture's industrialization led to increasingly segregated land uses, and the loss of productive synergies based on agrosilvopastoral integration. This resulted from certain land use types taking over others and the rupture of the previous ecosystemic equilibrium, in turn reflecting growing specialization trends. As we have seen, production efforts concentrated on cultivated lands, thus perpetuating the process of "agricolization" or the encouragement of agricultural use that had already been taking place since the nineteenth century. Livestock growth mainly relied on intensive landless farms, without any food ties to the land. In this way, close ties within agroecosystems between agricultural and livestock activities, not only in terms of food but also in terms of replenishing animal fertility and traction, were broken up into two almost distinct activities. The introduction of synthetic chemical fertilizers and mechanization rendered use segregation viable. The same phenomenon affected forest lands, dedicated to forestry or conservation policies, that often excluded or restricted other possible uses. In short, the trend towards production specialization and intensification has been a growing requirement of Spanish agriculture's growth model that has tended to impose specialized land uses on the territory based on market demands, soil attitude, or the endowment of natural resources especially water. The result is loss of geodiversity and spatial heterogeneity. Flows of energy and materials, once more local and contained, have ended up being globalized and of fossil origin.

At a farm scale, productive specialization entailed the following consequences: a strong tendency to suppress crop associations and polycultures; the simplification of rotations and their later suppression; the quasi elimination of fallows or their substantial reduction; and the fostering of crop alternatives governed by market demands. Agriculture shifted from crop and plant heterogeneity and their layout adjustments to monocultures, entailing significant reductions of genetic, structural and functional diversity (Gliessman 1998). If at the agroecosystem scale this phenomenon progressively reduced the capacity of agroecosystems of autonomous replacement fertility, at a farm level, the relative demand for fertilizers significantly increased. The spread of chemical fertilizers made this fundamental change possible. It sparked, in turn, the desire to suppress fallow and legume sowing. Cultivation of legumes was reduced to less than half since 1900. At that time, it represented 1.76% of total agricultural production in fresh matter and was a habitual part of wheat and barley crop rotations. In 2008, it represented only 0.25% and had disappeared from many cereal productions. As we will see in Chap. 5, this led to a reduction in the flow of nitrogen from symbiotic fixation and the ability of agroecosystems to replace soil fertility by themselves.

At the crop level, very important transformations affected both plants' morphologies and their uses. This evolution also applies to different livestock species. At the beginning of the century, when agriculture and livestock were still organic, cultivated plants and livestock species each served different purposes. Livestock provided meat and milk, but also carried out agricultural tasks or transported goods. In a previous study, we referred to Cantabrian Cornice cattle, of "mixed aptitude", i.e., both for agricultural and livestock use (Fernández Prieto 1992). High cereal stalks were aimed at producing large amounts of straw, the basis of horse stockfeed. Precisely because of progress in crops, a type of livestock was developed that could feed on croplands without competing with human food.

The aim of farmers was to maximize the harvestable parts of plants, especially those with highest commercial value; in turn, the aim of ranchers was to select the species and breeds generating the largest quantity of meat and milk yields. Thus, in the case of phytomass, seeds were selected and improved to concentrate photosynthetic capacity in the marketable part of cultivated plants, for example, by modifying cereals' morphology to concentrate more biomass in grains and less in straw. The process was applied to most arable crops and led to the considerable growth in grain yields per hectare. The "Green Revolution" was largely based on the genetic change that produced crop types with a lower "waste" weight. The phenomenon is visible in the changing relationships between grain and straw in cereals and legumes (Graph 2.15). Traditional cereal and legume varieties had high straw stems, therefore less grains, and were an essential part of animal feed. With the industrialization of agriculture, these were replaced with varieties that produced more grains and less straw (Graph 2.15).

In the case of woody plants, breeding and changes to management practices consisted of translocating stem and leaf biomass to the harvestable fruit. Multifunctional usage has been superseded by a preference for the commercial use of fruits. The case of olive-growing is paradigmatic: from producing firewood, foliage and olive kernel oil for livestock, table olives, domestic lighting, and edible oil, olive groves now almost exclusively produce oil, leading to changes in grove management and tree morphology (Graph 2.16) (Infante-Amate 2014; Infante-Amate and González de Molina 2013).

All these changes explain the remarkable increase of some crop yields (620% in the case of corn between 1900 and 2008 or 332% in the case of wheat), a much higher growth than that of NPP per hectare in cultivated areas (50%) or Domestic Extraction per cultivated ha (102%). In other words, the land's biophysical productivity has grown at a much lower rate than grain or fruit yields per hectare. The reason is the concentration of biomass in grain or harvestable fruit and, therefore, the reduction of harvest residues. Another factor is the reduction of grass (e.g. weeds) that accom-

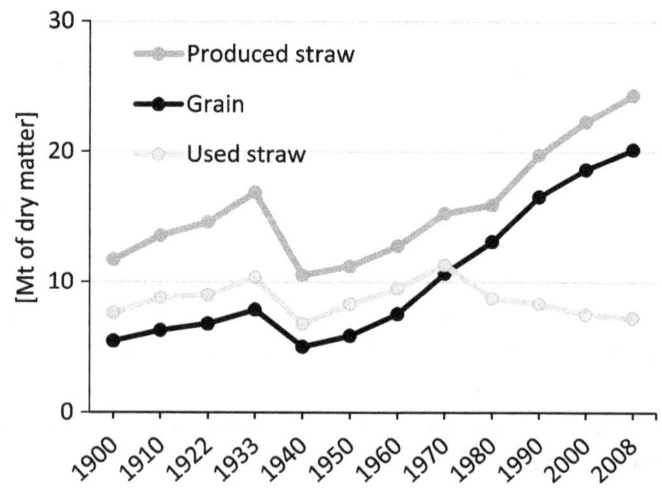

Graph 2.15 Evolution of grain and straw extracted from cereals, in Mt of dry matter. *Source* Agrarian Statistics

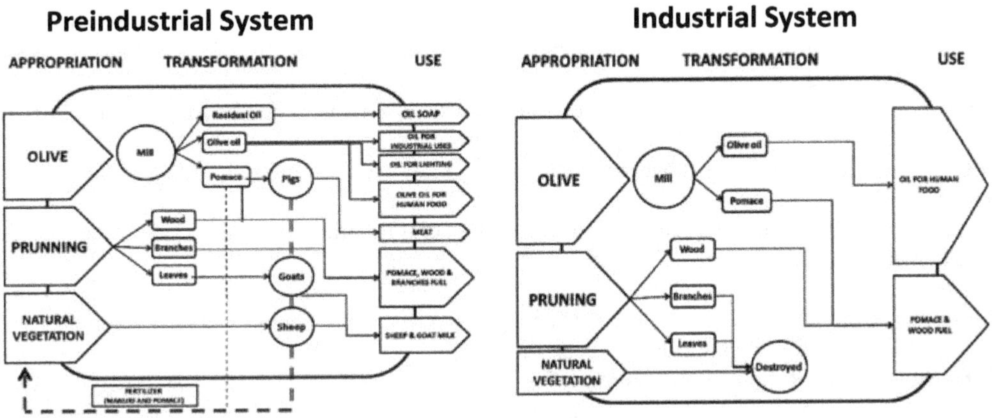

Graph 2.16 Changes in the multiple use of olive trees. *Source* Infante-Amate (2014)

panies crops based on mechanical or chemical means. Moreover, the introduction of improved and hybrid seeds, in the quest for maximum grain yields or the selection of woody plants varieties with high yields have led to a substantial loss of genetic diversity. This process has favored a more frequent and intensive use of fertilizers. The result has been the abandoning of seed varieties that were better adapted to soil and climate conditions and that, as far as we know, demanded less nutrients (Carranza et al. 2018).

In the case of livestock, a radical transformation has taken place. As livestock and production grew, extensive stock farming became less important and resorted more often to off-farm feeding. The bulk of livestock has lost its land ties and is stabled in intensive farms. Feed from abroad is often used, and breeds specialized in the production of meat or milk are imported from other countries. Labor livestock has almost completely disappeared and thanks to cheaper feed, monogastric livestock now have an unprecedented role. Ever bigger quantities of straw have been either burned or left on the farm. What remains of extensive stock farming is today the refuge of the traditional livestock breeds.

References

Agnoletti M, Cargnello G, Gardin L, Santoro A, Bazzoffi P, Sansone L et al. (2011) Traditional landscape and rural development: comparative study in three terraced areas in northern, central and southern Italy to evaluate the efficacy of GAEC standard 4.4 of cross compliance. Ital J Agron 6(1s), e16

Allen RC (2003) Progress and poverty in early modern Europe. The Econ Hist Rev 56(3):403–443

Barciela C (1986) Introducción. In: Garrabou R, Barciela C, Jiménez Blanco YJI (eds) Historia agraria de la España contemporánea. El fin de la agricultura tradicional (1900–1960). Editorial Crítica, Barcelona, pp 91–141

Barciela C, López Ortíz I (2003) El fracaso político del primer Franquismo, 1939–1959. Veinte años perdidos para la agricultura española. Barciela C (ed) Autarquía y mercado negro. El fracaso económico del primer Franquismo, 1939–1959. Editorial Crítica, Barcelona, pp 55–93

Barciela C, Giráldez J, Grupo de Estudios de Historia Rural, López I (2005) Sector agrario y pesca. In: Carreras de Odriozola A, Tafunell Sambola X (eds) Estadísticas Históricas de España, siglos XIX y XX. Fundación BBVA, Madrid, pp 245–356

Boserup E (1965) The conditions of agricultural growth: the economics of agrarian change under population pressure. Aldine, Chicago, Ill

Boserup E (1981) Population and technological change. A study of long-term trends. University of Chicago Press, Chicago, USA

Bringas MA (2000) La productividad de los factores en la agricultura española (1752–1935). Banco de España, Madrid

Carranza-Gallego G, Guzmán G, García-Ruíz R, González de Molina M, Aguilera E (2018) Contribution of old wheat varieties to climate change mitigation under contrasting managements and rainfed Mediterranean conditions. J Clean Prod 195:111–121

Cazcarro I, Duarte, R, Martín-Retortillo M, Pinilla V, Serrano A (2014) Water scarcity and agricultural growth in spain: from curse to blessing? Documento de Trabajo de la Asociación Española de Historia Económica, nº 14–19

Cazcarro I, Duarte R, Martín-Retortillo Pinilla V, Serrano A (2015) How sustainable is the increase in the water footprint of the spanish agricultural sector? A provincial analysis between 1955 and 2005–2010. Sustainability 7(5):5094–5119

Clar E (2005) Del cereal alimento al cereal pienso. Historia y balance de un intento de autosuficiencia ganadera, 1967-1972. Historia Agraria 37:513–544

Clar E, Martín-Retortillo M, Pinilla V (2016) Agricultura y desarrollo económico en España, 1870–2000. In: Gallego D, Germán L, Pinilla V (eds) Estudios sobre el desarrollo económico español. Dedicados al profesor Eloy Fernández Clemente. Prensas de la Universidad de Zaragoza, Zaragoza, pp 165–209

Cussó X (2005) El estado nutritivo de la población española 1900-1970. Análisis de las necesidades y las disponibilidades de nutrientes. Historia Agraria 36:329–358

Domínguez Martín R (2001a) La ganadería española: del franquismo a la CEE. Balance de un sector olvidado, Historia Agraria 23:29–52

Domínguez Martín R (2001b): Las trasformaciones del sector ganadero en España (1940–1985). Ager 1:47–83

Duarte R, Pinilla V, Serrano A (2014) The water foot print of the Spanish agricultural sector: 1860–2010. Ecol Econ 108:200–207

Fernández Prieto L (1992) Labregos con ciencia. Estado e sociedade e innovación tecnolóxica na agricultura galega, 1850–1936. Xerais, Vigo

Fernández Prieto L (2007) El apagón tecnológico del Franquismo. Estado e innovación en la agricultura española del siglo XX. Tirant Lo Blanch, Valencia

Gallego D (2001) Historia de un desarrollo pausado: integración mercantil y transformaciones productivas de la agricultura española. In: Pujol J et al. (eds) El pozo de todos los males. Sobre el atraso de la agricultura española contemporánea. Crítica, Barcelona, pp 147–214

Gallego Martínez D (1993) Pautas regionales de cambio técnico en el sector agrario español (1900–1930). Cuadernos Aragoneses de Economía, 3 (2):241–276

García Sanz A (1991) La ganadería española entre 1750–1865: los efectos de la reforma agraria liberal. Agricultura y Sociedad 72:81–120

Garrabou R, González de Molina M (2010) La reposición de la fertilidad en los sistemas agrarios tradicionales. Icaria, Barcelona

GEHR (Grupo de Estudios de Historia Rural) (1978) Contribución al análisis histórico de la ganadería española, 1865–1929. Agricultura y Sociedad 8:129–182

GEHR (Grupo de Estudios de Historia Rural) (1979) Contribución al análisis histórico de la ganadería española, 1865–1929. Agricultura y Sociedad 10:105–169

GEHR (Grupo de Estudios de Historia Rural) (1991) Estadísticas Históricas de la producción agraria española, 1859–1935. Ministerio de Agricultura Pesca y Alimentación, Madrid

Gliessman SR (1998) Agroecology. Ecological processes in sustainable agriculture. Lewis Publishers (CRC Press), Boca Ratón

González de Molina M, Casero Rodríguez F (1992) Mito y realidades de los incendios forestales en Andalucía. In: Alcantud JAG, Rey MJB (eds) El fuego: mitos, ritos y realidades. Barcelona, Anthropos

González de Molina M, Soto D, Infante-Amate J, Aguilera E (2013) ¿Una o varias transiciones? Nuevos datos sobre el consumo alimentario en España (1900–2008). In: XIV Congreso Internacional de Historia Agraria (Badajoz, 7–9 de noviembre 2013). https://doi.org/10.13140/2.1.1823.5684

González de Molina M, Soto, D, Aguilera E, Infante-Amate J (2014) Crecimiento agrario en España y cambios en la oferta alimentaria, 1900–1933. Hist Soc 80:157–183

González de Molina M, Soto D, Infante-Amate J, Aguilera E, Vila Traver J, Guzmán GI (2017) Decoupling food from land: the evolution of spanish agriculture from 1960–2010. Sustainability 9(12):23–48

Infante-Amate J (2014) ¿Quién levantó los olivos? Historia de la especialización olivarera en el sur de España (s. XVIII–XX). Ministerio de Agricultura, Alimentación y Medio Ambiente, Madrid

Infante-Amate J, González de Molina M (2013) 'Sustainable de-growth' in agriculture and food: an agro-ecological perspective on Spain's agri-food system (year 2000). J Clean Prod 38:27–35

Infante-Amate, J, Soto D, Aguilera E, García Ruiz R, Guzmán, G, Cid A, González de Molina M (2015) The Spanish transition to industrial metabolism long-term material flow analysis (1860–2010). J Ind Ecol 19(5):866–876. https://doi.org/10.1111/jiec.12261

Infante-Amate J Aguilera E, Palmeri F, Guzmán G, Soto D, García-Ruiz R, González de Molina (2018) Land embodied in Spain's biomass trade and consumption (1900–2008): Historical changes, drivers and impacts. Land Use Policy 78:493–502

Infante-Amate J, Parcerisas L (2013) El carácter de la especialización agraria en el Mediterráneo español. El caso de la viña y el olivar en perspectiva comparada (1850–1935). Comunicación al XIV Congreso Internacional de Historia Agraria (Badajoz, 7–9 de noviembre de 2013)

Infante-Amate J, Aguilera E, González de Molina M (2014) La gran transformación del sector agroalimentario español. Un análisis desde la perspectiva energética (1960–2010). Documento de Trabajo de la Sociedad de Estudios de Historia Agraria, nº 14–03. https://ideas.repec.org/p/seh/wpaper/1403.html

Infante-Amate J, Soto D, Iriarte Goñi I, Aguilera E, Cid A, Guzmán G, García Ruiz R, González de Molina M (2014b) La producción de leña en España y sus implicaciones en la transición energética. Una serie a escala provincial (1900–2000). Sociedad Española de Historia Agraria, Documento de Trabajo nº 14–16. http://econpapers.repec.org/paper/ahedtaehe/1416.htm

Kaplan JO, Krumhardt KM, Zimmermann N (2009) The prehistoric and preindustrial deforestation of Europe. Quat Sci Rev 28(27–28):3016–3034

Lassaletta L, Billen, G, Grizzetti B, Garnier J, Leach AM, Galloway JN (2014a): Food and feed trade as a driver in the global nitrogen cycle: 50-year trends. Biogeochemistry 118(1–3), 225–241.

Lassaletta L, Billen, G, Romero E, Garnier J, Aguilera E (2014b) How changes in diet and trade patterns have shaped the N cycle at the national scale: Spain (1961–2009). Reg Environ Change 14(2):785–797.

Lambin EF, Meyfroidt P (2011) Global land use change, economic globalization, and the looming land scarcity. Proc Natl Acad Sci 108(9):3465–3472

MAPAMA (Ministerio de Agricultura, Peca, Alimentación y Medio Ambiente) (2016) Anuario de las Producciones Agrarias. Madrid

Martín-Retortillo M, Pinilla V (2015) Patterns and causes of the growth of European agricultural production, 1950 to 2005. Agric Hist Rev 63(1):132–159

McNeill JR (2001) Algo nuevo bajo el sol: Historia ambiental del mundo en el siglo XX. Alianza Editorial, Madrid

Meyfroidt P Rudel TK, y Lambin EF (2010) Forest transitions, trade, and the global displacement of land use. In: Proc Natl Acad Sci 107(49):20917–20922

MF (Ministerio de Fomento) (1892) La ganadería en España. Avance sobre la riqueza pecuaria en 1891, formada por la Junta Consultiva Agronómica, conforme a las memorias reglamentarias que en el citado año han redactado los ingenieros del Servicio Agronómico. Ministerio de Fomento, Madrid

MF (Ministerio de Fomento) (1920) Estudio de la ganadería en España. Resumen hecho por la Junta Consultiva Agronómica de las memorias de 1917, remitidas por los ingenieros del servicio agronómico provincial. Imprenta de los Hijos de M.G. Hernández, Madrid

Parrotta JA, Trosper RL (2012) Traditional forest-related knowledge: sustaining communities, ecosystems and biocultural diversity. Springer, New York

Pinilla V, Serrano R (2008) The agricultural and food trade in the first globalization: spanish table wine exports, 1871–1935. A case study. J Wine Econ 3(2):132–148

Prados de la Escosura L (2003) El progreso económico de España (1850–2000). Bilbao, Fundación BBVA

Radkau J (2008) Nature and power: a global history of the environment. Cambridge University Press, Cambridge

Radkau J (2011) Wood. A History. Polity Press, Cambridge, UK

Simpson J (1997) La agricultura española (1765–1965): la larga siesta. Alianza, Madrid

Soto Fernández D (2006) Historia dunha agricultura sustentábel. Transformación sprodutivas na agricultura galega contemporánea. Santiago de Compostela. Xunta de Galicia, Consellería do Medio Rural

Warde P (2006) Fear of wood shortage and the reality of the woodland in Europe, c. 1450–1850. Hist Workshop J 62(1):28–57

Williams M (2003) Deforesting the earth: from prehistory to global crisis. University of Chicago Press, Chicago

Williams M (1990) Forests. In: Turner BL, Clark WC, Kates RW, Richard JF, Matthews JT, Meyers WB (eds) The Earth as transformed by human action, global and regional changes in the biosphere over the past 300 years. Cambridge University Press, Cambridge, pp 179–201

3

Quantification of Inputs in Spanish Agriculture

The socio-metabolic transition to industrial society led to displacing agriculture as the main source of energy and materials and generalizing the use of fossil fuels and minerals (Krausmann and Haberl 2002; Fischer-Kowalski and Haberl 2007; Krausmann et al. 2008; Kuskova et al. 2008; Infante-Amate et al. 2015). This transition also affected the agricultural sector itself, which underwent big quantitative and qualitative changes in its technical means of production. These technological changes are usually associated with corresponding increases in agricultural production (increase in land productivity) and in the reduction of human labor (increase labor productivity) (Boserup 1981; Giampietro et al. 1999; Fischer-Kowalski et al. 2014). As we have seen in Chap. 2, Spanish agroecosystems experienced significant intensification and productive specialization. Given that land uses were segregated, that many connections between them broke down, and technological innovation orientations (Fernández Prieto 2001), agroecosystems increasingly needed inputs imported from outside the agricultural sector. Organic inputs produced in the farms themselves and in the local environment, such as manure and animal traction, were replaced by large quantities of inorganic inputs powered and manufactured with fossil fuels, including synthetic fertilizers and pesticides, machinery, fuel and electricity (Guzmán Casado and González de Molina 2009).

In this chapter, we quantify the inputs used in Spanish agriculture since the beginning of the twentieth century. The aim is to examine the evolution of the social fund element, i.e., the technical means of production (TMP), during the industrialization process. TMP eventually not only had a decisive role in maintaining productive activity but productive activity ended up being dependent on it. The objective is not that of examining technical changes in agriculture, which goes beyond the scope of this research, but to evaluate the magnitude of the changes and their consequences for the functioning of Agrarian metabolism (AM). Although we have been expressing flows in tons of materials, in this chapter, we chose to express them in units of energy to give a fuller account of the developments. In Chap. 6, where we analyze the structure and functioning of agrarian metabolism by relating both the funds and the flows from which they originate, we will express inputs in tons to maintain metric consistency.

3.1 Comments on Methodology

According to our methodological proposal, inputs from outside the agricultural sector constitute the Imports (I) in the MEFA methodology. These Imports added to Domestic Extraction (DE) determine domestic consumption (DC) after deduction of Exports (E). Although we considered both, human work and animal traction, to calculate the EROIs, they are not taken into account here because they can be defined as flows that circulate within the agroecosystem and do not, therefore, come from outside. We thus developed a time series on the use of external inputs from 1900 to 2008. Their type and quantity were obtained mainly from agricultural statistics supplemented with technical reports as well as research studies and were based on the assumption that growth rates were constant during the years for which we lacked data. Yearbook data included fertilizers since 1933, tractors and other agricultural machines since 1955, fuels since 1960, pesticides since 1933 and greenhouses and tunnels from 1975 onwards. Pesticide data from 1950 to 1980 were expressed in monetary terms, and we converted them into weights using deflation data from Carreras and Tafunell (2005). Fertilizers in 1900–1922 were estimated from the data compiled by Gallego Martinez (1986) and Mateu Tortosa (2013). Fuel consumption data for 1950 and 1990–2008 were retrieved from Spanish statistics (MI 1961a, b; MINETUR 2015) and from FAOSTAT (FAO 2016) for the years 1970–1980. Fuel consumption from 1900 to 1940 was estimated based on the machinery's installed capacity. Electricity consumption data for 1950 were obtained from the INE (1960), from Carpintero and Naredo (2006) for 1960, from FAOSTAT (FAO 2016) for 1970–1980 and from MINETUR (2015) for 1990 onwards. The consumption of electricity before 1950 was estimated assuming that agricultural electricity represented the same proportion of total electricity consumption in Spain as in 1950. Corominas data (2010) was used to take into account upstream electricity consumption in the irrigation. The surface areas represented by each type of irrigation were taken from MAGRAMA (2015a). The official machinery data in the first half of the twentieth century was complemented by data from Martinez-Ruiz (2000). We considered that 97% of the greenhouses belonged to the "'Almería vineyard' 'type and 3% were of the' 'Glass greenhouse'" type (MAGRAMA 2008).

The data are presented at selected decadal intervals, that is, 1900, 1910, 1922, 1933, 1940, 1950, 1960, 1970, 1980, 1990, 2000, and 2008, as we did for biomass production. Where possible, the decennial values were defined based on 5-year averages. The series combines raw data taken directly from historical sources, estimates made by other researchers and our own estimates. Since official statistics are published annually, we show long-term trends for some years as we believe they are of interest. We divided the inputs into industrial (chemical fertilizers, machinery, etc.) and non-industrial (biomass) inputs. The energy we calculated for industrial inputs is embodied energy, that is, the sum of gross energy of the input plus the energy needed for its production and distribution. The embodied energy of industrial inputs evolved over time, following varying trends in production energy efficiency and the delivery of inputs. We developed a working document that describes the embodied energy (and its components) in agricultural inputs over the 1900–2010 period, together with theoretical and methodological considerations (Aguilera et al. 2015). Given that the data in this working document are provided in a 10-year format and do not always correspond to data for Spain, missing data were estimated by linear interpolation. To estimate the machinery's embodied energy, we took into account the installed power (MW) of the machinery, the years of manufacture of the "mix" of machinery and the actual replacement rate in the year studied (estimated according to records and annual census). Estimated embodied energy in the electricity was based on the Spanish electric mix and thermal power plant efficiency (Bartolomé-Rodríguez 2007; UNESA 2005; MINETUR 2016; REE 2012), supplemented with data on the embodied energy of fuels, taken from Aguilera et al. (2015). In terms of non-industrial inputs, the energy contained in the net imported biomass (seeds and feeds) was the gross energy of the different products, calculated using conversion factors included in Guzmán et al. (2014). The energy cost of transport taken from Aguilera et al. was added to it (2015). The energy required to produce the biomass was not taken into consideration, to avoid double accounting problems, since this cost must be attributed to the agroecosystems of origin.

The embodied energy metric would be equivalent to the concept of "cumulative energy demand" used in life cycle assessments as well as to the concept of "energy intensity" used in some energy studies. All embodied energy components are expressed in terms of gross energy. Energy requirements refer to the energy used in the production of a given input. They are divided into direct and indirect energy requirements. The direct energy requirements refer to the gross energy of the fuels used directly in the production process. Indirect energy requirements include all the remaining processes necessary for input production and its use on the farm, including the production and transport of fuel, production, and transport of raw materials, energy integrated into buildings and equipment and transportation of finished products to the farm. Worthy of note, only physical processes were included. For the analysis of the agricultural systems' energy inputs, we considered it more appropriate to use gross energy (GE), rather than net energy, since the former reflects total energy contained in the input. In addition, agricultural energy products were almost always expressed in gross energy values, as in our review of the energy content of biomass products and residues (Guzmán et al. 2014). Therefore, we also used GE values in our analysis of the embodied energy of agricultural inputs. On the other hand, we did not apply any quality correction factor to the calorific value of the different fuels.

Changes in energy efficiency in the manufacturing and operation of inputs were taken into account, especially industrial ones. To our knowledge, however, these changes in efficiency have barely been considered in historical analyses of agrarian systems. Only studies based on monetary data systematically take these changes into account (e.g., Cleveland 1995; Cao et al. 2010). The studies by Pelletier et al. (2014) on egg production in the USA and Pellegrini and Fernandez (2018) on global energy use in agriculture are two of the few examples of studies accounting for temporary changes in the energy efficiency of agricultural inputs from a life cycle analysis perspective.

The general trend has conisted in significant technological improvements in energy efficiency in the production of most agricultural inputs, such as nitrogen and phosphate fertilizers, steel for the production of machinery, etc. (Smil 1999, 2013; Jenssen and Kongshaug 2003; Ramirez and Worrell 2006; Dahmus 2014). Over some periods, such as during the 1980s energy crisis, this trend intensified due to the increase in energy prices and concerns about the security of energy supplies (Bhat et al. 1994). However, some agricultural inputs, which in the beginning required little energy due to the ease of extraction, have ended up demanding higher consumption levels for their extraction and refining (Meadows et al. 1972; Gutowski et al. 2013). Despite technological improvements, the energy efficiency of raw material production may have decreased. This is the case, for example, of oil and gas production in the USA (Hall et al. 2009, 2014), and generally worldwide (Gagnon et al. 2009; Hall et al. 2014), where energy return on investment (EROI) is already decreasing. An exhaustive historical compilation of embodied energy coefficients has been carried out for the main agricultural inputs (Aguilera et al. 2015). The calculations described in this chapter are based on this latter compilation.

3.2 Mechanical Traction

In our categorization, traction includes the energy used to perform mechanical work contained in agricultural tasks, excluding the pumping of water for irrigation, and also exluding the work performed by draft animals, which is based on internal energy from the agroecosystem. Therefore, it includes the work accomplished by self-propelled machines in the field, such as tillage, sowing or harvesting, but also the tasks that were performed outside the farm, such as threshing. Quantifying the total energy and environmental impacts of mechanical traction requires determining the amount of fuels used for direct and indirect energy consumption as well as the embodied energy of manufacturing and machinery maintenance. Additional data on animal traction and greenhouse gas emissions from animal and mechanical traction is provided in Aguilera et al. (2019a).

3.2.1 Machinery

In this section, we reconstruct the time series of the most important parameters to estimate energy and environmental impacts associated to farm machinery production and maintenance in Spain. In our case, this means reconstructing, for each type of farm machine, the annual census, the number of annual registrations in the census and removals from it, the average useful life, and the weight of new, removed and average machines of the census. The machine types studied are locomobiles, threshers, tractors, harvesters, other motors (static), tillage machinery and other farm implements.

The number of motorized farm machines is reported yearly by official statistics since 1970 (MA 1973; MAPA 1997; MAGRAMA 2012), and each 5 years since 1955 (MA 1973), although Martinez-Ruiz (2000) provides an annual series since 1955. We also have machine numbers in 1932 (MAIC 1932) and 1947 (Martinez-Ruiz 2000). Martinez-Ruiz (2000) provides threshers and locomotive numbers in 1900. We assumed that the first tractors were registered in 1902 and the first self-powered harvesters and 1-axis tractors (rototillers) in 1940. We estimated values in the missing years assuming constant growth rates between the two closest years with available data. We also assumed that there was no growth during the Civil War period (1936–1939).

The number of machine registrations each year is provided since 1970 (MAPA 1997; MAGRAMA 2012). There is no available information regarding yearly removals of farm machinery from the census. Therefore, we estimated this value as the census of a given year minus the sum of registrations of that year plus the census of the previous year. We estimated registrations and removals in years previous to 1970 assuming the same removal rate, with respect to the census, as the average of the first 20 years with available data (1970–1989).

In some specific years, the census and the yearly registration values do not match, i.e., when the change in the census between 2 years is higher than reported registrations in that year. This means that registrations are too low even assuming that there were no removals in that year. In those cases, we have assumed that there were no removals, and the number of registrations was equaled to the difference between the census of the previous and the current year. Moreover, in 1980 there was an unusual peak in the census, followed by the continuation of a fairly stable trend. We corrected the census value in that year assuming that it followed a constant growth rate.

There is no information on the total number of harvesters registered each year, only on the number of cereal harvesters. Likewise, the number of total harvesters is reported in 1990 for the first time. We used the ratios of total harvesters to cereal harvesters from the census, in those years where data is available, to estimate the total number of harvesters registered (in the whole series) and censed (in years previous to 1990). Total harvesters represented 105% of cereal harvesters in 1990 and 115% in 2011. The ratio of 1990 was used for the pre-1990 series.

The census of "other engines" (excluding irrigation) is only available for a few years along the whole studied period: 1932 (MAIC 1932), 1962 (INE 1963), 1973 (MA 1973), 1980, 1985 and 1988–1993 (MAPA 1997). We assumed constant growth rates in the missing years. In 1900, we assumed the same number of other engines as threshers, in line with the proportion between these two types of machines in the 1932 census.

The number of threshers is provided in 1900, 1947, 1955–1975 (Martinez-Ruiz 2000), 1932 (MAIC 1932) and 1976–1981, 1983 and 1986 (INE 1997). We estimated the missing years assuming constant growth rates.

There is high uncertainty in the evolution in the number of machines during the first half of the twentieth century (Graph 3.1a). However, we can observe a substantial growth between 1900 and 1932. From 308 locomobiles, 300 threshers and 300 "other engines" at the beginning of the century, to 538 locomobiles, 5062 threshers, 5312

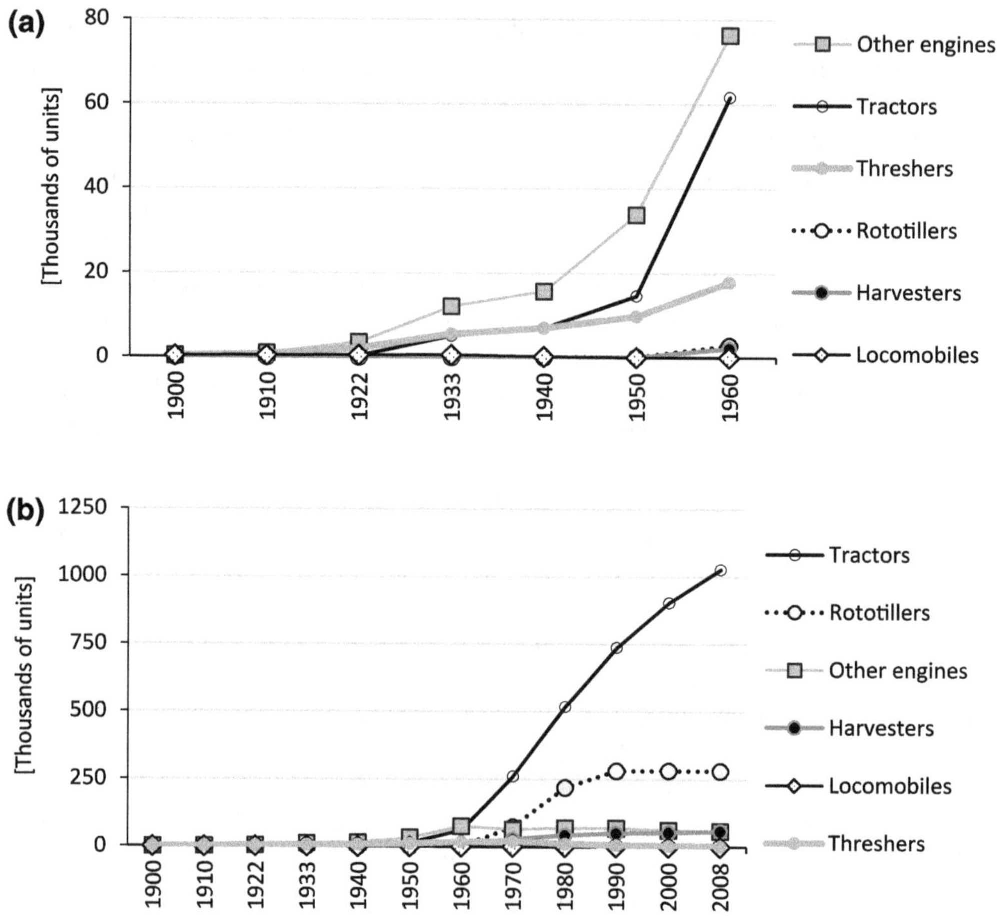

Graph 3.1 Historical evolution of the number of motorized machines in Spanish agriculture, 1900–1960 (**a**) and 1900–2008 (**b**), in thousands of units

"Other engines" and 4084 tractors in 1932. Despite the impact of the Civil War and the post-war period, by 1947 the number of tractors had doubled, and in 1955 there were 27,671 units. Thus, the data shows that locomobiles never grew too high above the levels of the nineteenth century, while most of the growth in fam machinery until almost 1930 was dominated by threshers and "other engines", followed by tractors, which became dominant in the 1950s. The number of tractors and harvesters kept growing until the end of the period in 2010, up to 1,049,950 and 60,263, respectively (Graph 3.1b). As we will see later, this is not due to ever-increasing mechanization of Spanish agriculture but to the lack of removals from the census.

The number of registrations peaked in the late 1970s in the case of tractors and rototillers (Graph 3.2a and b, respectively), and in the late 1960s in the case of harvesters (Graph 3.2c). However, removals from the census never matched registrations, implying the continuation of the growth in census numbers (Graph 3.2c) even decades after the peak in annual registrations. This implies that the average age of registered farm machinery was also growing.

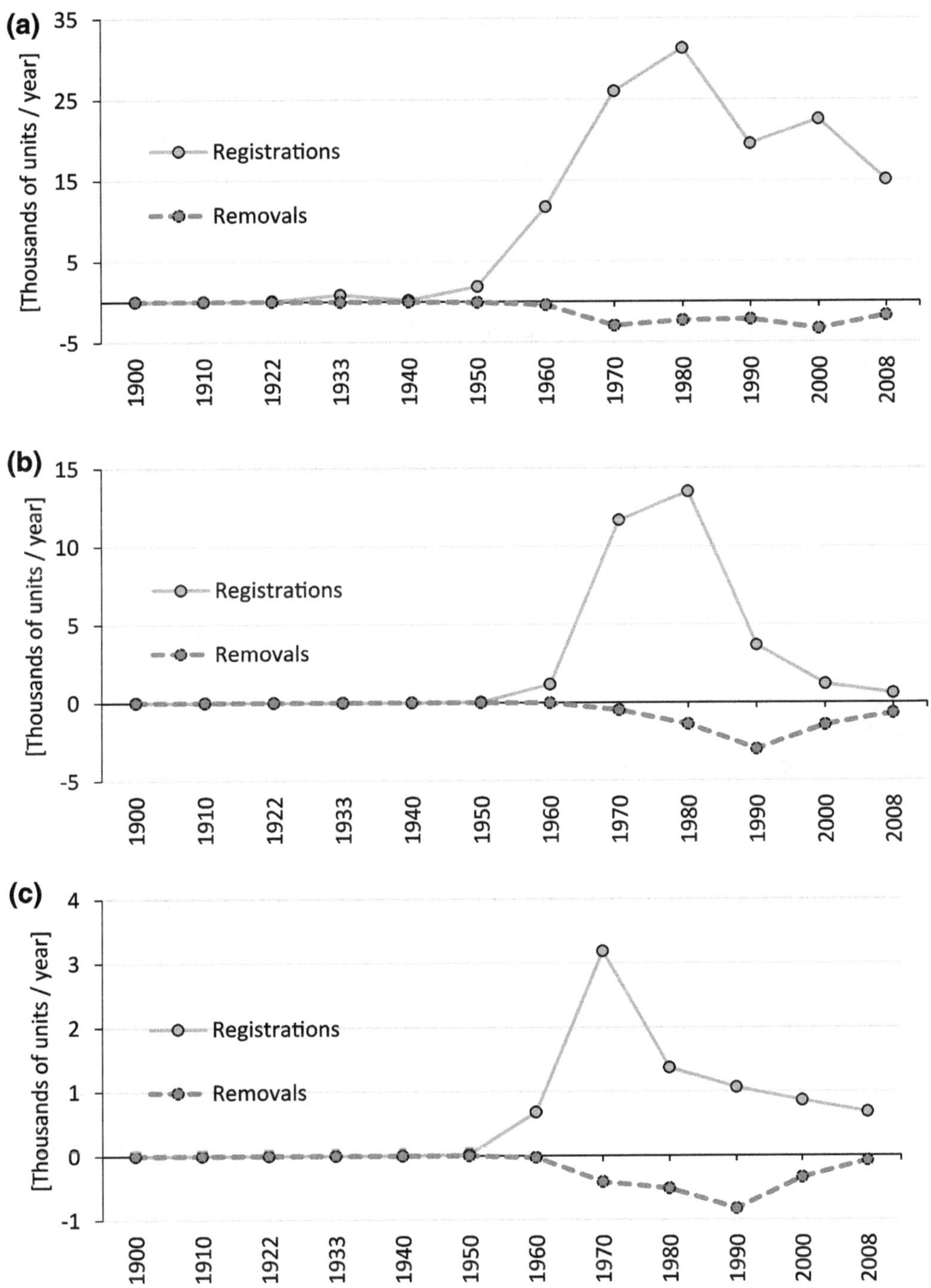

Graph 3.2 Historical evolution of the annual number of registrations and removals of tractors (**a**), rototillers (**b**) and harvesters (**c**) from the agricultural census, 1900–2008, in thousands of units per year (three-year moving average)

We estimated the average life expectancy of machinery assuming that the machines removed in a given year were the oldest ones in the census (Graph 3.1). In the period previous to 1970, when there were no registrations data, and thus the uncertainty of removals was very high, we assumed a constant life expectancy equal to the average of the first 20 years with available data (1970–1989). In the case of threshers, locomotives and other engines we do not have registrations and removals data during the whole studied period. In those cases, we assumed a constant life expectancy of 20 years.

Graph 3.3 shows a large increase in the estimated life expectancy of farm machinery in Spanish agriculture in the period (1970–2010). Some reports support this assumption showing that 31% of harvesters were more than 15 years old in 1990 (Pérez-Minguijón 1992), 55% of the tractors were more than 15 years old in 2006 (ANSEMAT 2006), and 47% of them were more than 20 years old in the same year (MAPA 2007). The age of the machines is typically inversely related to their yearly use time. Thus, harvesters less than 5 years old were used 38 days per year, while those older than 15 years were used 22 days per year on average in Spain in 1990 (Pérez-Minguijón 1992). Thus, the increase in apparent life expectancy does not necessarily mean that the machines are used for an increasingly longer period. Instead, the most probable cause is that old tractors with very little use or even not used anymore are not removed from the census (Pérez-Minguijón 1992, 1999). This possibility is supported by the fact that diesel fuel consumption has not grown significantly since the mid-1970s (Sect. 3.2.2), despite installed tractor power has quadrupled in the same period and specific fuel consumption of new tractors has only decreased by 33% in that period (Aguilera et al. 2015). Moreover, specific reports which have studied the real use of farm machines further support this argument. For example, a study by MAPA (2007) found that 14.5% of registered tractors were not being used. Thus, the overestimation of the census data implies an underestimation of removals and thus an increase in apparent life expectancy according to our calculations. (Graphs 3.4, 3.5 and 3.6).

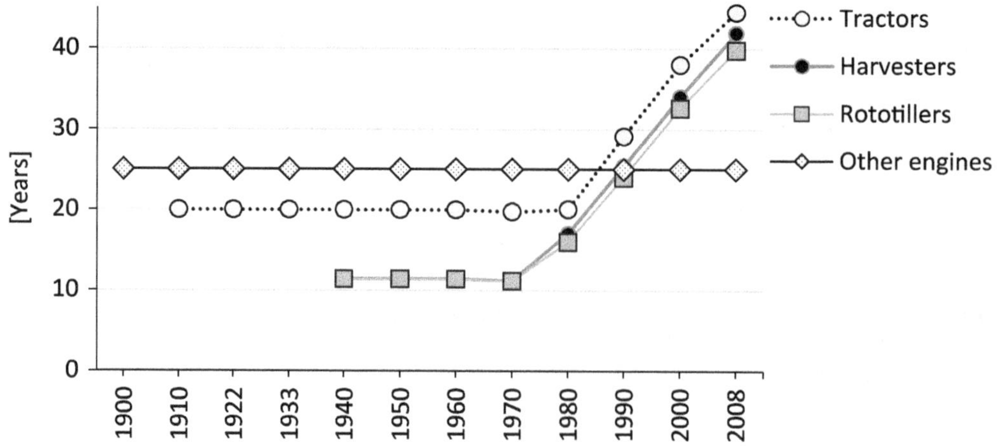

Graph 3.3 Historical evolution of the life expectancy of tractors, rototillers, and harvesters, 1970–2010, in years

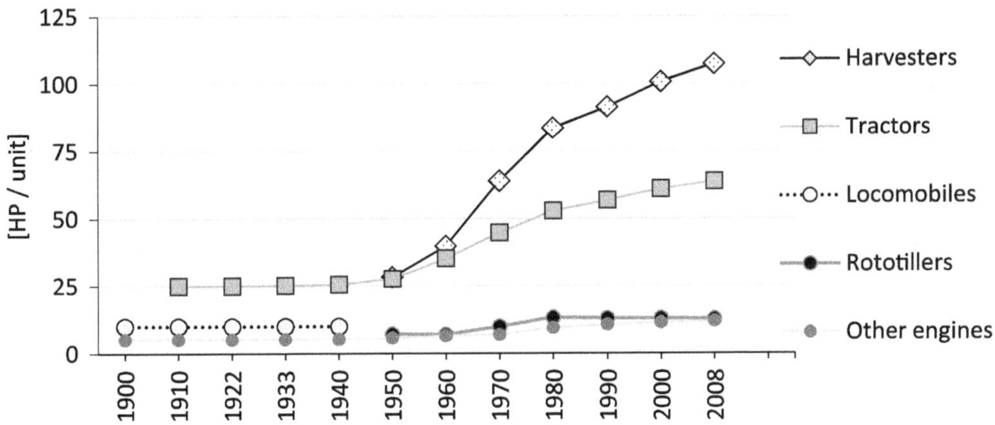

Graph 3.4 Historical evolution of the average rated power per machinery unit, 1900–2010, in HP/unit

3.2.2 Fuels

Direct energy consumed in traction includes all agriculture fuels reported by official agricultural statistics except those employed in irrigation and in modern livestock production (particularly for heating). We have estimated fuel energy employed in irrigation in Sect. 3.3. Regarding fuel consumption in livestock production, we have assumed that all non-liquid fuels, except coal during the first half of the twentieth century (which was used by threshers and locomobiles), were used in this activity. In the first part of this section, we show our reconstruction of total fuel consumption in Spanish agriculture. Then, we segregate traction, irrigation, and livestock fuel consumption.

The first official published data on fuel consumption in Spanish agriculture refers to 1950 and 1951 (MI 1961a, b). In 1978, the *Anuario de Estadística Agraria* report (MA 1978) included an annual series of fuel consumption in Spanish agriculture starting in 1958. Another series, also by MA, includes gasoline, diesel, and oil fuel consumption from 1960 to 1977 (MA 1966, 1970, 1975a, b, 1976, 1977). FAOSTAT also reports agricultural fuel consumption data in Spain, starting in 1970 (FAO 2016). The last official series is released by the Spanish Ministry of Industry and starts in 1990 (MINETUR 2015). On the other hand, scholars studying energy balances have also compiled historical sources to estimate fuel consumption in Spain. The first of these works, by Naredo and Campos (1980), reports fuel consumption data, distinguishing diesel fuel and other fuels, in 1950, 1951, 1977, and 1978. A more recent work (Carpintero and Naredo 2006) reports total fuel consumption from 1960 to 2000, based on OECD energy balances. The latter reference is also the source of FAOSTAT data.

The studies mentioned above employ different units for reporting fuel consumption, including kcal, toe (Mg oil equivalent), tce (Mg coal equivalents), liters, and joules. Moreover, some of them express the data as net energy and other as gross energy. We have attempted to harmonize the data expressing all of them as TJ gross

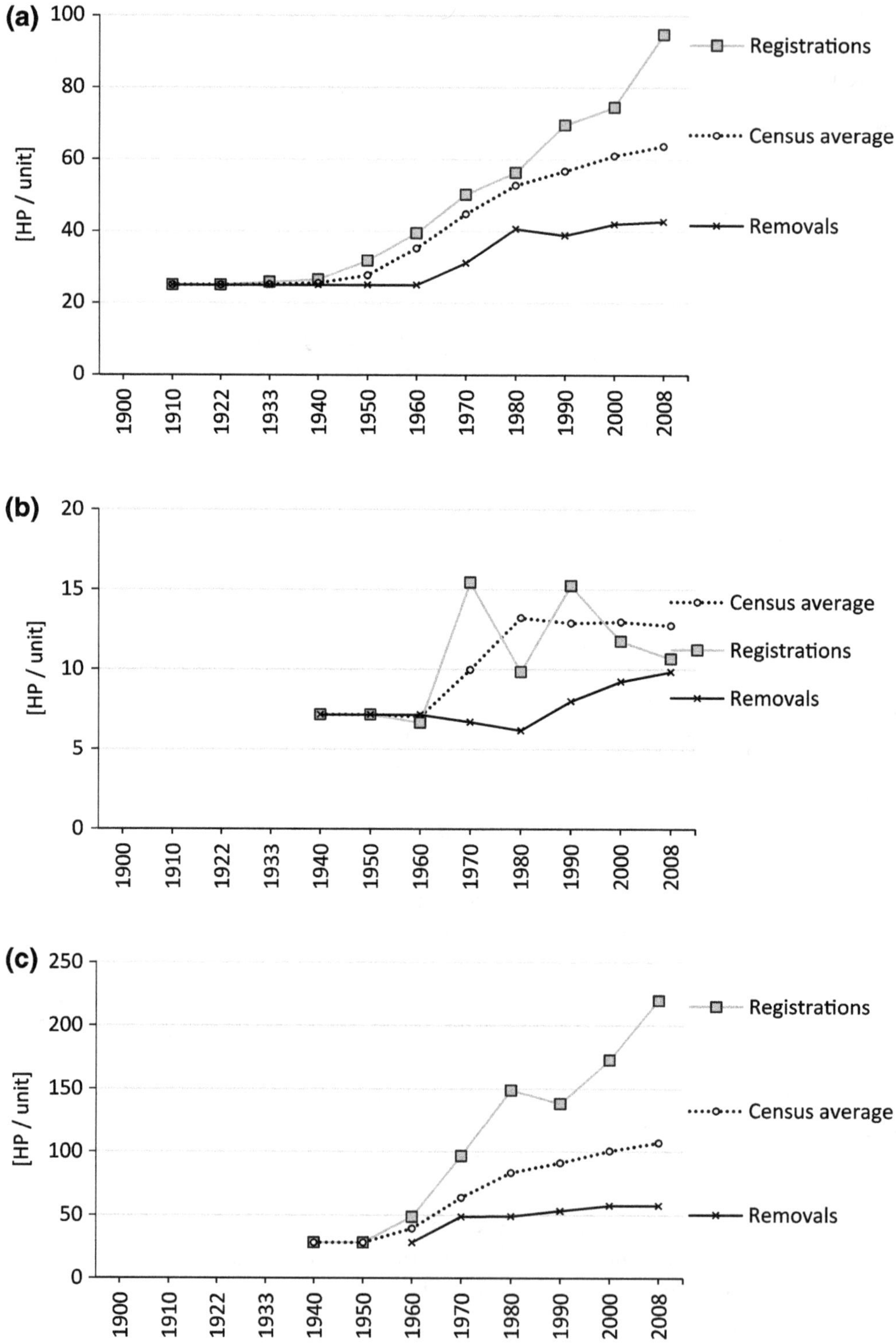

Graph 3.5 Historical evolution of the rated power of the average machines, registered machines and removed machines, including tractors (**a**), rototillers (**b**), harvesters, (**c**) and other engines (**d**), in HP/unit (three-year moving average)

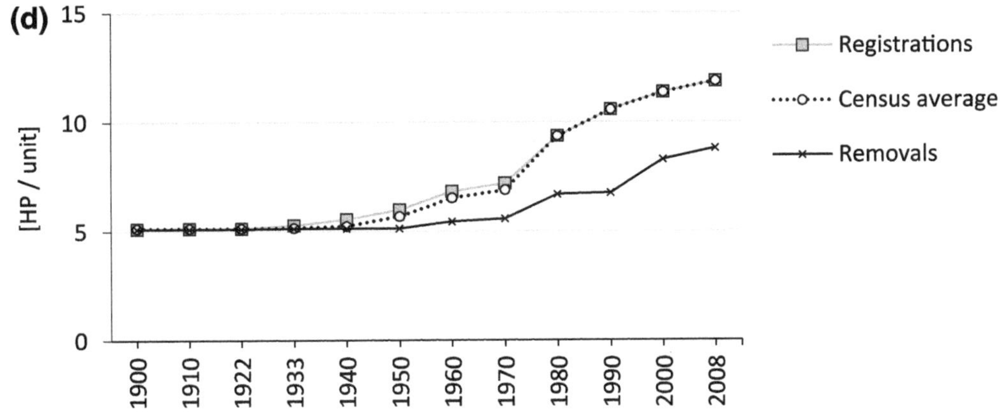

Graph 3.5 (continued)

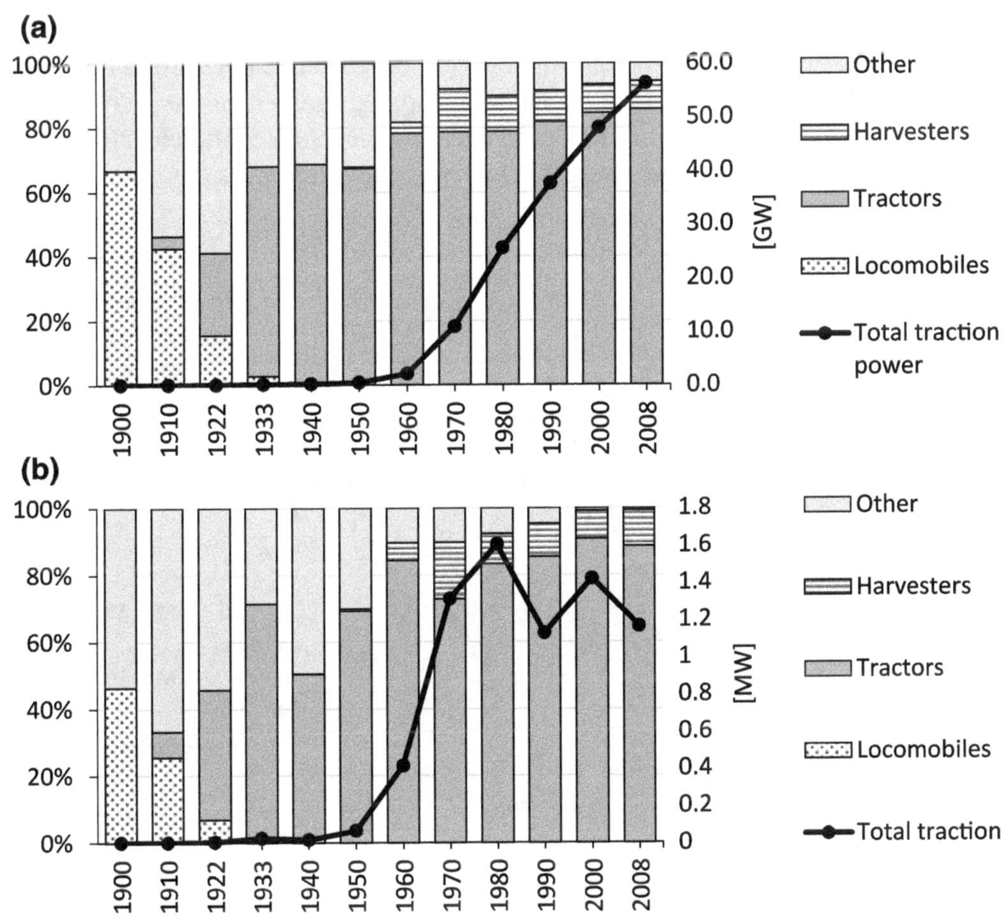

Graph 3.6 Historical evolution of the installed traction power in Spanish agriculture, 1900–2010. Total installed power, in GW (**a**) and gross annual registrations, in MW/year (**b**)

Table 3.1 Energy conversion units of fossil fuels

	Density kg/m^3	Litres/ton	GCV GJ/t	NCV GJ/t	GCV GJ/litro	NCV GJ/litro
Kerosene	802.6	1246	46.2	43.9	37.1	35.3
Motor gasoline	740.7	1350.1	47.1	44.8	34.9	33.1
Gas /dieseloil	843.9	1185	45.7	43.4	38.5	36.6
Naphta	690.6	1448	47.7	45.3	33	31.3
LPG	522.2	1915	50.1	46.2	26.2	24.1
Natural Gas	799.6	1250.6	50.4	45.4	40	36.3

Source IEA (2015)

energy. Energy conversion units employed are shown in Table 3.1. The results of the comparison are shown in Graph 3.7.

In Graph 3.7 we can observe varying levels of coincidence in reported fuel consumption in Spanish agriculture between the compared sources. Some of the differences might be explained by differences in energy conversion coefficients (between different units and net to gross energy). However, the differences between sources

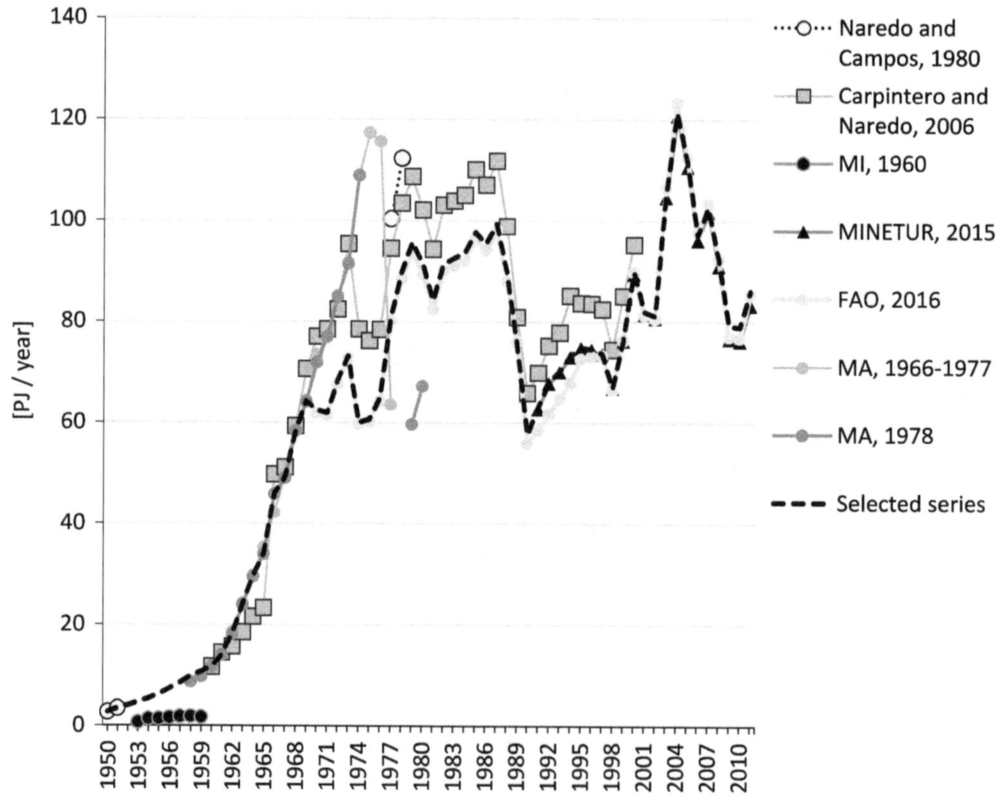

Graph 3.7 Historical evolution of fuel consumption in Spanish agriculture according to different sources, 1950–201, in petajoules

are often not fixed, implying that there are other factors responsible for the observed differences. We constructed an own series (dotted line in Graph 3.7) prioritizing the national official sources, followed by international official sources. We completed the series in some periods with data of missing fuel types from secondary sources (research articles and monographs). We have tried to include the least number of conversions between energy units and types, and we have discarded the data not coherent with the other series. In particular, FAOSTAT (FAO 2016) data from 1970 to 1989 has been complemented with data from Spanish reports. Our series starts with 1951 and 1952 data from Naredo and Campos. Then, we assume a constant growth rate in fuel consumption until 1958, when data from MA (1978) starts. From 1970 to 1974 we take diesel and LPG data from FAOSTAT, but gasoline and other liquids data from MA (1978). From 1975 to 1979 we take all data from FAOSTAT, but estimate other liquids consumption by interpolating MA (1978) value in 1969 with FAOSTAT value in 1980. From 1980 to 1989, all values are from FAOSTAT. From 1990 to 2013, all values are from MINETUR (2015), the official Spanish statistics, which is very similar to FAOSTAT for most fuel types, although there are significant differences in some specific fuels types and periods, such as fuel oil from 1990 to 2003 (Graph 3.8). Moreover, FAO data does not include renewable fuels reported by

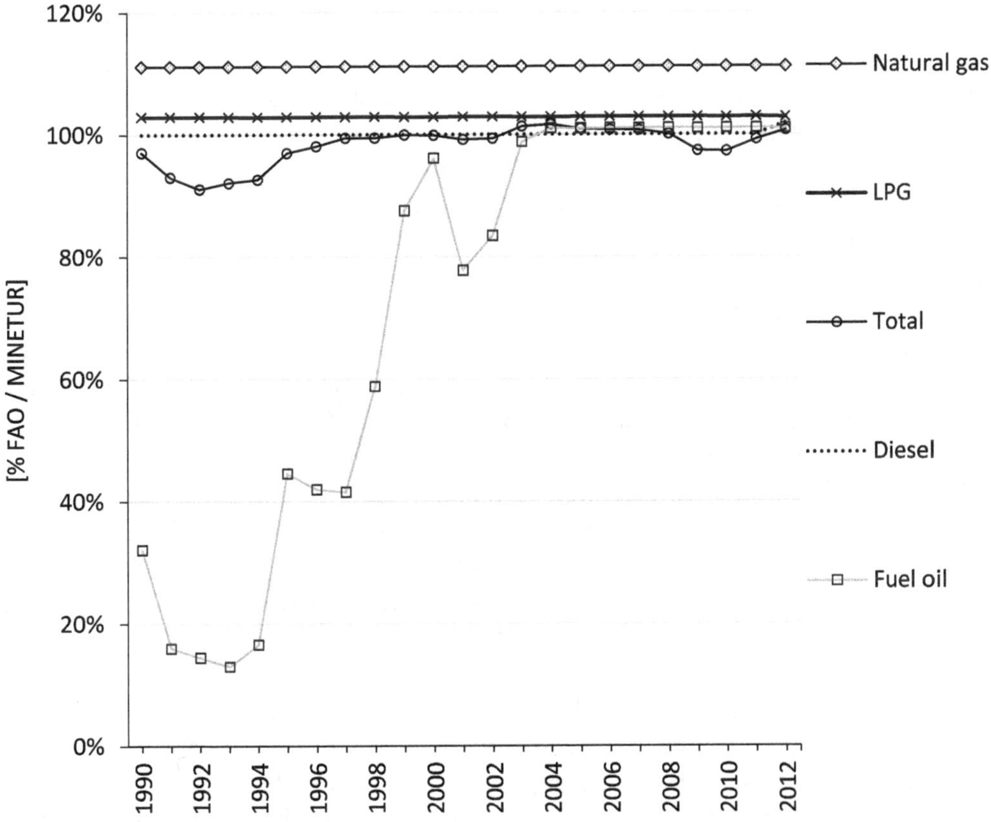

Graph 3.8 Comparison of FAO (2016) and MINETUR (2015) data of fuel consumption in Spanish agriculture, by type of fuel, 1990–2012 (FAO values expressed as % of MINETUR values)

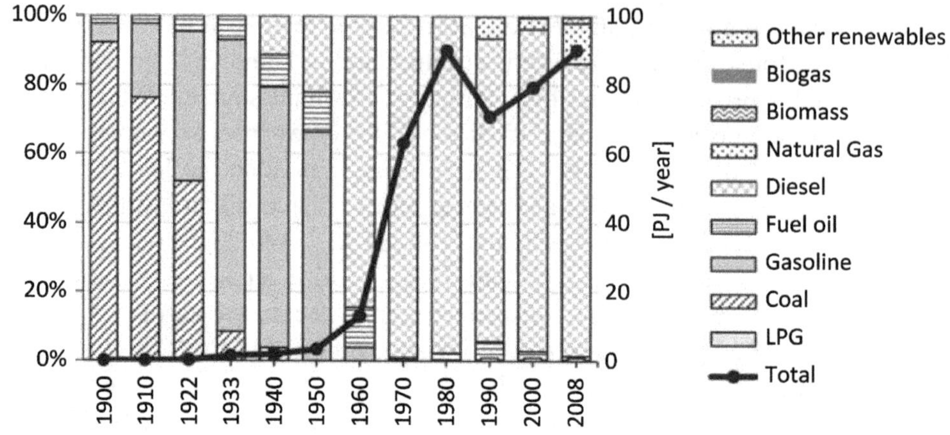

Graph 3.9 Historical evolution of fuel consumption in Spanish agriculture, by fuel type, 1900–2008, in petajoules/year

MINETUR (2015), such as biomass and biogas, nor renewable thermal energy such as solar thermal and geothermal. In Graph 3.9, we show the selected annual series of fuel consumption from 1958 to 2012.

For constructing the full 1900–2008 decadal series, we had to estimate fuel consumption in the pre-1950 period. To our knowledge, there is a lack of published official statistics or estimations made by scholars on fuel consumption in Spanish agriculture before 1950. In our estimation, we assume that fuel consumption is proportional to the installed power of machinery, taking 1950–1951 data as a reference for fuel consumption. We distinguished the relative proportions of the different types of fuel based on qualitative information. For example, Naredo and Campos report that in 1948 diesel engines still represented only 20.6% of total installed capacity of agricultural machinery. According to Martinez-Ruiz (2000), 40% of the tractors still had petrol engines in 1955, and the majority of them in previous periods. We assumed that all new tractors from 1955 onwards used diesel fuel. This means that in 1960 gasoline would represent about 8% of fuel consumption in agriculture. This figure somewhat contradicts the annual data reported by MA (1978), which indicates that gasoline consumption from 1958 to 1961 was 0, while fuel oil represented 10–25% and diesel fuel 75–90%. In fact, null or almost null gasoline consumption values during the whole annual series from different sources in the 1958–2011 period also contradict the fact that many types of common agricultural small machines (such as chainsaws or weeding machines) employ gasoline. Therefore, we do not know whether gasoline amount is included within other types of fuel, or it is just not reported (the most probable).

This reveals two transitions in the major types of fuels used during the studied period, from coal to gasoline, during mainly the 1920s, and from gasoline to diesel, during the 1950s. It also shows the appearance in the last decades of gas products employed for heating, which significantly contributed to the increase in total fuel consumption in Spanish agriculture. The distinction between the different uses of fuels is shown in Graph 3.10.

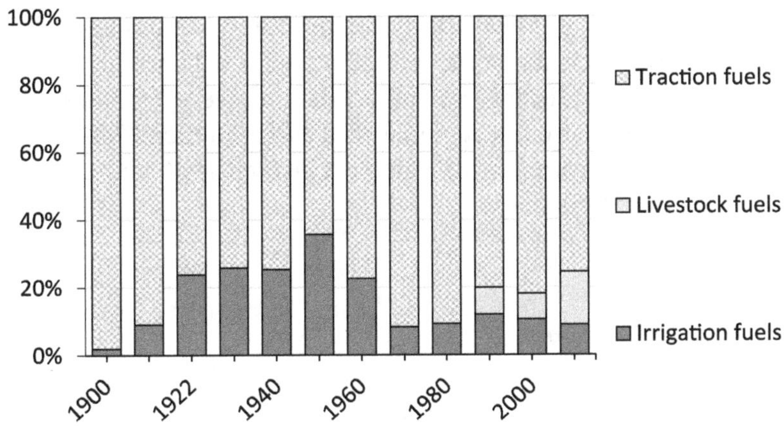

Graph 3.10 Historical evolution of fuel consumption in Spanish agriculture in selected time steps, by activity type, 1900–2008, expressed as %. Own estimation (see text)

3.3 Irrigation

Irrigation inputs include infrastructure, on-farm energy use (energy consumption by water pumps), off-farm energy use (energy consumed in desalination and diversion channels) and indirect energy use (energy required for the production of fuels and electricity). Additional information on irrigation inputs and their C footprint is provided in Aguilera et al. (2019b).

In the last decades, starting with the "Tajo-Segura" water diversion in 1979 and continuing with other diversions and desalination projects in the last two decades, there has been an increase of high-energy consuming water sources (Graph 3.11).

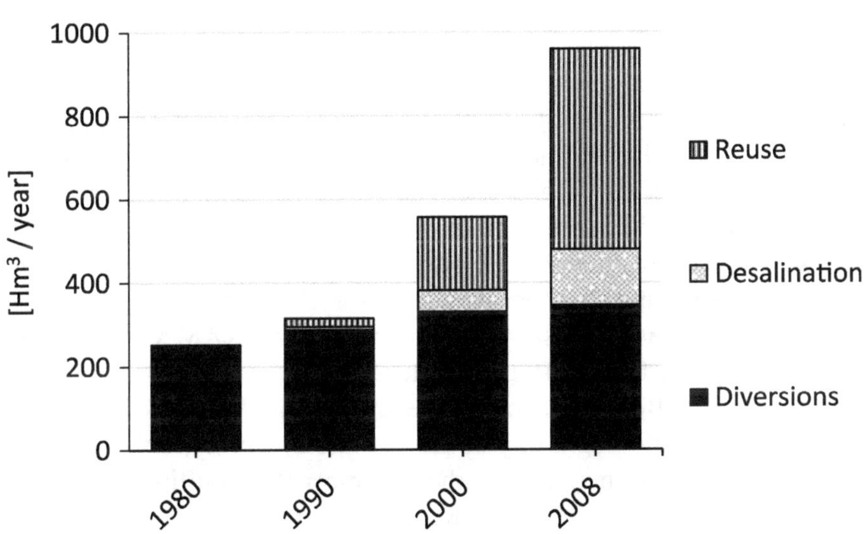

Graph 3.11 Historical evolution of agricultural water use from unconventional sources, 1900–2008, in hm^3/year. *Source* Corominas (2010) and own estimations (see text)

Thus, all the increase in water availability for irrigation has implied higher energy requirements and environmental costs, a trend which is enhanced by the increase in the average depth of wells. There is a lack of data of average values of this factor at the national level in Spain, but regional data of Comunidad Valenciana in East Spain shows an increase from 23 to 87 meters between 1940 and 1970 (Calatayud and Martínez-Carrión 2005).

3.3.1 Irrigation Systems

We have estimated the surface area of each type of irrigation, including surface, sprinkle and trickle irrigation systems. The official statistics include an additional category named "intermittent irrigation" ("riego eventual"), which includes surface irrigation systems where water is only applied on certain occasions. This category has been equaled to one-third of surface irrigation systems for the estimation of the effectively irrigated surface area. Our estimation of the evolution of the surface area of each type of irrigation at the beginning of the twentieth century was based on two government reports (MAICOP 1904; MF 1918). We have used MAICOP (1904) data for our 1900 time-step, MF (1918) data for our 1922-time step and the average between the two of them for the 1910-time step. We have found no other data distinguishing each type of irrigation until the 1962 Agrarian Census (INE 1963). Thus, the values of the intermediate time steps (1933, 1940 and 1950) have been linearly interpolated. Up to 1960-time step, the only two irrigation categories were surface and intermittent surface irrigation. As we have data of total irrigation area for all time steps, we estimated intermittent surface irrigation values as explained above, while surface irrigation area values were calculated as the difference between total irrigation and intermittent irrigation. Sprinkle irrigation data for the 1960–1990 period was obtained from Calatayud and Martínez-Carrión (2005). We assumed that intermittent surface irrigation disappeared from 1970 onwards. Drip irrigation area data at the national level starts with 1989 data from Calatayud and Martínez-Carrión (2005). We used that value for our 1990-time step and assumed that drip irrigation area in 1980 was 1/10 of the area in 1989, based on the approximate growth rate of Murcia region from 1975 to 1992 (Calatayud and Martínez-Carrión 2005). Data for 2000-time step was taken from 2003 report by the Spanish Ministry of Agriculture (MAPA 2003), based on a 2002 survey. Data for 2008-time step was taken from 2015 report by the Spanish Ministry of Agriculture (MAGRAMA 2015a), based the average of 5 surveys performed from 2006 to 2010.

We can observe different stages in the evolution of irrigated area in Spain (Graph 3.12). During the beginning of the twentieth century, only surface irrigation systems existed. The majority of them, almost 0.89 million hectares (Mha), had constant irrigation, while 0.33 Mha were irrigated only intermittently. Thus, our corrected irrigated area series starts at just 1.00 Mha, and grows significantly until 1922 (1.24 million hectares). Then, the share of intermittent irrigation starts growing, making the corrected series to drop even if the sum of constant and intermittent

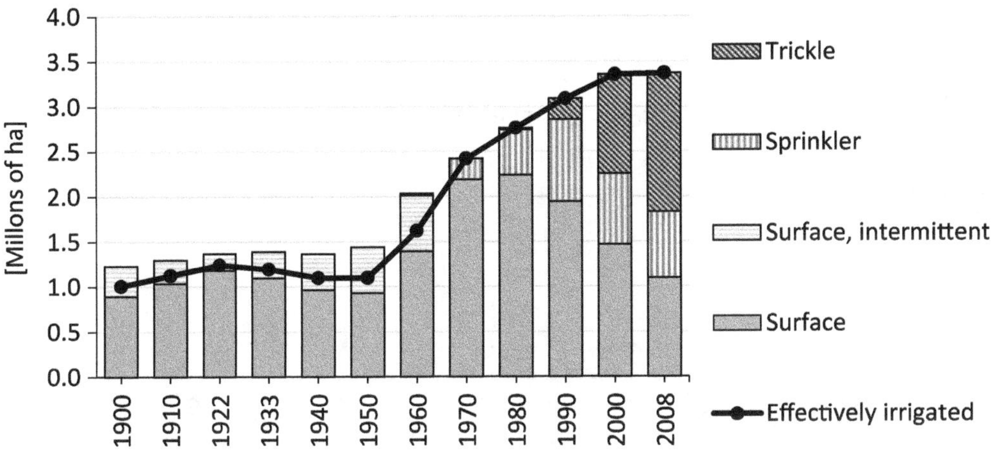

Graph 3.12 Historical evolution of irrigated area in Spain, 1900–2008, in hectares. Own estimation (see text)

surface irrigation areas still grows until 1940. The Autarky period (1940–1950) was associated to a reduction of both the gross and the corrected total irrigated area, the latter reaching 1.12 Mha. In 1950 starts a strong growth trend. The corrected irrigated area doubles in just 2 decades, and reaches 3.18 Mha in 1990. Growth continues until the present, although the growth rate has slowed down in the last decades. The first stage of the growth period was based on the expansion of constant surface irrigation area, while sprinkler irrigation expanded from 1960 to 1990 and trickle irrigation from 1980 onwards. In 1980, the surface irrigation area peaks at 2.30 Mha, and in the last time step it has dropped to half of this value (1.09 Mha), while trickle irrigation area reaches 1.54 Mha. This trend has continued until 2014, when surface irrigation dropped below 1 Mha and trickle irrigation covered 1.76 Mha.

3.3.2 Installed Mechanical Power

Data on the number of irrigation engines was gathered from different official publications, namely MF (1918) for 1916, MA (1933) for 1932, and MAPA (1980) for 1955–1980 (Graph 3.13). For the estimation of the number of irrigation engines in our selected time steps (Graph 3.14), we assumed a constant growth rate between 1900 and 1933, based on the observed growth rate between 1916 and 1932. For the period after 1980, in the case of combustion engines, we assumed the same growth rate as in the 1970–1980 period, while the number of electric engines was based on the evolution of electricity consumption in agriculture.

The number of installed engines (Graph 3.13) grew from about 1000 units in 1916 (JCA 1918) to 264,000 units in 1995 (MAPA 1997), reaching nearly 279,000 units in 2008. Most of the irrigation water in the early twentieth century was directly supplied by gravity from rivers, springs, and reservoirs through channels and *acequias*. Water

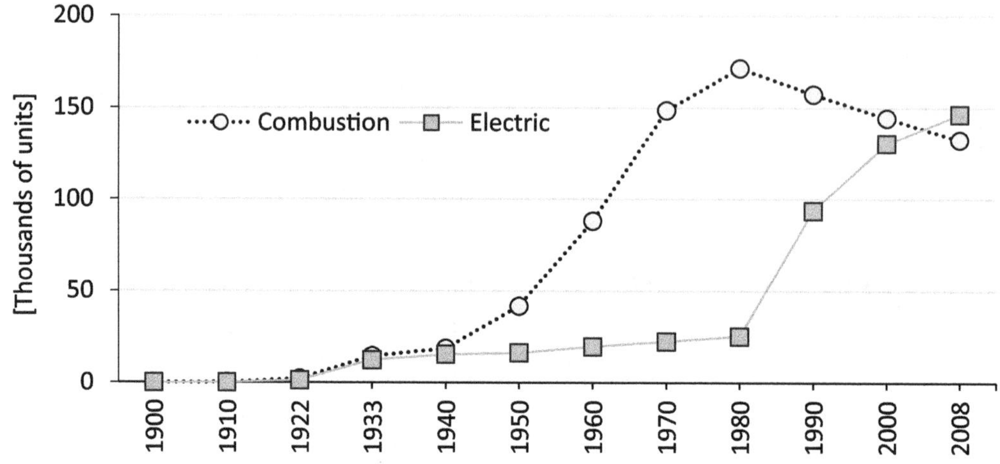

Graph 3.13 Historical evolution of the number of irrigation engines, 1900–2008, in thousands of units. Own estimation

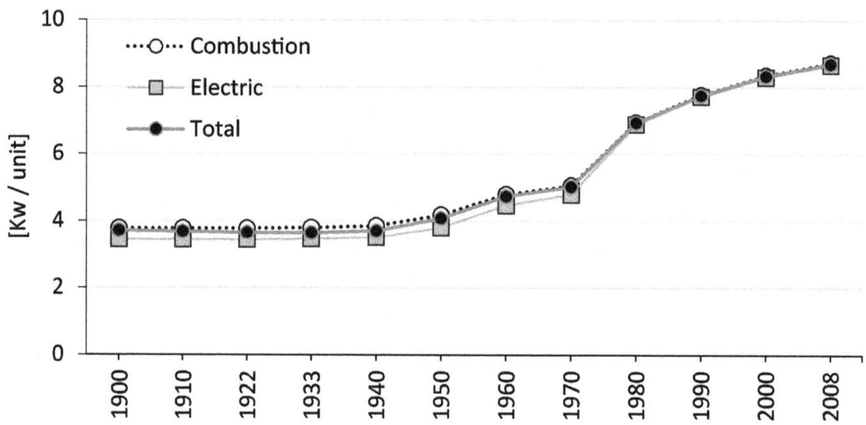

Graph 3.14 Historical evolution of the average rated power of irrigation engines, 1900–2008, in KW /unit. Own estimation (see text)

elevation was only required in about 8 and 9% of the irrigated surface in 1904 and 1918, respectively (MAICOP 1904; MF 1918, respectively). In 1916, pump engines were used in about 31% of this surface requiring water elevation, while manual pumps (13,000 units) were used in 2% of this surface, water pumping windmills (3,000 units) in 3%, and water wheels (48,000 units) in 64% (MF, 1918). The number of water pumping windmills doubled up to 7000 units in 1932 (MAIC 1932), but there is no data on the continuation of the trend. The number of waterwheels had grown to 73,000 units in 1932 (MAIC 1932), and to 85,000 in 1962 (INE 1963), and there is no more data after that year.

There is information on the installed power of irrigation engines from 1955 to 1979 (MAPA 1980). The evolution of average engine power in the previous and later periods was estimated based on the evolution of the rated power of average traction

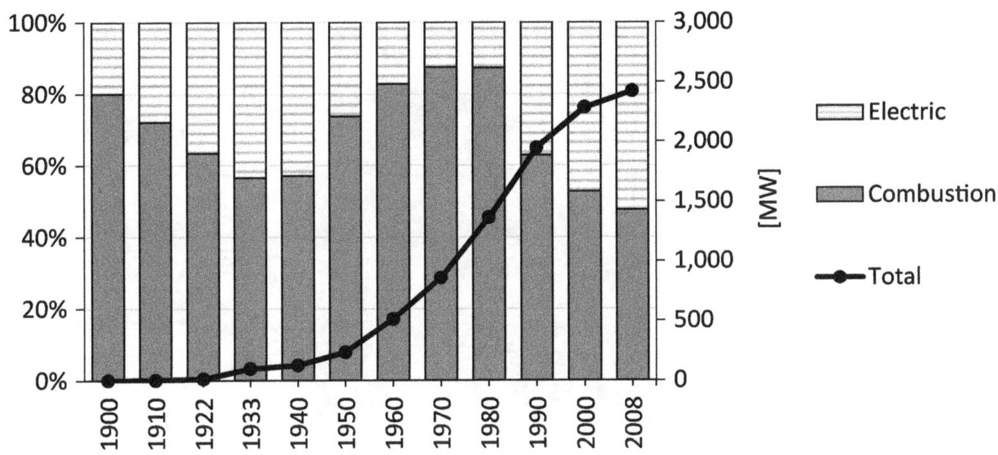

Graph 3.15 Historical evolution of the total installed power of irrigation engines, 1922–2008, in GW. Own estimation (see text)

machinery (see Sect. 2) (Graph 3.14). The total installed power of irrigation engines (Graph 3.15) was calculated by multiplying the number of engines by their average rated power.

The rated power of irrigation engines shows significant growth, more than doubling during the studied period (Graph 3.14). Most of the growth is observed during the 1950–1980 period when official statistics are available. These numbers are substantially lower than those of the Comunidad Valenciana region of Spain in the 1940–1970 period, where the average power of irrigation pumps was 21 KW/unit in 1940 (Calatayud and Martínez-Carrión 2005). But, according to the same source, the average power had grown up to 57 KW/unit in 1970, thus showing a similar growth pattern as in Spain as a whole.

The installed power for irrigation grew from 14 to 2496 MW in the 1922–2008 period (Graph 3.15). Electric engines represented around half of the installed power up to 1940, when heat engines started increasing their share up to nearly 90% in 1980. After that year trend reversed, and electric engines represented over 50% of the installed power in 2008.

3.3.3 *Fuels*

Fuel consumption in irrigation (Graph 3.16) was estimated based on the number of irrigation combustion engines, assuming a fixed fuel consumption rate per engine installed power of 7.0 GJ/KW, which is the consumption rate in 1995, the only year with available data on fuel consumption in irrigation (Corominas 2010). The distribution between coal and liquid fuels was based on the relative share of steam engines and internal combustion engines. In 1916 and 1932 the historical sources (respectively, MF 1918; MA 1933) provide specific data of engine fuel types. We

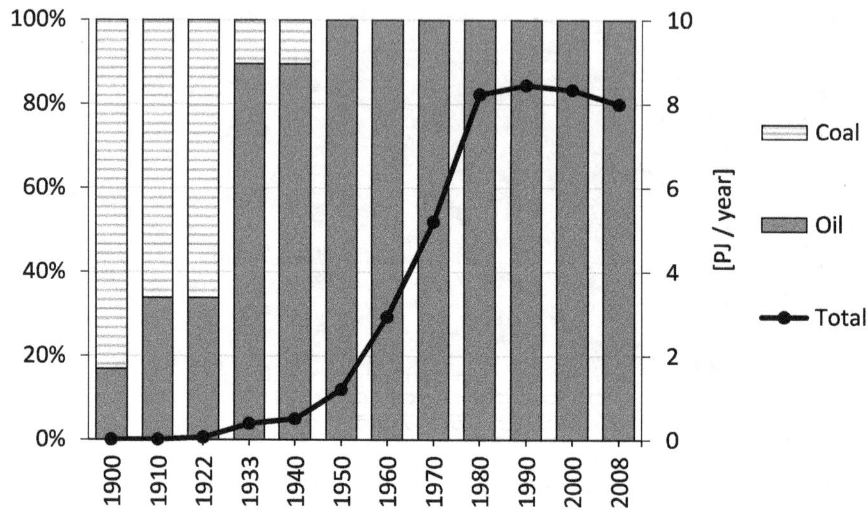

Graph 3.16 Historical evolution of direct fuel consumption in irrigation, 1900–2008, in petajoules. Own estimation (see text)

took the 1916 value for 1910 and 1922, and the 1932 value for 1933 and 1940. We assumed that the relative share of oil fuels in 1900 was half of that in 1916. Engines using coke oven gas, which were relatively common in the early twentieth century, have been included within the "oil" category. After 1950, we assumed steam engines to be phased out.

Total fuel consumption increased from 0.5 TJ in 1900 up to a peak 8450 TJ in 1990 and slowly declined after that year, reaching 7993 TJ in 2008. Coal dominates fuel consumption in the early twentieth century. Despite our estimation coal consumption remains practically flat from 1922 to 1940, its relative share drops dramatically from 1922 to 1932, due to the sharp increase in internal combustion engines.

3.3.4 Electricity

We found various sources reporting agriculture electricity use in different periods. The National Statistics Institute (INE) reports electricity consumption in agriculture from 1938 to 1958 (INE 1960), while another official study from the Industry Ministry (MI) provides a series from 1945 to 1959 (MI 1961a, b). The data in INE report is expressed as MWh, while the data in MI report is expressed as tons of coal equivalents (tce). If we express them as TJ, using the coefficient reported in MI document of 29.3 GJ/tce, we get much higher values from MI than for INE (Graph 3.9). However, the trends are identical from 1945 to 1955. On the other hand, INE data is similar to 1950 data reported by Naredo and Campos (1980), but show a drop in electricity consumption in 1956–1958 that is not supported by historical data, and which does not fit well with the following data series starting in 1960 (Carpintero and Naredo

2006). Thus, we made the exercise of recalculating MI values using a modified GJ/tce coefficient, equaling its data with those of INE (1960) in 1945. The results are shown in Graph 3.16. We can observe that the modified MI data series (MI-modified) matches well all overlapping INE series except the period 1956–1958. It also matches well Carpintero and Naredo (2006) series starting in 1960, which in turn matches FAOSTAT data starting in 1970. Therefore, we used INE (1960) data from 1939 to 1945, the modified MI data series from 1945 to 1959, Carpintero and Naredo (2006) data from 1960 to 1969, FAOSTAT (FAO 2016) data from 1970 to 1989, and IDAE (2015) for 1990–2013.

Graph 3.17a shows the disparity between different published statistics of electricity consumption in Spanish agriculture during the mid-twentieth century, which is solved by our corrected estimation of MI data. Graph 3.17b shows our estimation of

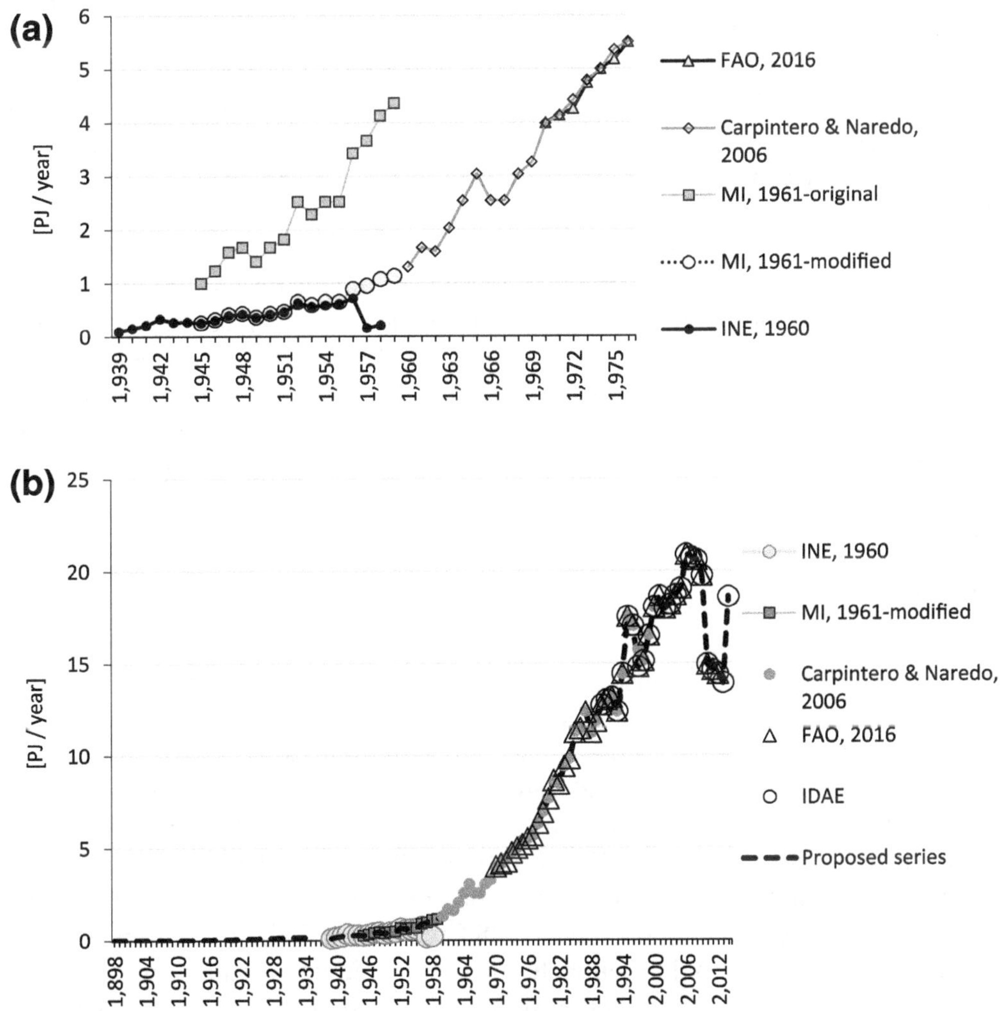

Graph 3.17 Historical evolution of on-farm electricity use in Spanish agriculture according to various sources, 1939–1976 (**a**) y 1898–2014 (**b**), in petajoules/year

annual electricity consumption from 1898 to 2013. The estimated direct electricity consumption in agriculture grew 10-fold in two decades, from 7 TJ at the beginning of the twentieth century to 70 TJ in 1922. By 1955, it had grown ten-fold again, at 656 TJ, and again in 1980, at 7646 TJ. It peaked in 2007 at 20,279 TJ and dropped down to 13,924 in 2013, to grow again to 18.6 PJ in 2016.

We have also estimated off-farm ("upstream") electricity consumption for agriculture in our selected time steps (Graph 3.18b and c). This electricity is used for obtaining unconventional water sources, including desalination and water pumping in channels and diversions. We multiplied the quantity of water coming from these sources calculated in Sect. 3.1 by energy consumption coefficients from Corominas (2010), of 1.2, 3.7 and 0.25 KWh/m^3 for diversions, desalination and reuse, respectively. The results are shown in Graph 3.18a.

The electricity employed to obtain irrigation water from unconventional sources (Graph 3.18a) rose from 1092 TJ after the construction of Tajo-Segura diversion in 1979, to 3716 TJ in 2008 when other diversions and many desalination plants were operating. Irrigation electricity grew from 7 TJ in 1900 to 223 TJ in 1933, 473 TJ in 1950 and 23,099 TJ in 2008. A large share of the growth in the last decades has been due to the rise in off-farm electricity use.

Overall, direct energy consumption in irrigation in Spain grew from 8 TJ in 1900 to 31,091 TJ in 2008. Despite most energy consumption was electricity at the beginning of the twentieth century, oil fuels consumption rapidly expanded since 1933, and dominated the use of energy in irrigation during the middle decades of the century. Fuel consumption, however, peaked in 1980, while electricity continued growing until the end of the studied period, in line with the growth in electric engines numbers and installed power. In addition, off-farm electricity consumption was also boosted in the last decades, with the expansion of technologies such as desalination and diversion channels.

3.4 Fertilizers

We have reconstructed the consumption of mineral fertilizers in Spain in terms of nutrients on an annual basis, including synthetic fertilizers, saltpeter, and guano. This estimation is based on historical data expressed in total fertilizers amount for the period 1900–1940 and expressed as nutrients during the 1945–2010 period. The first period up to 1940 is composed mainly by statistical data for all types of fertilizers in some single years (namely 1907, 1908, 1919, 1920, 1928, 1930–1935, 1939 and 1940) and a continuous data series of the apparent consumption of total fertilizers from 1898 to 1940 (excluding the 1936–1938 Civil War period). In addition, we have continuous series of ammonium sulphate and sodium nitrate consumption from 1900 to 1914. Our main data source is the work by Gallego (1986), who compiled and systematized historical statistics of fertilizers consumption in Spain during the first half of the twentieth century, including data from Alonso de Illera for 1907 and 1908, data from Junta Consultiva Agronómica government agency from 1919 and 1920,

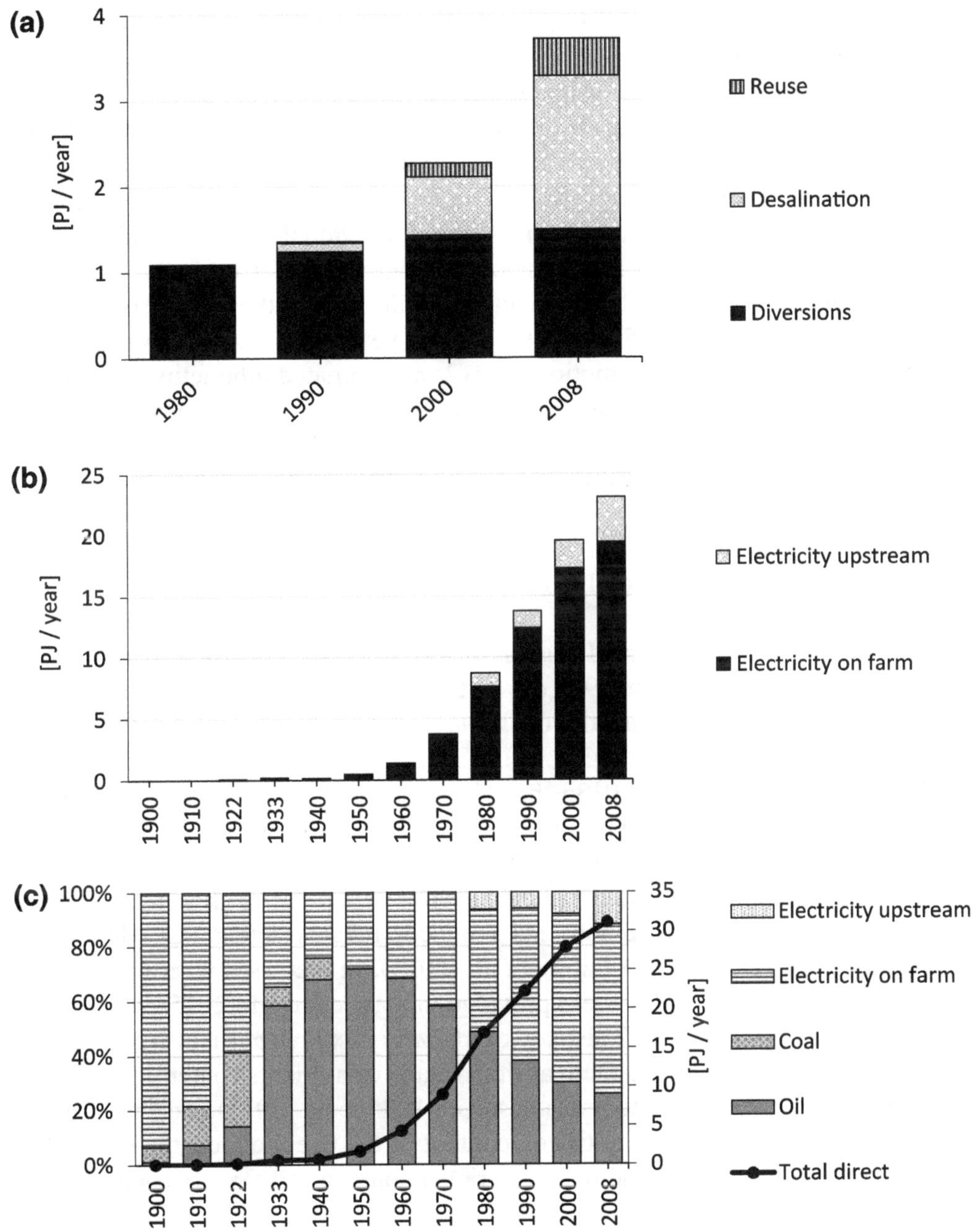

Graph 3.18 Direct energy consumption for irrigation in Spanish agriculture including upstream electricity consumption from unconventional water sources 1980–2008 (**a**), total electricity consumption 1900–2008 (**b**) and total direct energy use 1900–2008 (**c**), in petajoules/year

and from the *Anuario de Estadística Agraria* reports from 1928 to 1940. In addition, he compiled the foreign trade fertilizer data from the nineteenth century up to the Spanish Civil War (1936). These statistics include information of different items with various levels of aggregation, but the only item that could be studied separately during the whole series is ammonium sulphate. In addition, Mateu (2013) reconstructed sodium nitrate (saltpeter) imports (which can be equaled to apparent consumption) from 1901 to 1914. Therefore, the missing gaps of individual types of fertilizers up to 1914 were estimated using total apparent consumption of fertilizers in the given year, minus ammonium sulphate and sodium nitrate, multiplied by the average of the share of each individual type of fertilizer in the closest years where data is available. For example, superphosphate consumption in 1912 was estimated subtracting ammonium sulphate and sodium nitrate consumption to total fertilizer apparent consumption in 1912 and multiplying the result by the average of the percentage represented by superphosphate in total fertilizer consumption minus ammonium sulphate and sodium nitrate in 1907, 1908, 1919 and 1920. Potassium sulphate and potassium chloride consumption values in 1920 were much higher than the values of close years, so we considered them outliers and assumed that production was the same as in 1919.

Data for the 1941–1944 period was estimated by exponential interpolation using 1940 and 1945 data. Data from 1945 onwards were taken directly from the selected Spanish government official statistics (*Anuario de Estadística Agraria* reports, MAIC 1928, 1930, 1931; MA 1933, 1934, 1935, 1939, 1940, 1973; MAPA 1990, 1999; MARM 2010; MAGRAMA 2013a, b, c). We only altered this official series in 1957–1964, when we added potassium sulphate data gathered from a National Statistics Institute publication (INE 1965).

As we mentioned above, the data up to 1940 is expressed as gross tons of fertilizers. Therefore, we calculated nutrients using coefficients (Table 3.2). Most of the coefficients were taken from Aguilera et al. (2015). In the cases where no data, or ranges instead of single values, were provided in Aguilera et al. (2015), we took the values from Gallego (1986). We also took fertilizers mixtures nutrient contents from Gallego (1986), who estimated them as the average of nutrients contents. Moreover, we estimated guano nutrient contents based on Wikipedia (2015) and Smil (2001), taking into account that guano nitrogen content in the twentieth century should be in the lower side of its historical range, as richer sources had already been depleted (Table 3.2).

Graph 3.19a shows the evolution of the amounts of nitrogen consumed. We can observe an upward trend during the pre-Civil War period (1898–1935, Graph 3.19a), when consumption grew 26-fold. This growth was only briefly interrupted in 1918 by the effect of the First World War. The dominant N fertilizer during this period was ammonium sulphate, which was obtained mainly through recovery from coke oven gases. Imported Chilean nitrate (and lately also calcium nitrate) also played a role in the supply of mineral N in Spain, while guano was already almost exhausted by the beginning of the century.

Table 3.2 Nutrient contents of fertilizers employed in Spain, percentage of nutrient

	N (%)	P_2O_5 (%)	K_2O	Source
Ammonia	82.0	0.0	0.0	Aguilera et al. (2015)
Ammonium Nitrate	35.0	0.0	0.0	Aguilera et al. (2015)
Ammonium Sulfate	21.0	0.0	0.0	Aguilera et al. (2015)
Calcium-Ammonium Nitrate	25.0	0.0	0.0	Aguilera et al. (2015)
Calcium Nitrate	16.0	0.0	0.0	Aguilera et al. (2015)
Urea	46.0	0.0	0.0	Aguilera et al. (2015)
Potassium Nitrate	13.5	0.0	45.0	Gallego (1986)
Complex NPK fertilizers	15.0	15.0	15.0	Aguilera et al. (2015)
Mono Ammonium Phosphate	11.0	52.0	0.0	Aguilera et al. (2015)
Di Ammonium Phosphate	18.0	46.0	0.0	Aguilera et al. (2015)
Ammonium phosphate*	14.5	49.0	0.0	Aguilera et al. (2015)
Phosphate rock	0.0	32.0	0.0	Aguilera et al. (2015)
Triple Superphosphate	0.0	48.0	0.0	Aguilera et al. (2015)
Single superphosphate	0.0	18.0	0.0	Gallego (1986)
Slag	0.0	16.7	0.0	Gallego (1986)
Complex PK fertiizers	0.0	22.0	22.0	Aguilera et al. (2015)
Muriate of potash (potassium chloride)	0.0	0.0	60.0	Aguilera et al. (2015)
Sulfate of potash	0.0	0.0	50.0	Aguilera et al. (2015), INE (1965)
Sodiumnitrate	15.0	0.0	0.0	Smil (2001)
Mixtures	Dynamic coefficient			Gallego (1986)
Cianamida de cal	20.0	0.0	0.0	Smil (2001)
Kainita	0.0	0.0	18.0	Gallego (1986)
Silvinita	0.0	0.0	18.0	Gallego (1986
Guano	10.0	10.0	2.5	Wikipedia (2015), Smil (2001)

The post-Civil War period (Graph 3.19b) started with stagnant use of N fertilizers, which only surpassed pre-war levels by the mid-1950s. This stage corresponds to the Autarky period of Franco's dictatorship and was followed by an exponential growth up to 1979. Growth slowed down onwards, with significant variability between years, but still continued up to 2000, when N fertilizer consumption peaked at 1279 Gg N. After that year, N fertilizer consumption fell to a minimum of 740 Gg N in 2008, and partially recovered afterwards, reaching 962 Gg in 2013. The N fertilizer mix was dominated by ammonium sulphate until the late-1960s, when it was surpassed by calcium-ammonium nitrate. Urea and complex fertilizers became major N sources in the 1970s and now represent the majority of the N fertilizer mix, together with calcium-ammonium nitrate.

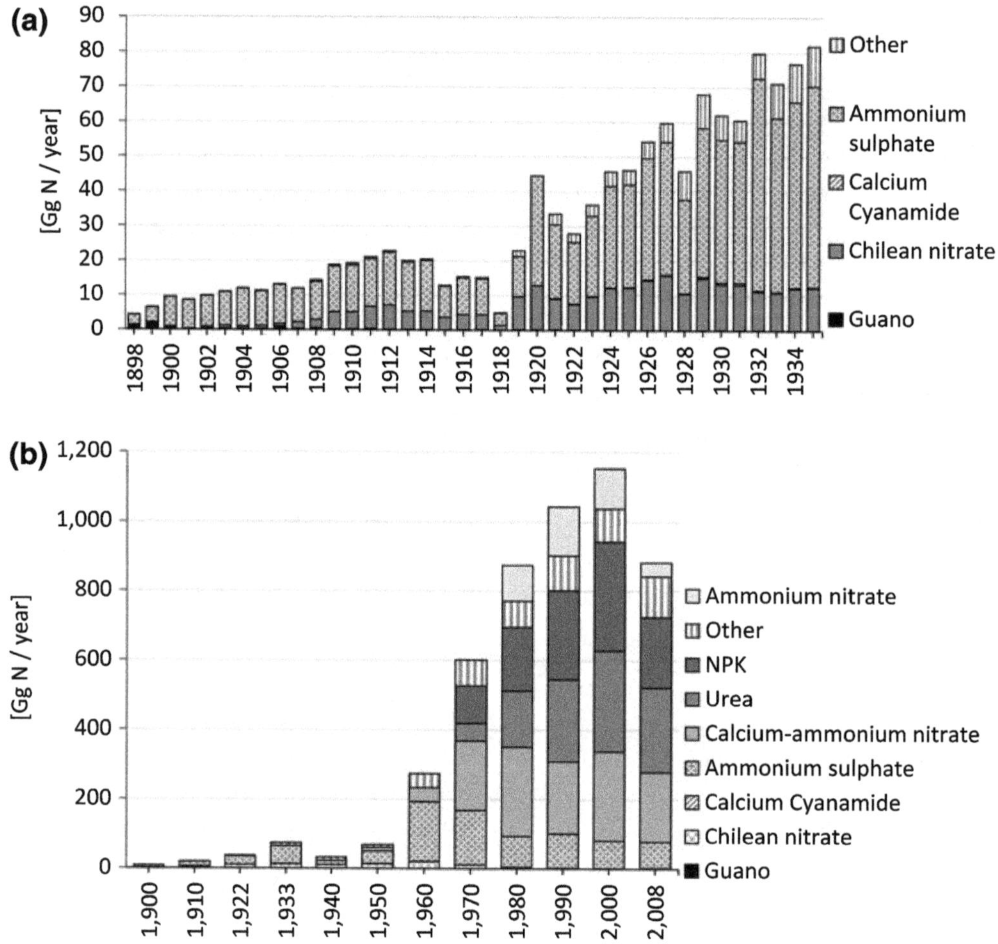

Graph 3.19 Nitrogen fertilizer consumption in Spain, 1898–1935 (**a**) and 1900–2013 (**b**), in Gg N/year

Phosphate fertilizer consumption (Graph 3.20a) show a similar pattern than N fertilizer consumption. Phosphate consumption grows by one order of magnitude during the first third of the twentieth century (Graph 3.20b). The relative growth is not as high as for N, but initial levels of P consumption are higher. The drop in 1918 is also very evident in this case. Phosphate fertilizer consumption is completely dominated by single superphosphate during this period, with minor amounts of rock phosphate, Thomas meal, mixtures and other (Graph 3.21).

Potassium fertilizer consumption trends resemble those of the other nutrients. Growth rates were significant during pre-Civil War period and slowed down during the 1940s and 1950s. Potassium chloride, followed by potassium sulphate, was the major K fertilizers until the mid-1960s. Then, the arrival of compound NPK fertilizers meant a boost in potassium fertilizer consumption growth. After slowing down in the 1970s due to a decrease in the use of potassium chloride, total potassium consumption grew fast again in the 1980s and 1990s, peaking in 1998 at 356 Mg K_2O.

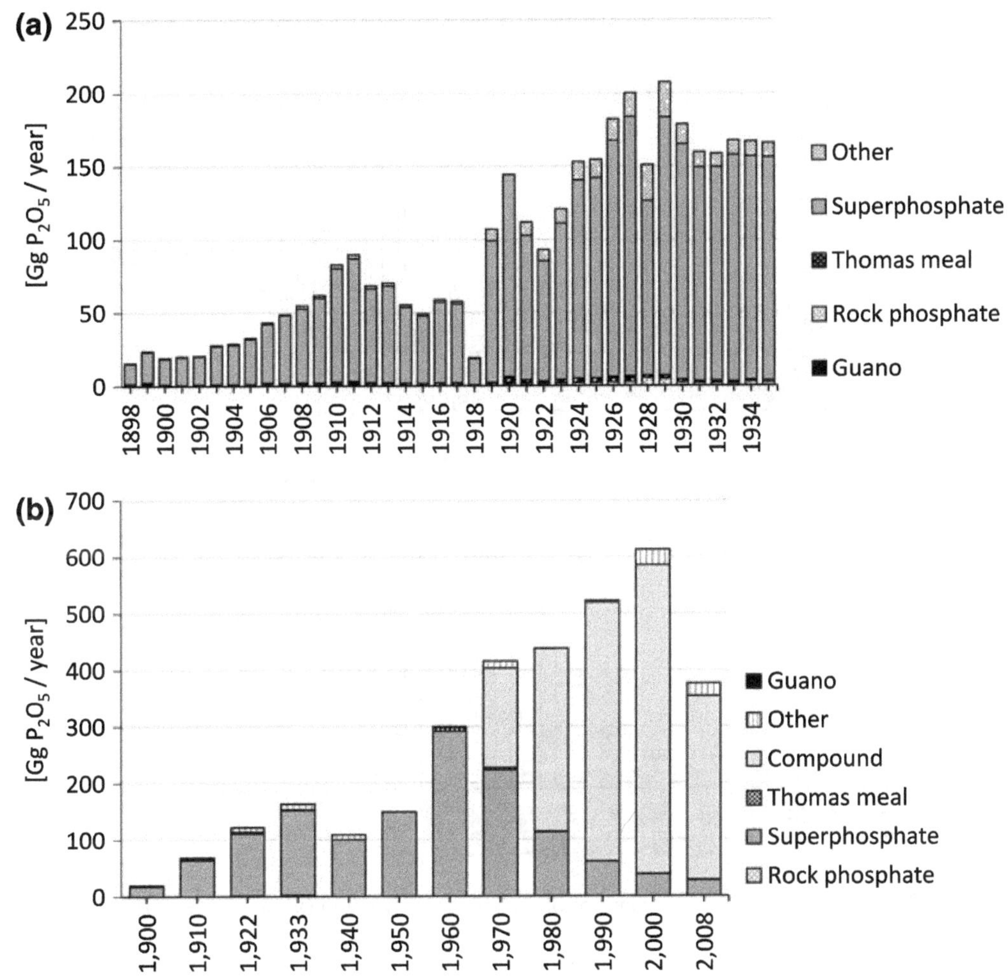

Graph 3.20 Phosphate fertilizer consumption in Spain, 1898–1935 (**a**) and 1939–2013 (**b**), in Gg P_2O_2 /year

3.5 Crop Protection

3.5.1 Pesticides

There is very limited information on the amounts of pesticides employed in Spain during the studied period. The available statistics are fragmentary both temporally and in terms of a full accounting of pesticide products. Moreover, only recently they have been expressed as active matter weight. The first available data, from the *Anuario de Estadística Agraria* (MA 1930, 1931, 1933, 1934, 1935, 1939, 1940) reported sulphur, copper sulphate and iron sulphate. An annual series expressed as monetary value of main pesticide categories (including insecticides and acaricides, fumigants, nematicides, fungicides, herbicides, veterinary products, and other) and covering the 1950–1967 period is reported by the 1973 *Anuario de Estadística Agraria* report (MA

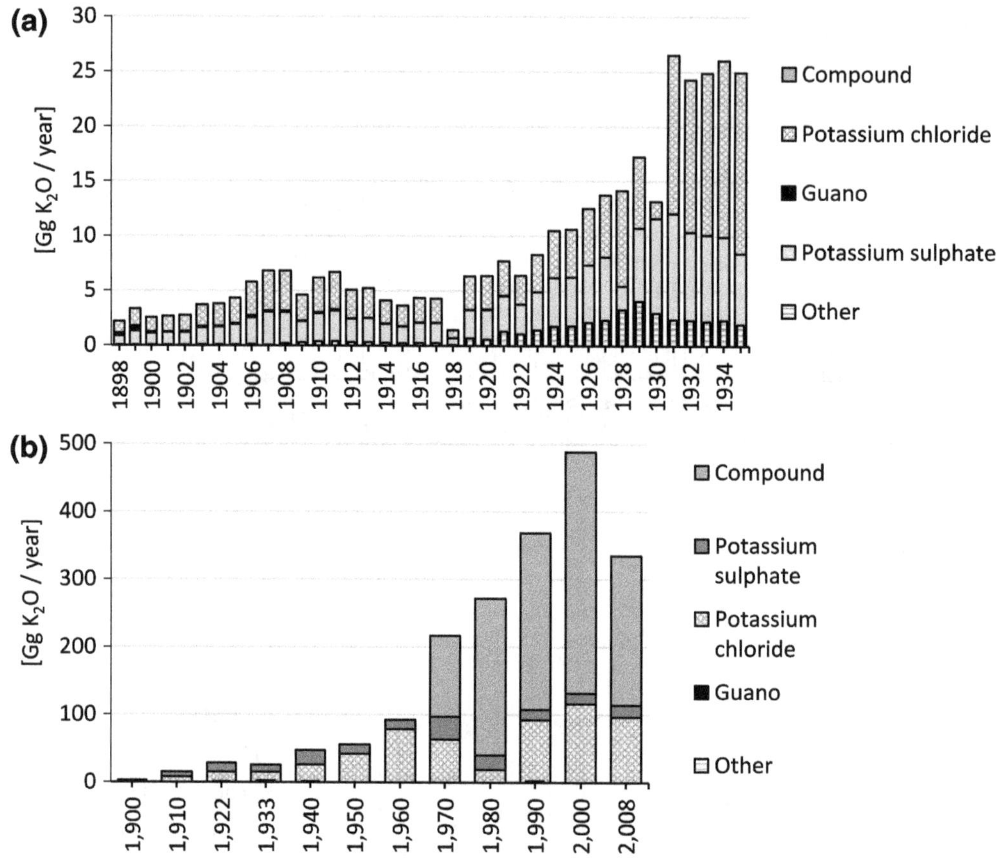

Graph 3.21 Potassium fertilizer consumption in Spain, 1898–1935 (**a**) and 1939–2013 (**b**), in Gg K_2O /year

1973), and this series is extended up to 1993 by the 1994 edition of that report (MAPA 1994). In 1955, starts another series including the quantities of pesticide products grouped by active matter (including DDT, lindane, malathion, nitro compounds, arsenic compounds, cyanuric compounds, and many other) but reported as total weight, not active matter weight. This series finishes in 1968 and is published in three reports (INE 1960, 1965, 1970). Last, the available data in FAOSTAT (FAO 2016) starts in 1990, groups pesticides by main categories (including insecticides, fungicides herbicides and other and it is expressed as active matter weight.

We estimated the trends in the use of pesticides (sulfur and copper pesticides) from 1900 to 1920 taking the 1933 value as a reference and assuming that the rate of change was similar to that of total synthetic fertilizers during each interval (see Fertilizers section). In 1933 and 1940 we took the reported values of the official statistics, assuming that only sulfur and copper pesticides were used. There is very high uncertainty in the estimation of pesticide use from 1940 to 1990. The conversion of the monetary value series to physical units is hindered by the lack of information of pesticide prices. On the other hand, the conversion of the total weight series to active matter weight requires to know the average richness of each type of pesticide. If we estimate pesticide use based on 1990 inflation-adjusted (inflation data from Carreras de Odriozola and Tafunell Sambola 2006) monetary value series we get much lower pesticide consumption values than using the total weight series taking Naredo and Campos (1980) approach of assuming 40% active matter weight (Graph 3.22a). In fact, this assumption is really uncertain, as active matter content of commercial pesticides can vary from less than 1% to almost 100%, with very large variations between different commercial products even within a single active matter compound. Moreover, this series shows a sudden drop in the late 1960s that is not supported by the historical sources, and if we examine the trends of the specific compounds, we

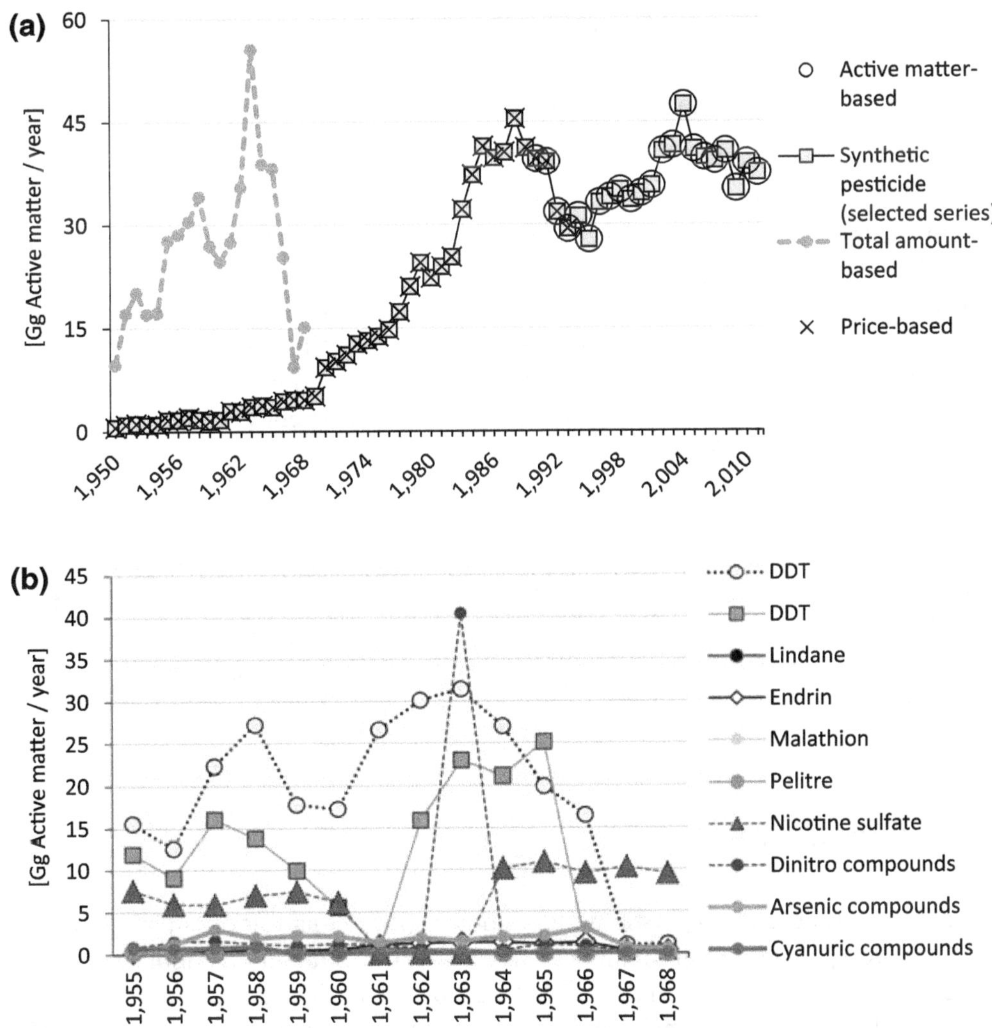

Graph 3.22 Historical evolution of synthetic pesticide use in Spain (excluding sulfur and copper pesticides), according to different estimations, 1950–2011, Gg Active matter /year. Price-based series is calculated from MA (1973) and MAPA (1994); total amount-based series is calculated from INE (1960, 1965, 1970), assuming 40% average active matter; active matter series is from FAOSTAT (FAO 2016)

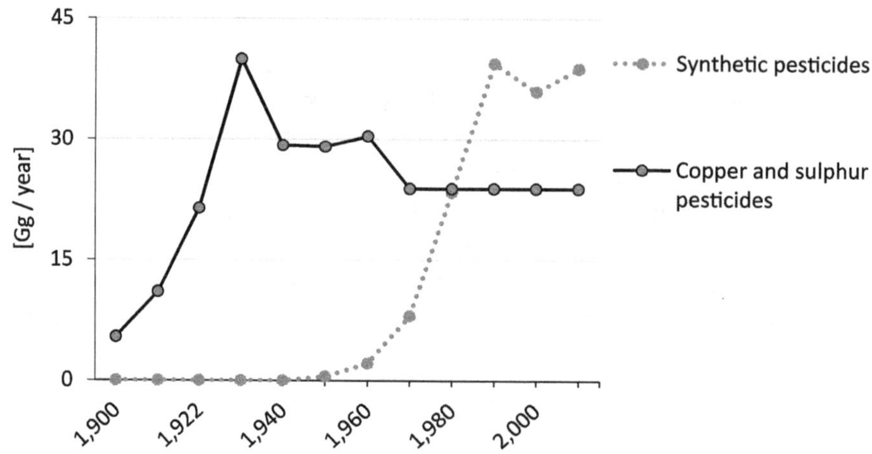

Graph 3.23 Historical evolution of pesticide use in Spain, 1900–2008, Gg /year

find very unrealistic oscillations that suggest that those statistics are not very accurate (Graph 3.22b). Therefore, we have selected the estimation based on prices instead of that based on total weight.

The selected series in Graph 3.22a suggests that most growth in synthetic pesticide use took place during the 1970s and 1980s decades, being more or less stagnant afterwards. Graph 3.22b shows the erratic behavior of individual pesticides use during the 1955–1968 period according to INE. This lack of consistency has led us to discard this data source (Graph 3.23).

Our reconstruction of pesticide use during the twentieth and early twenty-first centuries shows two major growth periods, one during the first third of the century, with the growth in copper and sulfur pesticides, and another one after the autarky period of Franco's dictatorship, with the growth in synthetic pesticide use. However, the copper and Sulphur series is only based on published statistics during the 1933–1970 period. Therefore, there is a high uncertainty in this series before and after that period.

3.5.2 *Greenhouses*

Greenhouses, tunnels, and plastic mulches now represent an important share of horticultural crop production in Spain. The first published data on greenhouse and protected crop areas in Spain is in the 1975 *Anuario de Estadística Agraria* report (MA 1975a, b). Thereafter, the official statistics offer data on some of these items in 1981 (MAPA 1981), and on all of them in 1984 and 1986 (MAPA 1984, 1986). We estimated area values in the middle years of those periods, and in years previous to 1975, assuming constant growth rates. After 1986, protected crops data were published on an annual basis. In some of these years, we have data from the previous year or years. We selected preferably the latest published data, which sometimes had mod-

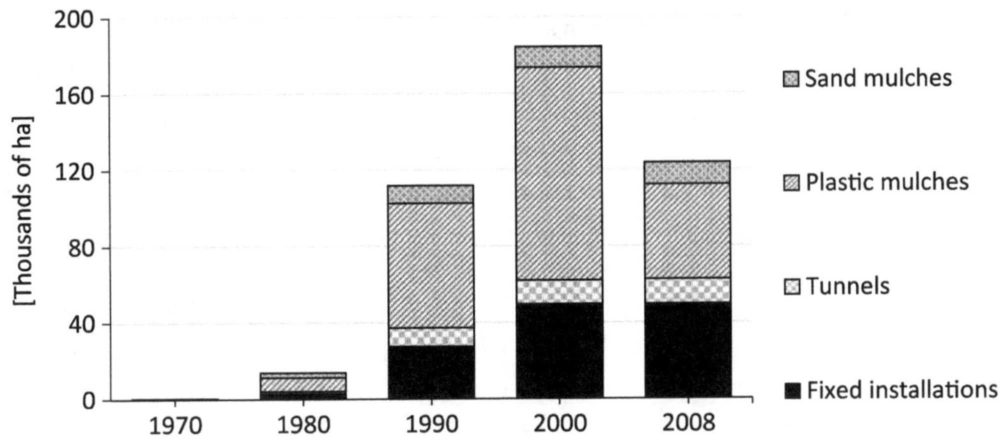

Graph 3.24 Historical evolution of greenhouses, tunnels and mulches surface areas in Spain, 1970–2012, in thousands of hectares

ifications of previous reports (MAPA 1988, 1989, 1990, 1992, 1993, 1997, 1999, 2001, 2002, 2004, 2006, MARM 2008, 2010, MAGRAMA 2012, 2013a, b, c). The complete series is shown in Graph 3.24a while Graph 3.24b shows 5-year average data in our selected time steps. According to an official 2008 survey, about 97% of fixed installations are plastic-covered and 3% glass-covered (MARM 2008a, b, c).

The reported surface areas in 1975 were 310 ha of fixed installation, 45 ha of tunnels and 120 ha of plastic mulches. Maximum surface area of fixed greenhouse installations was reached in 2006, with 52,867 ha, and in 2012 it had dropped to 48,206 ha. Maximum surface area of tunnels was reached in 2006, with 14,621 ha, and maximum area of plastic mulches was reached in 2002, with 116,172 ha. The latter dropped heavily in the following years, down to 44,827 ha in 2012.

Thus, the growth in this input was one of the latest among the agricultural inputs employed in Spanish agriculture. It was only possible when the technologies used in the construction of plastic-covered greenhouses and other crop protection techniques used in Spain were mature enough to be economically applied at the large scale. Moreover, greenhouse crop production is largely devoted to off-season and fresh produce for export, which are dependent on well-developed distribution chains that were only present in Spain relatively late.

3.6 Use of Inputs in the Agricultural Sector (Imports)

In the previous sections, we saw how the growth of external inputs was unstoppable, especially since 1960, making it possible to intensify and specialize production as described in the previous chapter. Table 3.3 shows all the inputs used from 1900 to 2008, distinguishing between industrial and biological inputs, as well as their different behavior throughout the period. The use of inputs generally increased by two

Table 3.3 External inputs used by Spanish agriculture (TJ), 1900–2008

	1900	1930	1950	1970	1990	2008
Feed	1269.9	2046.8	1119.7	53,402.6	51,822.0	187,841.6
Seeds	314.6	183.6	578.2	0.0	0.0	5335.6
Total non-industrial inputs	1584.5	2230.4	1697.9	53,402.6	51,822.0	193,177.0
Traction	528.9	2131.5	5681.3	81,695.6	103,296.4	129,665.6
Irrigation	995.3	2132.2	5096.0	20,712.0	58,565.5	75,406.9
Chemical fertilizers	1884.1	17,865.7	15,142.6	75,240.8	104,987.8	77,618.1
Crop protection	25.8	794.9	653.5	2384.8	24,369.3	31,135.1
Total industrial inputs	3432.2	22,924.2	26.573.5	180,033.3	291,219.0	313,825.7
Total external inputs	5018.7	25,154.6	28,271.3	233,436.0	343,041.0	507,002.7
1900 = 100						
Feed	100	161	88	4205	4081	14792
Seeds	100	58	184	0	0	1696
Total non-industrial inputs	100	141	107	3370	3271	12192
Traction	100	403	1074	15,446	19,530	24,516
Irrigation	100	214	512	2081	5884	7576
Chemical fertilizers	100	948	804	3003	5572	4120
Crop protection	100	3081	2533	9243	94,455	120,679
Total industrial inputs	100	668	774	5245	8485	9144
Total external inputs	100	501	653	4651	6835	10102

Source Author's own compilation based on statistical sources and the use of coefficients included in Aguilera et al. (2015)

orders of magnitude in the period between those two years, two thirds of which were industrial inputs. Industrial inputs and their energy costs grew strongly in the second half of the 20th century, first with the use of chemical fertilizers and mechanization, then by irrigation and crop protection. As a result, yields per unit area increased, especially in irrigated areas and in farms using new seed varieties, both hybrid and improved. In fact, the use of seeds particularly increased between 1980 and 2008 with the arrival of industrial germplasm from specialized global companies. The most immediate effect of the application of this land-saving technology was the use of varieties that were more productive than traditional ones under optimum nutrient and moisture supply conditions. But that was not the only effect. It also broke the necessary rotations of traditional management to adapt to the shortage of both nutrients and moisture. Thus, there was an expansion of monocultures and crop rotations determined not by agronomic rationality but by agricultural market demands. Resulting biodiversity reductions favored the appearance of plant plagues and diseases and additional use of phytosanitary products that had until then been quite limited. These kinds of chemical remedies generated a vicious circle in which the breaking of trophic chains (the use of insecticides led to the disappearance of beneficial insects that control insect plagues), along with the progression of crops and homogeneous varieties over large stretches of land, made it necessary to increasingly use these substances to control pests and diseases. This item visibly grew the most over time, despite not exceeding 6% of total energy expenditure in 2008.

In the next chapter, we will see the effects of mechanization on human work, including the elimination of animal labor. The use of mechanical traction became widespread between the sixties and seventies and never stopped growing, even in the midst of the economic-financial crisis. Irrigation followed a similar course: at first, irrigation was linked to large hydraulic works and later to so-called "irrigation modernization", consisting in the ever-increasing role of groundwater elevation and pressurized irrigation networks requiring a high energy use. To finish, we have to mention that coinciding with the livestock production process described above, feed grew spectacularly, multiplying by a factor of 121 since 1900. In 2008, it accounted for 38% of the energy value of all inputs used in the agricultural sector. This phenomenal growth of inputs generally explains Spanish agriculture's loss of efficiency, as we will see in Chap. 5. Graph 3.25 clearly shows the enormous amount of energy originating from outside the agricultural sector that was necessary to inject into agroecosystems to maintain the continuous growth of agricultural production. While biomass DE grew by 38%, the use of external energy multiplied by a hundred. In 1900, industrial inputs from outside the agroecosystem and used for production represented only 14.5% of total invested energy; in 2008, that percentage had risen to 62%, and if we add the feed from Latin America, it reached 99.4% of total energy invested in agricultural production; that is, practically all the energy invested came from outside the agricultural sector. The socio-economic consequences of this are described in the following chapter.

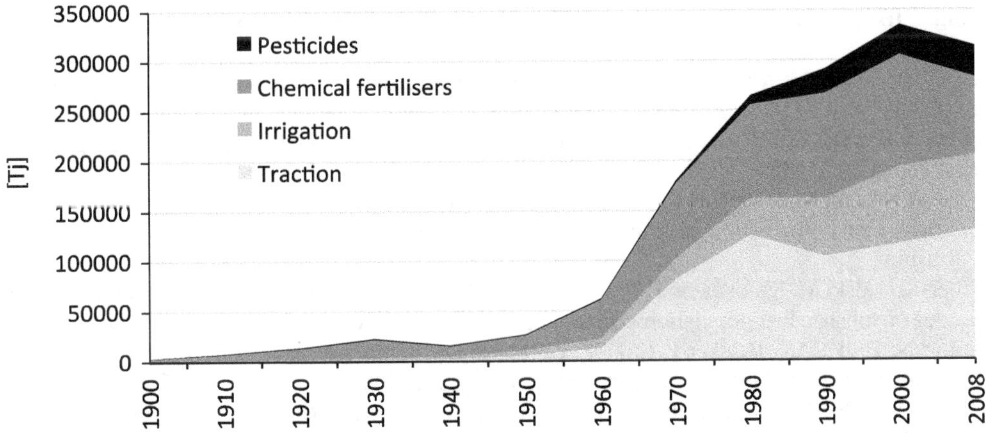

Graph 3.25 Energy value of the inputs used in Spanish agriculture, in TJ. *Source* see Table 3.3

References

Aguilera E, Guzmán GI, Infante-Amate J, Soto D, García-Ruiz R, Herrera A, Villa I, Torremocha E, Carranza G, González de Molina M (2015) Embodied energy in agricultural inputs. Incorporating a historical perspective. Sociedad Española de Historia Agraria. DT-SEHA 1507

Aguilera E, Guzmán GI, González de Molina M, Soto D, Infante-Amate J (2019a) From animals to machines. The impact of mechanization on the carbon footprint of traction in Spanish agriculture, 1900–2014. J Clean Prod 221:295–305

Aguilera E, Vila-Traver J, Deemer BR, Infante-Amate J, Guzmán GI, González de Molina M (2019b) Methane Emissions from Artificial Waterbodies Dominate the Carbon Footprint of Irrigation: A Study of Transitions in the Food-Energy-Water-Climate Nexus (Spain, 1900–2014), Environ Sci Technol 53:5091–5101

ANSEMAT (Asociación Nacional de Maquinaria Agropecuaria Forestal y de Espacios Verdes) (2007) Estudio de la situación del parque nacional de maquinaria agrícola Balance 2006. Madrid

Bartolomé Rodríguez I (2007) La industria eléctrica en España (1890–1936). Madrid. Banco de España, Estudios de Historia Económica, no. 50

Bhat MG, English BC, Turhollow AF, Nyangito HO (1994) Energy in synthetic fertilizers and pesticides: revisited. Final project report. Knoxville, TN, USA. Tennessee University Press, Department of Agricultural Economics and Rural Sociology

Boserup E (1981) Population and technological change. A study of long-term trends. University of Chicago Press, Chicago, USA

Calatayud S, Martínez-Carrión JM (2005) El cambio tecnológico en el uso de las aguas subterráneas en la España del siglo XX. Un enfoque regional. Revista de Historia Industrial 28:81–114

Cao S, Xie G, Zhen L (2010) Total embodied energy requirements and its decomposition in China's agricultural sector. Ecol Econ 69:1396–1404

Carpintero O, Naredo JM (2006) Sobre la evolución de los balances energéticos de la agricultura española, 1950–2000. Hist Agrar 40:531–554

Carreras de Odriozola A, Tafunell Sambola X (2005) Estadísticas históricas de España, siglos XIX–XX. Madrid. Fundación BBVA

Cleveland CJ (1995) The direct and indirect energy use of fossil-fuels and electricity in USA agriculture, 1910–1990. Agric, Ecosyst Environ 55:111–121

Corominas J (2010) Agua y energía en el riego, en la época de la sostenibilidad. Ingeniería del Agua 17:219–233

Dahmus JB (2014) Can efficiency improvements reduce resource consumption? J Ind Ecol 18:883–897

FAO (2016) Database for food and agriculture. Rome. Available in: http://faostat3.fao.org/browse/R/*/E

Fernández Prieto L (2001) El cambio tecnológico en la historia agraria de la España contemporánea. Hist Agrar 24:59–86

Fischer-Kowalski M, Haberl H (2007) Socioecological transitions and global change: Trajectories of social metabolism and land use. Institute of Social Ecology. Vienna (Austria). Edward Elgar Publishing

Fischer-Kowalski M, Krausmann F, Pallua I (2014) A sociometabolic reading of the Anthropocene: modes of subsistence, population size and human impact on earth. Anthropocene Rev 1(1):8–33

Gagnon N, Hall CAS, Brinker L (2009) A preliminary investigation of energy return on energy investment for global oil and gas production. Energies 2:490–503

Gallego D (1986) La producción agraria de Álava, Navarra y La Rioja desde mediados del siglo XIX a 1935. Universidad Complutense de Madrid, Madrid

Giampietro M, Bukkens SGF, Pimentel D (1999) General trends of technological changes in agriculture. CritAl Rev Plat Sci 18:261–282

Gutowski TG, Sahni S, Allwood JM, Ashby MF, Worrell E (2013) The energy required to produce materials: constraints on energy-intensity improvements, parameters of demand. Philos Trans R Soc A-Math Phys Eng Sci 371:1–14

Guzmán casado GI, González de Molina M (2009) Preindustrial agriculture versus organic agriculture: the land cost of sustainability. Land Use Policy 26:502–510

Guzmán GI, Aguilera E, Soto D, Cid A, Infante-Amate J, García-Ruiz R, Herrera A, Villa I, González de Molina M (2014) Methodology and conversión factors to estimate the net primary productivity of 112 historical and contemporary agro-ecosystems (I). Documento de Trabajo de la Sociedad Española de Historia Agraria, no 14–06. Disponible en: www.seha.info

Hall CAS, Balogh S, Murphy DJR (2009) What is the Minimum EROI that a sustainable society must have? Energies 2:25–47

Hall CAS, Lambert JG, Balogh SB (2014) EROI of different fuels and the implications for society. Energy Policy 64:141–152

IDAE (Instituto para la Diversificación y Ahorro de la Energía) (2015) Balances del consumo de energía final: Serie histórica 1990–2013. Ministerio de Energía, Industria y Turismo, Madrid

IEA (2015) Energy Statistics and Balances of Non-OECD Countries and Energy Statistics of OECD Countries, and United Nations, Energy Statistics Yearbook. In: Agency IE (Ed)

INE (Instituto Nacional de Estadística) (1960) Anuario de estadística 1960. Madrid

INE (Instituto Nacional de Estadística) (1963) Censo agrario 1962. Madrid

INE (Instituto Nacional de Estadística) (1965) Anuario de estadística 1965. Madrid

INE (Instituto Nacional de Estadística) (1970) Anuario de estadística 1970. Madrid

INE (Instituto Nacional de Estadística) (1997) Anuario de estadística 1997. Madrid

Infante-Amate J, Soto D, Aguilera E, García Ruiz R, Guzmán G, Cid A, González de Molina M (2015) The spanish transition to industrial metabolism long-term material flow analysis (1860–2010). J Ind Ecol 19(5):866–876. Available in: https://doi.org/10.1111/jiec.12261

JCA (Junta Consultiva Agronómica) (1918) Medios que se utilizan para suministrar el riego a las tierras y distribución de los cultivos en la zona regable. Resumen hecho por la Junta Consultiva Agronómica de las Memorias de 1916, remitidas por losingenieros del Servicio Agronómico provincial. Ministerio de Fomento. Dirección General de Agricultura, Minas y Montes, Madrid, Spain

Jenssen TK, Kongshaug G (2003) Energy consumption and greenhouse gas emissions in fertilizer production. International Fertiliser Society Meeting London, London, UK

Krausmann F, Haberl H (2002) The process of industrialization from the perspective of energetic metabolism: socioeconomic energy flows in Austria 1830–1995. Ecol Econ 41:177–201

Krausmann F, Erb KE, Gringrich S, Lauk C, Haberl H (2008) Global patterns of socioeconomic biomass flows in the year 2000: a comprehensive assessment of supply, consumption and constraints. Ecol Econ 65:471–487

MA (Ministerio de Agricultura) (1930) Anuario de Estadística Agraria 1930. MA, Madrid

MA (Ministerio de Agricultura) (1931) Anuario de Estadística Agraria 1931 MA, Madrid

MA (Ministerio de Agricultura) (1933) Anuario de Estadística Agraria 1933. Madrid

MA (Ministerio de Agricultura) (1934) Anuario de Estadística Agraria 1934. Madrid

MA (Ministerio de Agricultura) (1935) Anuario de Estadística Agraria 1935. Madrid

MA (Ministerio de Agricultura) (1939) Anuario de Estadística Agraria 1939. Madrid

MA (Ministerio de Agricultura) (1940) Anuario de Estadística Agraria 1940. Madrid

MA (Ministerio de Agricultura) (1966) La agricultura, la pesca y la alimentación en España, año 1966. Madrid

MA (Ministerio de Agricultura) (1970) La agricultura, la pesca y la alimentación en España, año 1970. Madrid

MA (Ministerio de Agricultura) (1973) Anuario de Estadística Agraria 1973. Madrid

MA (Ministerio de Agricultura) (1975) Anuario de Estadística Agraria 1975. Madrid. Ministerio de Agricultura

MA (Ministerio de Agricultura) (1975) La agricultura, la pesca y la alimentación en España, año 1975. Madrid

MA (Ministerio de Agricultura) (1976) La agricultura, la pesca y la alimentación en España, año 1976. Madrid

MA (Ministerio de Agricultura) (1977) La agricultura, la pesca y la alimentación en España, año 1977. Madrid

MA (Ministerio de Agricultura) (1978) Anuario de Estadística Agraria 1978. Madrid

MAGRAMA (Ministerio de Agricultura Alimentación y Medio Ambiente) (2008) Encuesta sobre superficie y rendimiento de cultivos. Resultados 2008. Madrid.

MAGRAMA (Ministerio de Agricultura, Alimentación y Medio Ambiente) (2012) Anuario de Estadística Agraria 2012. Madrid

MAGRAMA (Ministerio de Agricultura, Alimentación y Medio Ambiente) (2013a) Anuario de Estadística Agraria 2013. Madrid

MAGRAMA (Ministerio de Agricultura, Alimentación y Medio Ambiente) (2013b) Balance del nitrógeno en la agricultura española. Año 2011. Madrid

MAGRAMA (Ministerio de Agricultura, Alimentación y Medio Ambiente) (2013c) Ganado Porcino de Ciclo Cerrado en Aragón: estudios de Costes y Rentas de las Explotaciones Agrarias. Resultados Técnico-Económicos. Madrid

MAGRAMA (Ministerio de Agricultura, Alimentación y Medio Ambiente) (2015a) Encuesta sobre superficies y rendimientos de cultivo. Informe sobre regadíos en España. Madrid

MAGRAMA (Ministerio de Agricultura, Alimentación y Medio Ambiente) (2015b) *Inventario de presas y embalses.* Madrid

MAGRAMA (Ministerio de Agricultura, Alimentación y Medio Ambiente) (2016) El libro digital del agua. Madrid

MAIC (Ministerio de Agricultura, Industria y Comercio) (1928) Anuario de Estadística Agraria 1928. Madrid

MAIC (Ministerio de Agricultura, Industria y Comercio) (1930) Anuario de Estadística Agraria 1930. Madrid

MAIC (Ministerio de Agricultura, Industria y Comercio) (1931) Anuario de Estadística Agraria 1931. Madrid

MAIC (Ministerio de Agricultura, Industria y Comercio) (1932) Anuario de Estadística Agraria 1932. Madrid

Maicop (Ministerio de Agricultura, Industria, Comercio y Obras Públicas) (1904) El regadío en España. Resumen hecho por la Junta Consultiva Agronómica de las memorias sobre riegos remitidas por los ingenieros del Servicio Agronómico provincial. Madrid

MAPA (Ministerio de Agricultura, Pesca y Alimentación) (1980) Anuario de Estadística Agraria 1980. Madrid

MAPA (Ministerio de Agricultura, Pesca y Alimentación) (1981) Anuario de Estadística Agraria 1981. Madrid

MAPA (Ministerio de Agricultura, Pesca y Alimentación) (1984) Anuario de Estadística Agraria 1984. Madrid

MAPA (Ministerio de Agricultura, Pesca y Alimentación) (1986) Anuario de Estadística Agraria 1986. Madrid

MAPA (Ministerio de Agricultura, Pesca y Alimentación) (1988) Anuario de Estadística Agraria 1988. Madrid

MAPA (Ministerio de Agricultura, Pesca y Alimentación) (1989) Anuario de Estadística Agraria 1989. Madrid

MAPA (Ministerio de Agricultura, Pesca y Alimentación) (1990) Anuario de Estadística Agraria 1990. Madrid

MAPA (Ministerio de Agricultura, Pesca y Alimentación) (1992) Anuario de Estadística Agraria 1992. Madrid

MAPA (Ministerio de Agricultura, Pesca y Alimentación) (1993) Anuario de Estadística Agraria 1993. Madrid

MAPA (Ministerio de Agricultura, Pesca y Alimentación) (1994) Anuario de Estadística Agraria 1994. Madrid

MAPA (Ministerio de Agricultura, Pesca y Alimentación) (1997) Anuario de Estadística Agraria 1997. Madrid

MAPA (Ministerio de Agricultura, Pesca y Alimentación) (1999) Anuario de Estadística Agraria 1999. Madrid

MAPA (Ministerio de Agricultura, Pesca y Alimentación) (2001) Anuario de Estadística Agraria 2001. Madrid

MAPA (Ministerio de Agricultura, Pesca y Alimentación) (2002) Anuario de Estadística Agraria 2002. Madrid

MAPA (Ministerio de Agricultura, Pesca y Alimentación) (2003) Encuesta sobre superficies y rendimientos de cultivos del año 2002. Memoria. Madrid

MAPA (Ministerio de Agricultura, Pesca y Alimentación) (2004) Anuario de Estadística Agraria 2004. Madrid

MAPA (Ministerio de Agricultura, Pesca y Alimentación) (2006) Anuario de Estadística Agraria 2006. Madrid

MAPA (Ministerio de Agricultura, Pesca y Alimentación) (2007) Análisis del parque nacional de tractores agrícolas en 2005–2006. Madrid

MARM (Ministerio de Medio Ambiente, Medio Rural y Marino) (2008) Inventario de emisiones de gases de efecto invernadero de España 1990–2006 Madrid. Ministerio de Medio Ambiente, Medio Rural y Marino

MARM (Ministerio de Medio Ambiente, Medio Rural y Marino) (2008) Encuesta sobre superficies y rendimientos de cultivos. Madrid

MARM (Ministerio de Medio Ambiente, Medio Rural y Marino) (2010) Anuario de Estadística Agraria 2010. Madrid

MARM (Ministerio de Medio Ambiente, Medio Rural y Marino) (2008) Anuario de Estadística Agraria 2008. Madrid

Martinez-Ruiz JI (2000) Trilladoras y tractores: energía, tecnología e industria en la mecanización de la agricultura española (1862–1967). Universidad de Sevilla, Sevilla, Spain

Mateu Tortosa E (2013) Agriculture and propaganda: chilean nitrate fertilizers in Spain. Hist Agrar 59:95–123

Meadows DH, Meadows DL, Randers JY, Behrens III WW (1972) The Limits to growth: a report for the Club of Rome's project on the predicament of mankind. Universe Books, New York

MF (Ministerio de Fomento) (1918) Medios que se utilizan para suministrar el riego a las tierras y distribución de los cultivos en la zona regable. Resumen hecho por la Junta Consultiva Agronómica de las memorias de 1916, remitidas por los ingenieros al Servicio Agronómico provincial. Madrid

MI (Ministerio de Industria) (1961a) *Estadística de la industria de la energía eléctrica. Resumen del año 1960.* Madrid

MI (Ministerio de Industria) (1961b) *La energía en España. Evolución y perspectivas (1945–1975).* Madrid

MI (Ministerio de Industria) (1972) Estadística de la industria de energía eléctrica 1970. Madrid

MIE (Ministerio de Industria y Energía) (1981) Estadística de la industria de energía eléctrica 1980. Madrid

MIE (Ministerio de Industria y Energía) (1991) Estadística de la industria de energía eléctrica 1990. Madrid

MIE (Ministerio de Industria y Energía) (2003) Estadística de la industria de energía eléctrica 2002. Madrid

MINETUR (Ministerio de Energía, Industria y Turismo) (2015) Balances de energía final (1990–2013). Madrid

MINETUR (Ministerio de Energía Industria y Turismo) (2016) Estadísticas eléctricas anuales (1958–2009). http://www.minetur.gob.es/energia/balances/Publicaciones/ElectricasAnuales/Paginas/ElectricasAnuales.aspx. Accessed 12 Feb 2016. MINETUR, Madrid.

MITYC (Ministerio de Industria, Turismo y Comercio) (2009) Estadística de la industria de energía eléctrica 2008. Madrid

Naredo JM, Campos P (1980) Los balances energéticos de la agricultura española. Agricultura y Sociedad 15:163–255

Pelletier N, Ibarburu M, Xin H (2014) Comparison of the environmental footprint of the egg industry in the United States in 1960 and 2010. Poult Sci 93:241–255

Pellegrini P, Fernández RJ (2018) Crop intensification, land use, and on-farm energy-use efficiency during the worldwide spread of the green revolution. Proc Nat Acad Sci, USA 115: 2335–2340.

Pérez-Minguijón M (1992) Análisis del parque nacional de cosechadoras de cereales. Rev Estud Agro-Soc 159: 271–289

Pérez-Minguijón M (1999) El nuevo reglamento general de vehículos y maquinarias agrícolas. Agrotécnica 2:9–11

Ramírez CA, Worrell E (2006) Feeding fossil fuels to the soil: An analysis of energy embedded and technological learning in the fertilizer industry, Resour Conserv Recycl 46:75–93

REE (Red Eléctrica de España) (1998) El sistema eléctrico español 1997. Madrid

REE (Red Eléctrica de España) (2000) El sistema eléctrico español 1999. Madrid

REE (Red Eléctrica de España) (2005) El sistema eléctrico español 2004. Madrid

REE (Red Eléctrica de España) (2010) El sistema eléctrico español 2009. Madrid

REE (Red Eléctrica de España) (2012) El sistema eléctrico español (2011). Red Eléctrica de España, Madrid

REE (Red Eléctrica de España) (2015) Balances de energía eléctrica (1990–2014). Madrid

Smil V (1999) Energies: an illustrated guide to the biosphere and civilization. The MIT Press, Cambridge, MA, USA

Smil V (2001) Enriching the earth: Fritz Haber, Carl Bosch, and the transformation of world food production. The MIT Press, Cambridge, MA, USA

Smil V (2013) Harvesting the biosphere: what we have taken from nature. The MIT Press, London

UNESA (Asociación Española de la Industria Eléctrica) (2005) El sector eléctrico a través de UNESA (1944–2004). Madrid

Wikipedia (2015) Guano

4

Evolution of Agricultural Population

4.1 Introduction

In the previous chapter, we reviewed Spanish agroecosystems' means of production, that is, one of its two social fund elements. We did so by analyzing the evolution of inputs used. In this chapter, we describe the evolution of the agricultural population, the second social fund. The agricultural population emits a work flow that is measurable in terms of energy. It also originates an integrated information flow that supports its own structure and functioning. This flow is generated by households, whose "reproduction" is not only of a biological nature: it also relies on economic costs that have varied over time. Consequently, the concept of agrarian metabolism we put forward here takes into account not only the number of individuals engaged in agricultural work and the time spent on it but also their families or households and the paid or unpaid work time that is required to sustain them. The maintenance of a constant flow of human energy needed to manage agroecosystems depends on the reproduction of these agricultural groups. Reproduction costs must be covered by income from the sale of agricultural production or from wages obtained from the sale of labor power.

This second fund element is often characterized by paid work. But paid work is not fully operational in the case of agrarian metabolism because not all work going into agriculture is marketed, i.e., paid. Reproductive work and family care associated with biological reproduction constitute unpaid reproductive jobs though they are essential to reproduce agricultural work flows. We do not dispose of adequate sources to make a full-scale estimate of the time budget of all these reproductive tasks nor of all the other gainful activities (paid work outside the sector) that became ever more widespread as agricultural activity became less profitable. Though such estimates have been carried out for local case studies and provide us with valuable information for organic agriculture contexts (Marco 2017; Villa 2017), they go beyond the scope of this research for now. Statistical sources do not include any category enabling to quantify the number of households dedicated to agricultural activity anyway.

However, in the absence of a full-blown analysis of agricultural households, it is possible to estimate their reproduction costs, understood as the amount of goods and services necessary to maintain and reproduce the household measured in monetary terms.

It is interesting to examine for that matter whether Spanish agriculture was able to sustain its agricultural population during the twentieth century and what impact it had on other fund elements. To this end, this chapter attempts to quantify the monetary flow that farmers received in exchange for their marketed production and their work throughout the period under study. Our main hypothesis is that they perceived insufficient income to sustain and reproduce their household. They were driven to seek higher incomes by specializing in crops with the most profitable market outlets or by intensifying production. Thus, close links can be established between insufficient income and tendencies towards intensification and productive specialization. Insufficient income derived from the unequal distribution of income among farmers themselves. Small farmers were especially affected, and many eventually abandoned their activity. But the reasons for income inequality differed at the beginning and at the end of the period under study. During the first half of the century, unequal distribution of agricultural income (internal inequality) meant that a large share of farmers perceived insufficient income. During the second half of the century, however, the sector at large saw its income drop. This fall persisted over time and especially weighed on small farmers. Income decline resulted from the deterioration of terms of trade, giving rise to the transfer of income to other sectors of activity (external inequality). In this context, productive intensification and specialization became a common strategy for all farmers, regardless of their size.

The lack of data sources was a major problem when performing our analysis of the first 50 years of the twentieth century. For this reason, all the information available for this period has been brought together in the first section. We have attempted to estimate household income and expenditure and their impact on employment and farm structures. Nevertheless, we disposed of much more information on developments since the 1950s, when inequality became external and most farmers began to follow the intensification pattern. The following sections separately analyze income, household spending behavior, the extent of achieved expenditure coverage and impacts on employment and farm structures. The chapter ends with some general conclusions on the agricultural population fund element's evolution, its current situation, and its prospects.

4.2 The Agricultural Population During the First Half of the Twentieth Century

By the end of the nineteenth century, communal property and rights had been completely dismantled. Small plots of municipal land (called "*Propios*") distributed to day laborers and small landowners had been privatized during the 1855 General

Disentailment. From then on, day laborers began to depend almost exclusively on earned wages and small farmers, whether owners, tenants or sharecroppers, depended on the sale of their products. The commodification of their economies, that began when the Liberal Agrarian Reform was introduced at the beginning of the nineteenth century, had been completed (for a review, see Acosta et al. 2009). In this context, wages and prices became the central explanatory factors of farmers and farmworker decisions.

As we saw in Chap. 2, during the first third of the twentieth century, Spanish agriculture grew slowly initially and more vigorously later. This had consequences on the agricultural population's size and composition as well as on their reproduction strategies. The sources available for estimating the agricultural population during these years differ from the sources available after the 1950s, although all estimates are based directly or indirectly on population censuses led and published by the National Institute of Statistics or prior agencies. For the study period before 1950, we used the professional classifications established in the censuses to compile the information in Table 4.1 that reflects the evolution of the active population throughout those years. At least two distinct periods can be distinguished: that of falling active agricultural population between 1900 and the 1930s; and the forties and fifties when the population recovered and even exceeded its 1900 size.

In 1900, the active agricultural population exceeded 4.5 million male and female individuals. Female workers tend to be underestimated by official statistics, not to mention domestic and care work that largely remains outside the market still today, unreflected in national accounting. Women accounted for only 17% of total workers, and 1900 was also the year in which the largest amount of women's work appeared in the sources. Research carried out at a local level during these years suggests a much greater weight of women in agricultural work (Marco 2017; Villa 2017). The figure, as we shall see later, was not the biggest in the series in absolute terms, but it did represent the highest percentage of the active population (60.4%) and of the population generally (25%). For a sector that could still be defined as organic agriculture, human labor was essential, given its major role in agricultural work processes.

Table 4.1 Active agricultural population (thousand) and its percentage over total active population and total population

Year	Men	Women	Active agricultural pop	Total active pop	%	Total pop	%
1900	3782.6	775.7	4558.3	7547.0	60.4	18,830.6	24.2
1910	3861.1	359.4	4220.5	7581.5	55.7	20,360.3	20.7
1920	4232.8	322.8	4555.6	7962.4	57.2	22,012.6	20.7
1930	3777.3	263.4	4040.7	8772.5	46.1	24,026.5	16.8
1940	4518.9	262.1	4781.0	9360.9	51.1	26,386.8	18.1
1950	4827.6	409.5	5237.2	10.793.1	48.5	28,172.2	18.6

Source Population census. *INE*

Table 4.2 Evolution of cultivated land productivity (considering only primary crops and not including residues and labor) in biophysical and monetary terms

Year	t dm/ha	1900 = 100	Pts/ha	1900 = 100	t dm/active pop. member	1900 = 100	Pts/active pop. member	1900 = 100
1900	0.68	100	138	100	2.45	100	723.7	100
1910	0.71	105	127	92	2.99	122	783.8	108
1922	0.79	116	138	100	3.28	134	814.4	112
1933	0.82	121	166	120	4.13	168	1173.3	162

Source Compiled by the author. Monetary data was obtained from Simpson (1997, 58), based on 1910 pesetas. Only the main crops were included to get as close as possible to monetary values. Therefore, the rest of the cultivated lands' net primary productivity is not accounted for
Pts = pesetas; dm = dry matter

The period of evolution of the active agricultural population[1] can be divided in turn into two sub-periods: a first period during which the numbers of the active population remained more or less stable and a second, starting in the 1920s, when the numbers began to fall sharply. Labor productivity increase, measured in monetary terms (see Table 4.2), explains the 7.4% reduction reflected in the 1910 figures. This evolution pattern should have persisted. However, the numbers of the active population had recovered by the early 1920s and its percentage with respect to the total active population had barely dropped to 57.2%. We can, therefore, say that the size of the active population remained almost stable during the first twenty years of the century.

This behavior is consistent both with agriculture's growth rates and the technological changes implemented in those years. The agricultural sector grew significantly between 1900 and 1933, as shown both by the increase in production per hectare and per worker, as illustrated in Table 4.2. But between 1900 and 1922, productivity per hectare grew very little in physical terms and not at all in monetary terms; the increase in labor productivity was slightly higher for both measurements. In contrast, during the 1922–1933 period, land productivity grew in physical terms at a similar rate to that of previous decades, but labor productivity accelerated, as the number of workers dropped sharply. In 1930 the countryside had lost more than half a million of its active population compared to 1900. These active members of the population already represented less than half of the total active population and less than 17% of the Spanish population.

Productivity increases were due to Spanish agriculture's incipient processes of intensification and production specialization, as described in previous chapters. From a labor demand perspective, the dissemination of new irrigation technologies (Calatayud and Martínez Carrión 1999) and the use of chemical fertilizers (see Chap. 3) played an essential role. Increases in productivity per hectare could have

[1] The agricultural population is the population aged between 12 and 65 who can work in agriculture.

been even greater had the cultivated land not grown by 23.7%, thanks, among other things, to chemical fertilizers and new plows that carried out heavier-duty work. Increases in labor productivity rates were influenced, in turn, by the spread of new mechanical technologies. These technologies became more widespread during the twenties and were used more intensely, precisely as the active agricultural population began to decrease. Most technologies at the time were designed to save land and not labor force. Even as mechanization gained ground, it concerned traditional crops that required less labor than intensive crops, which are much more difficult to mechanize.

Similar developments were taking place in the rest of Europe, although the number of farm workers had begun to fall more rapidly. This relative divergence is due to multiple factors, but two issues proper to the case of Spain are worthy of mention: on the one hand, the structural nature of the differences in productivity, hard to overcome, and the new mechanical technology problems of adapting to Spanish agriculture's land and climate conditions. The fact is, despite agricultural growth, work processes continued to demand large numbers of workers on a punctual basis to realize concrete tasks: for example, the weeding and harvesting of cereals or the harvesting of olives. Olive grove harvesting would scarcely change until the 1980s (Infante-Amate 2011). Self-propelled machinery began to spread as early as the fifties. Until then, a relatively narrow relationship existed between each country's land and climate conditions and the possibilities of replacing human work with animal labor, leading to greater work productivity. Spain, with low net primary productivity, could not reach the same productivity levels per hectare or per worker achieved by other more northern European countries (González de Molina 2001). Productivity was conditioned by the employment offer that Spanish cereal rainfed lands could generate. These lands covered extensive stretches of territory and presented few alternatives of use. The employment offer was highly seasonal for crop and occasional in the case of fallows. This capacity was in stark contrast with that, for example, offered by irrigation, for which work demand existed almost throughout the year. It has been argued that the active agricultural population was in fact so abundant it discouraged the adoption of major measures of productivity improvement. However, this view, so widespread in Spanish economic historiography (Simpson 1997), obliviates that in reality there was very little that could be done to converge with other humid European countries.

Thus, during those years, agriculture relied on large numbers of individuals gathering in the fields to accomplish major tasks; these farm workers already relied exclusively on wages or received prices to survive. Wages were usually low and irregular. Agricultural unemployment did not consist in a percentage of workers without access to work, but in the number of days per year when there were no employment needs, unlike industry or services unemployment. This may explain why the different sector's administrations did not offer concrete data on agricultural unemployment during the first third of the twentieth century, unlike after the 1950s. Furthermore, the Spanish industry offered a limited amount of opportunities for alternative employment (Gallego 2001, 1986). Nevertheless, these general trends hide great territorial disparities: the number of primary sector employees dropped mainly in industrial areas, while in inland agricultural areas, these majority percentages remained largely unchanged. Precisely for this reason, unbalanced property distribution led to notorious peasant revolts giving rise to the so-called "agrarian question" (Acosta et al. 2009). We will return to this issue later.

Having described the size of the active agricultural population, we must now turn towards their income and expenditure requirements. Income is usually reflected in agricultural sector accounts and national accounts. These accounts are provided both by the Ministry of Agriculture and the National Institute of Statistics. However, the sector's accounts only began to be drawn up in the fifties, so we do not dispose of accounting information for the first half of the century. Reconstructing macromagnitudes over those years goes beyond the scope of this book. However, we do have some estimates on the value of total agricultural production (GEHR 1983) and final production (Simpson 1997) for the period before the civil war.

We also dispose of two estimates of the share of agriculture, fisheries and forestry in Gross Value Added (GVA): one carried out by Maluquer (2016) and another by Prados de la Escosura (2017). Although both estimations are based on the same final agricultural production size series provided by Prados, itself based on Simpson's final production calculations, we can find considerable differences between them. These differences are due to the various ways in which both authors related historical estimates with national accounting. Neither of them offers a series that is comparable with that offered by the sector's macromagnitudes, much less with an agricultural income series. In any case, we used both series as a proxy for the monetary remuneration that farmers would have received in exchange for their products. These series were calculated per farm worker, using our own employment series for the sector.

Although considerable differences can be observed between both series, both share a similar evolution that is generally consistent with that typically accepted by historiography. According to Prados, agricultural GVA per member of active agricultural population grew between 1900 and 1935 by 95.7%, and by 55.8% according to Maluquer. It dropped between 1935 and 1950 by 35.4% and 1.9%, respectively. Either way, there was an upward trend during the first third of the twentieth century in both series. Total GVA growth rates between 1900 and 1935 are similar in both series to those of GVA per member of active agricultural population (89.4% for Prados and 50.7% for Maluquer).

Also worthy of interest is its relationship with the rest of the economy. Graphs 4.1 and 4.2 show the evolution of GVA per worker in agriculture and in the economy as a whole throughout the first third of the twentieth century. Although agricultural productivity was between 40 and 50% lower than that of the economy as a whole, this ratio did not decrease significantly in the first third of the twentieth century and that is the most relevant point. It is much more difficult to know to what extent the agricultural sector's productivity improvements were able to meet the needs of the agricultural population. We do not dispose of a series on disposable income or a series on basic rural household consumption. To solve this problem, we made some calculations the result of which should be considered as merely indicative. The first household budget survey to be conducted provides data on household consumption

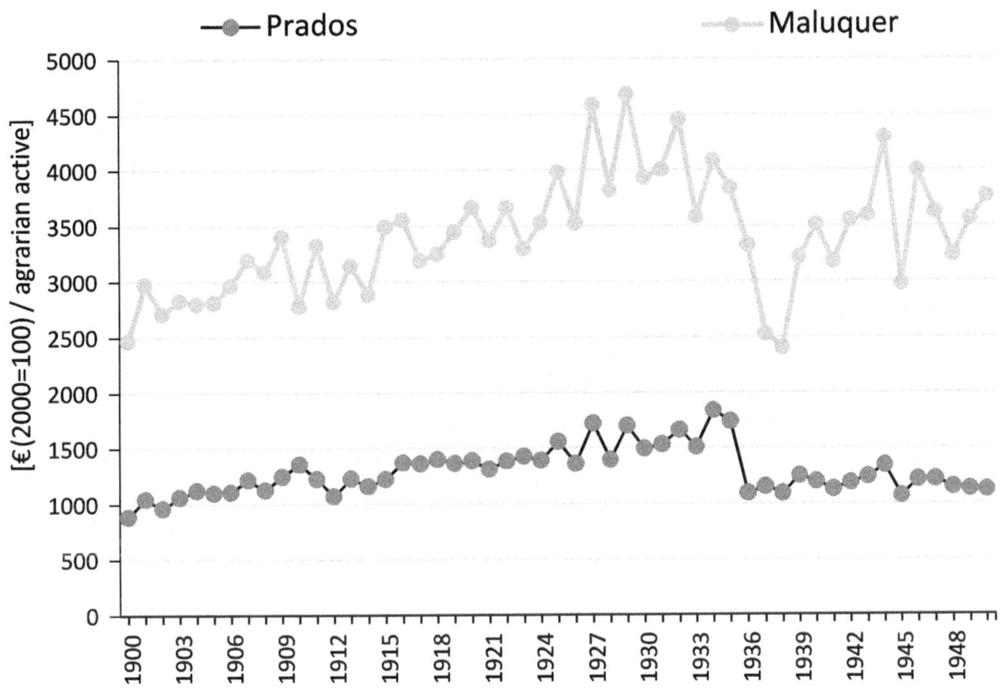

Graph 4.1 Gross value-added per active population member, in constant year 2000 euros. *Source* Maluquer 2016; Prados (2017) and our own active population data

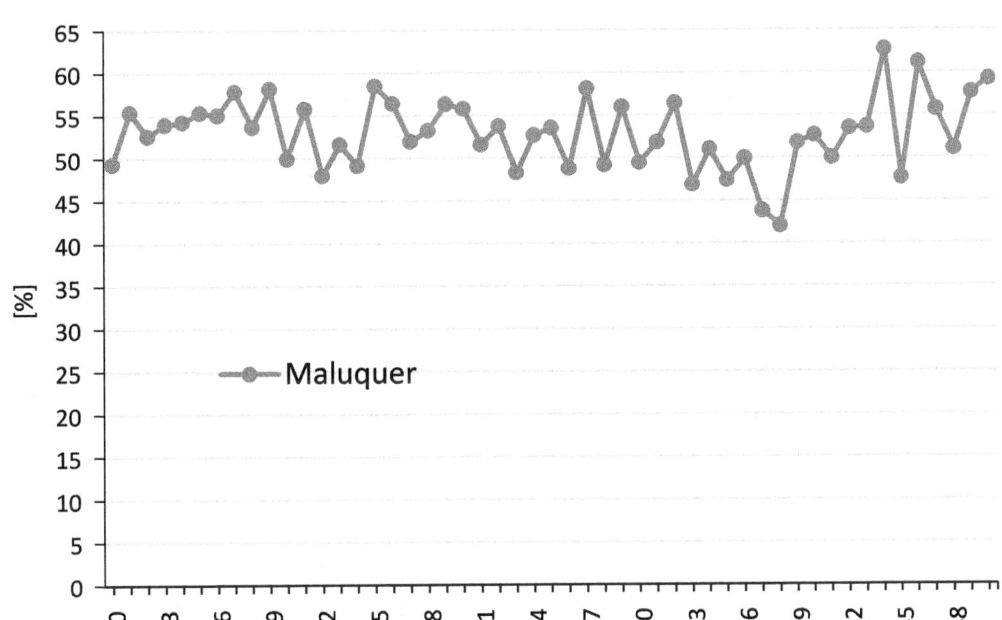

Graph 4.2 Share of GVA/member of active agricultural population compared to GVA/member of total active population. *Source* Maluquer 2016 and our own active population data (Given that the original series used by Prados and Maluquer is the same, the percentage ratio between agricultural and total GVA is identical. That's why we showed only one of the two series.)

in 1958, including households living in rural areas, described as "suburban" in the surveys. According to our own reconstruction of the Spanish economy's metabolic profile (Infante-Amate et al. 2015), exosomatic consumption levels were similar to those in the first half of the century and endosomatic consumption levels were recovering those prior to the civil war (González de Molina et al. 2014, 2017). However, there must have been differences between 1900 and 1958. To uncover these differences, we assumed that household budgets followed a similar trend to that of national private consumption, estimated by Prados (2003, 405–410; and Carreras and Tafunell 2005, 1284–1285). We, therefore, estimated household budgets since 1900 based on 1958 household consumption, applying the same variation rate as that of private consumption from 1901 to 1950.

Table 4.3 compares the GVA per active agricultural population member of both estimates with average household expenditure over the 1900 to 1950 period in current pesetas, which gives an idea of the extent of coverage provided by agricultural income.[2] Table 4.3 presents the data for three different periods: 1901–1905, 1931–1935, and 1946–1950. Discrepancies between Prados and Maluquer estimations can be observed, the latter indicating a wider coverage. In any case, at the beginning of the twentieth century, GVA per active agricultural population member covered 79.5% of household budgets according to Prados, while GVA was sufficient based on Maluquer's values, and there was even some remaining income. The relationship worsened in the thirties in both GVA estimates. At the end of the 1940s, the situation improved slightly according to Prados and clearly improved according to Maluquer. Real GVA values are probably situated between both estimates, therefore, close to the country's average household expenditure. Nevertheless, both estimates reveal a similar trend: a declining percentage of coverage throughout the first third of the century that is consistent with the progress of the country's economy. Despite agricultural production growth and the falling numbers of workers, household spending grew more rapidly and this reduced the extent of coverage of agricultural income. The evolution of apparent food consumption described in previous work corroborates this increase in household expenditure (González de Molina et al. 2014). The forties

Table 4.3 Comparison between estimated household budget and GVA per worker (current pesetas)

	Household budget	GVA/Active pop member	GVA/Active pop member	% coverage	% coverage
		Prados	Maluquer	Prados	Maluquer
1901–05	962.3	765.4	1114.8	79.5	115.8
1931–35	2881.6	1913.5	2745.4	66.4	95.3
1946–50	10,250.2	7104.9	11,314.5	69.3	110.4

Source 1958 survey of household budgets, Maluquer (2016), Prados (2003, 2017) and our active population data

[2]This calculation is merely an approximation given that GVA is not equivalent to income, but it provides a useful comparison framework.

would represent a decline in the agricultural sector's GVA in constant pesetas, both for Prados and for Maluquer.

Strategies adopted by agricultural households must have been conditioned by income from agricultural activity in the absence of other production alternatives, as we have seen. In that sense, we have already emphasized the major role of inequality to explain intensification and specialization processes of Spanish agroecosystems during the entire study period, as inequality was the driver of a range of farmer strategies. Inequality was of course not the only reason. Agricultural income was indeed unevenly distributed and this encouraged the adoption of distinct production strategies that converged in the specialization and intensification of production. The analysis of property and farm structures should allow us to discover whether the land was distributed fairly among farmers, though it is only possible to perform this analysis starting from the first agricultural census to be published in 1962. The only information we dispose of is that published by Pascual Carrión, who gathered the results of the preliminary land register (*Avance Catastral*) until 31 December 1930 (Carrión 1932 [1975]). This information describes the structure of property and not that of farms. In addition, the data provided refer to 27 provinces, i.e., 62.4% of the total area. More northern provinces of the Peninsula are missing, generating a data bias favorable to larger farms and an underestimation of small farms.

Table 4.4 shows the taxable income declared by owners included in the Rural Land Cadastre (*Catastro de Rústica*), based on plot sizes. Obviously, the table does not fully capture the real number of owners as it is based on all registered landowners omitting possible redundancies in different areas and municipalities. Table 4.5 shows the approximate number of owners calculated by Carrión. Consequently, it classifies

Table 4.4 Distribution of registered wealth (taxable income) of farms (plots) in 1930

	Pesetas	%	Owners	%	Pts/owner
Up to 10 ha	361,221,300	51.66	10,016,115	98.06	36.1
From 10 to 100 ha	134,484,324	19.24	169,472	1.66	793.5
From 100 to 250 ha	58,757,117	8.40	16,305	0.16	3,603.6
More than 250 ha	144,857,076	20.71	12,467	0.12	11,619.2
Total taxable income	699,118,386	100.00	10,214,359	100.00	68.4

Source Pascual Carrión (1932 [1975], 100–101)

Table 4.5 Distribution of agricultural wealth among owners in 1930

	Pesetas	%	Owners	%	Pts/owner
Small owners [>1000 pts]	228,431,436	32.68	1,699,585	94.94	134.4
Medium owners [1000–5000 pts]	176,711,520	25.28	73.092	4.09	2,417.7
Large owners[+5000 pts]	294,028,428	42.06	17.349	0.97	16,947.9
Total	699,118,386	100.0	1,790,026	100.0	390.6

Source Pascual Carrión (1932 [1975], 112–113)

the cadastral plots according to their owners. As mentioned, the table does not reflect the real number of leasing and sharecropping cases, so there must have been a smaller number of large farms and a higher number of small and medium farms.

A total of 98.06% of owners had plots under 10 hectares that generated barely more than half the taxable income. Average income per plot and owner was just over 36 pesetas, not including labor costs that counted as income for most farmers. Either way, we can assume that many small plots were grouped into larger farms via leases or other transfer agreements, given that 77.67% of them were below one hectare and 96% below five hectares.

Nevertheless, Carrión's data gives an idea of land access inequality and the unequal distribution of agricultural activity income. The data also confirms the overwhelming weight of small properties and dependence on work on self-owned farms or work for others. We can logically assume that the vast majority of these small landowners sought to maximize agricultural income either by specializing their production in more easily marketable crops or by intensifying the production of subsistence crops. Based on the data in Table 4.3, they succeeded. Both strategies have been observed among small farmers in the region of the Vega de Granada. At the turn of the century, they adopted beet crops and were the first to use chemical fertilizers (González de Molina and Guzmán Casado 2006); they have also been observed among a large part of small farmers, who turned to cultivating olive groves in Andalusia and managed them more intensively (Infante-Amate 2014).

For their part, employees, threatened by seasonal unemployment and the lack of alternative employment outside the sector, developed strategies to strengthen their role in the labor market. They found ways to receive better salaries and working conditions by constituting local unions at first, and later, by becoming members of national unions (*CNT* and *UGT*), or even political parties such as the *PSOE* and some republican groups. This resulted in notable increases in peasant protests that became especially intense during the so-called *Bolshevik Triennium* (1918–1920). Since then, an increasing number of agricultural conflicts were resolved through collective bargaining accompanied by wage increases. Interested readers may find a detailed description of this process in our research (Acosta et al. 2009) on the origins of the National Federation of Land Workers (*Federación Nacional de Trabajadores de la Tierra*). This pressure strategy and union mobilization of rural workers resulted in wage increases (Graph 4.3).

The specialization and intensification strategy came to be shared by farmers who depended on external labor. During the crisis at the end of the century, when agricultural prices were low, wages were relatively low and there were structural shortages of fertilizers (most of them were still organic), the most logical strategy was to replace soil fertility within the farm, without using external fertilizers. This meant extensifying production, while reducing labor and fertilization costs. This extensification strategy explains the rise in unemployment during those years and the early twentieth century's fierce social crisis (Acosta et al. 2009). The doctoral thesis by Inmaculada Villa (2017) regarding Montefrío (Granada), an inland Andalusian town, shows that this strategy was followed by medium and large landowners. Protests for more work and better wages were repressed by authorities almost throughout Spain:

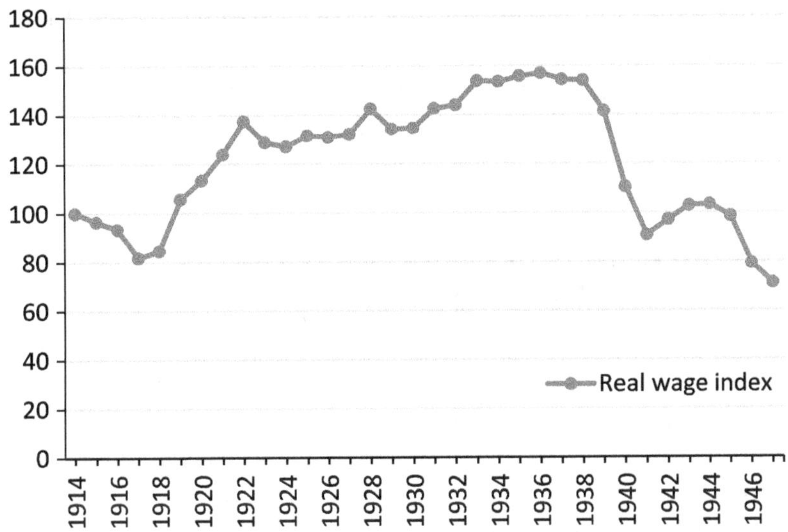

Graph 4.3 Index of real average wages of agricultural workers (men) 1900 = 100. *Source* Carreras and Tafunell (2005). Deflated by the CPI prepared by Maluquer (2013)

repression by the Spanish military police (*Guardia Civil*), by the army, strike bans, restrictions of association rights, etc. Things would change during the First World War. As described just above, wages rose to compensate for the increase in living expenses and working conditions improved thanks to rural workers' protests. The rise in wages should have led to a change in the strategy by farmers hiring wage labor. The most logical outcome would have been the substitution of labor by machines, given the bigger weight of labor costs on income. Meanwhile, the use of chemical fertilizers intensified. Intensification by increasing the yields per unit area, mainly possible thanks to chemical fertilizers, became an achievable means to compensate for the rise in labor costs. Although we cannot expect a direct correlation between wage increases and the use of mineral fertilizers, both variables visibly followed a similar trend during those years. In this way, practically all farmers gradually adopted intensification and specialization strategies.

Franco's dictatorship led to a "disorderly turnaround" back to organic agriculture, that was unbalanced, notoriously inefficient and now lacking agrosilvopastoral integration. Agricultural production dropped to levels recorded at the beginning of the century. In a context of nutrient deficits and scant machine use, the number of agricultural workers rose to unprecedented levels. As illustrated in Table 4.1, the 1940 active agricultural population exceeded that of 1900, accounting for over half the workers in a country that went back to being mainly agricultural. The number of workers reached its peak in 1950 with more than 5.2 million, i.e., 48.5% of total workers. The active agricultural population was still above that of 1900 at the start of the sixties, with 4.6 million individuals, 91% of whom were men. As we saw in Chap. 2, the Civil War led to considerable damage to the work animals and considerable efforts were necessary in the following years for its recovery. The lack of animal traction had to be compensated by greater use of human labor, subject to an almost punitive regime of labor relations and low wages (Naredo and Sumpsi 1984). Agricultural worker income followed the same trend, almost reverting, as shown in Table 4.3, to values at the turn of the century. Extensification and the punitive labor regime did not allow for input-based intensification to take place.

4.3 An Estimate of the Agricultural Sector's Macromagnitudes (1950–2008)

To compile farmers' incomes since the 1950s, we dispose of the Sector's Accounts included in the National Accounts. Despite this, the task was challenging given the variations in information quality, included items and the distribution of economic sectors throughout the period. The main problem is the lack of a consistent series. We dispose of three large accounting models with notable variations. Between 1950 and 1989, agricultural macromagnitudes were published in accordance with the FAO's methodology, which integrates agriculture, livestock and forestry accounts. With Spain's incorporation into the EU, there were changes and the SEC 79 Eurostat methodologies were adopted to draw up national accounts. The *Agrarian Statistics Yearbooks* (*Anuarios de Estadística Agraria*) include the 1986–1996 series based on this methodology. The White Paper on Agriculture and Rural Development (*Libro Blanco de la Agricultura y el Desarrollo Rural*, MAPA 2003) published the retrospective series between 1974 and 2000 for the Agrarian Branch based on this methodology only. A final conceptual change occurred with the SEC 1995 methodology, still in use today that makes considerable changes to national accounting (Maluquer 2009a). As mentioned, the FAO series includes agricultural, livestock, and forestry production, while the latter methodology separates forestry accounts (also published only until 2003). But the changes are not only conceptual, since agricultural production and income values are greater based on the SEC 95 methodology than on the two previous methodologies for years where matching figures can be found.

To solve the problem, we chose to build a uniform series on agricultural production, intermediate consumption and income using the retropolation technique from the SEC 95 series up until the two former series.[3] Given that the series is meant to provide approximate total monetary flows, we also included silvicultural production data. As information is lacking for this field since 2004, we assumed that the same percentage weight of the 1999–2003 average had been maintained in the 2004–2008 period.[4] The series offers the results both in current prices and in real values. To deflate the production series, the farmers' price index (available since 1953 in the

[3]This technique takes into account that new accounts incorporate items not included in previous ones and, therefore, maintains older series trends by increasing the size. We are aware of the limitations, but it is the only possible solution given the current information we dispose of. The problems and advantages of different linking alternatives are discussed in Maluquer (2009a, b) and Prados de la Escosura (2009).

[4]The forestry sector's contribution is not very significant anyway: 2.4% of production, 0.8% of Intermediate Consumption and 3.6% of income.

Agricultural Statistics Yearbooks) was used. The index of prices paid by farmers (available from the same source since 1957) was used for the intermediate consumption series and the Consumer Price Index (CPI) (found in Maluquer 2013) was used for agricultural income. Costs of hired labor were deducted from agricultural income to obtain farmers' income, which we consider a proxy of Farm Family Income. Thus, farmers' income enables us to analyze the evolution of their purchasing power. We will compare this latter purchasing power with the evolution of household expenditure in the following section. However, agricultural income is not necessarily the only type of income perceived by agricultural households. It is well known that over the last decades, unemployment insurance, pensions, and other public transfers have represented a significant source of income, added to income from other gainful activities. Either way, we assumed that the amount of income derived from agricultural activity largely explained farmers' production decisions.

Agriculture's confirmed loss of relative weight in the economy is not the end of the story. Over the last decades, the exchange ratio between agriculture and the rest of the economy has steadily declined, partly explaining agriculture's falling relative weight. This weight decline has led to a lopsided relationship that affects agriculture's economic viability and is one explanatory factor of its industrialization process. The best demonstration of this is the relationship between price indices. Graph 4.4 shows the evolution of the ratio between the prices received by farmers and the CPI on the one hand and between the prices received by farmers and the prices paid on the other. The figures show a regular decline in the received price/CPI ratio throughout the study period, with significant repercussions, as we will see later, on the evolution of farm income. Certain events had particularly strong impacts. For example, in the years after the Moncloa Pacts (1977), the decline accelerated. On the other hand,

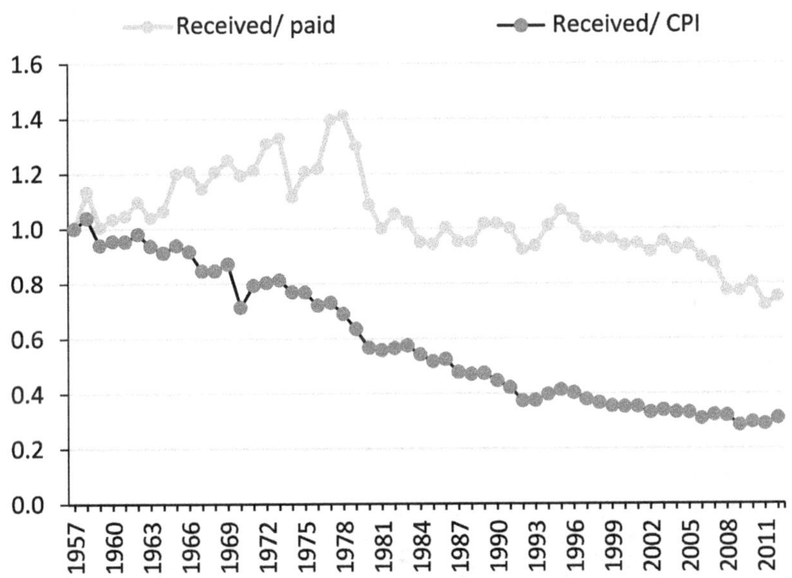

Graph 4.4 Relationship between price indices, "received", "paid" and "Consumer Price Index (CPI)". *Source* Agricultural Statistics Yearbooks and Maluquer 2013

terms of trade ceased to deteriorate between 1980 and 1983 and from 2007 onwards due to the effects of the crisis. The evolution of the ratio between prices received and prices paid was much more affected by the evolution of some basic products prices such as oil, but two distinct periods can be distinguished. The first period, between 1957 and 1978, shows a favorable trend for farmers, somewhat tinged by the oil crisis. This fact is striking since it coincides precisely with the acceleration period of agriculture's industrialization. However, as from the so-called Moncloa Pacts (1977) between the Government, employers' associations and trade unions, the trend reversed and the terms of trade between agricultural products and input prices became negative. The decline accelerated from the mid-nineties, especially since 2007 and constitutes a major mechanism of transfer of agricultural income to other economic sectors, accentuated by the increase in absolute terms of input use (Graph 4.6).

Main agricultural production trends, both in monetary and physical values, have been described in Chap. 2. Graph 4.5 summarizes Agricultural Production trends both in current values and in real terms. In both cases, growth has been notable, with annual cumulative growth rates of 8.4% in current values and 2.8% in constant values. Either way, we can confirm that production stagnated in our century. In Chap. 2 we showed how a large share of this growth was explained by the concentration of production in marketable biomass and not by the limited growth of net primary productivity per hectare.

Agricultural production growth was accompanied by ever greater reliance on external inputs, as seen in Chap. 3. Agriculture's intermediate consumption grew much more intensely (Graph 4.6), with annual cumulative growth rates of 10.8% in current values and 4.7% in constant values. But several phases can be clearly distinguished even during the period of almost constant growth until the beginning of the century: a first phase of intense growth, coinciding with agriculture's industri-

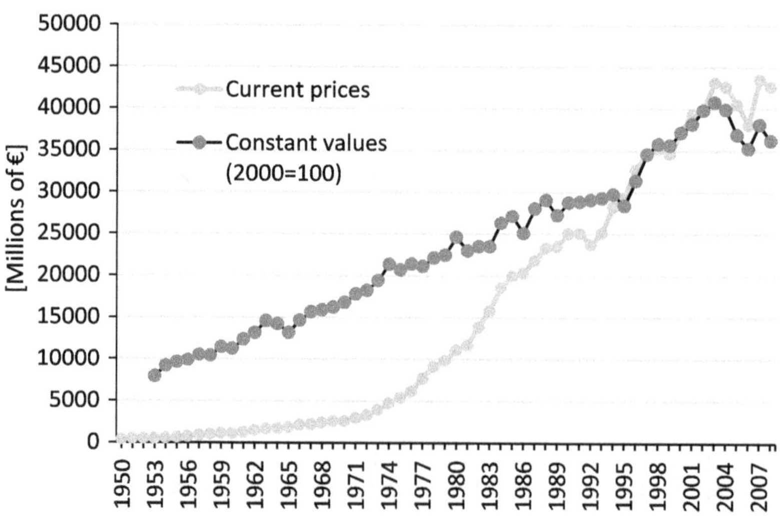

Graph 4.5 Agricultural production. *Source* Agricultural Statistics Yearbooks (see justification in the text)

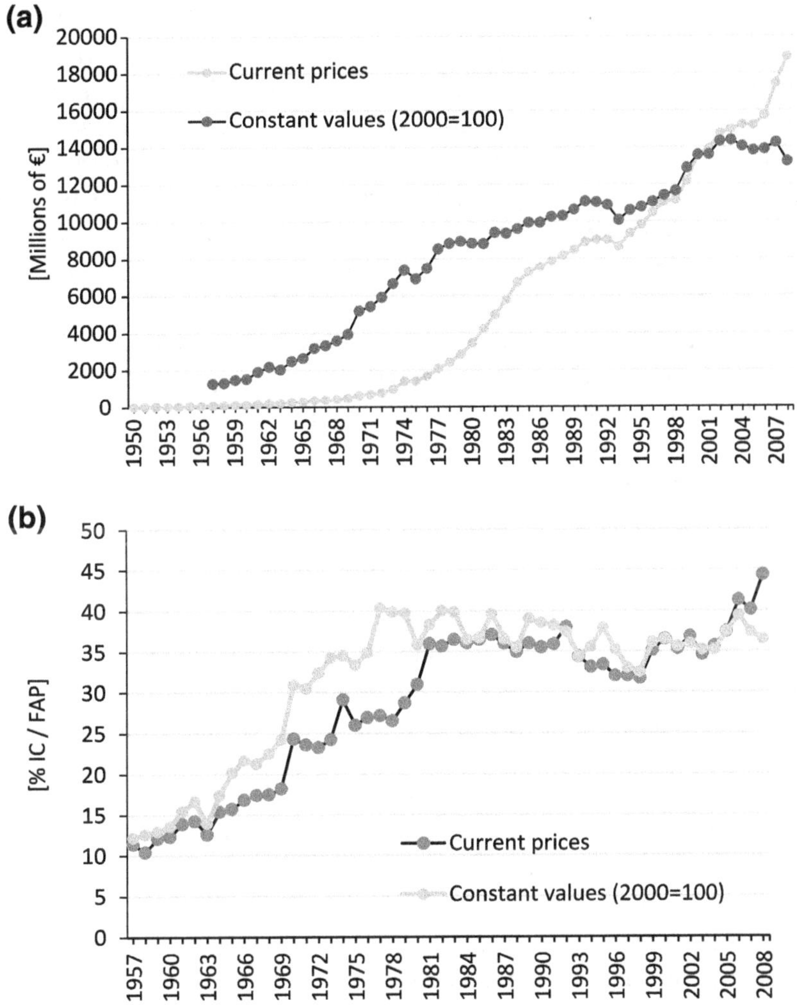

Graph 4.6 Evolution of intermediate consumption, millions of year 2000 euros (**a**) and in percentage of Final Agricultural Production (**b**) *Source* Agricultural Statistics Yearbooks (see justification in the text)

alization process, until 1978 (annual growth rate of 9.6%); a phase of much more moderate growth until 1993 (0.89% per annum) coinciding with Spain's accession process to the EU; and a short period of 4.4% annual growth from 1993 until the turn of the century. To finish, intermediate consumption began to drop in real terms since the year 2000.

Despite recent drops in intermediate consumption, after 2010, price dynamics have prevented this trend from being beneficial for farmers. Farmers have been forced to continue to dedicate an increasing share of their agricultural production value to covering intermediate consumption, notably affecting their income. Graph 4.6 shows the production percentage represented by expenditure outside the sector in constant and current values. These two indicators express different meanings. The first indicator (when deflating the production series with the received prices and intermediate consumption with the prices paid) measures the variation in external input requirements per unit of output. In this sense, the evolution corresponds to the trend commented above. During the agricultural industrialization period, intermediate consumption reached 40% of production, and has remained close to that figure ever since. The percentage between intermediate consumption (hereon IC) and Final Agricultural Production (hereon FACP) in current values measures the production share that farmers dedicated to paying for intermediate consumption. This indicator is strongly dependent on price dynamics; it is much better suited to explain agricultural income dynamics and reveals a somewhat different evolution. Until 1981, the percentages were lower than in constant values (not exceeding 35%), because of farmers' favorable price ratio as mentioned earlier, although from that date onwards it stabilizes, growing strongly at the turn of the century until reaching 44.3% in 2008. The sharp rise in prices paid in the first years of the twenty-first century has meant that farmers have had to allocate an increasing part of their production to satisfy inputs needs from outside the sector.

Intermediate consumption growth is traditionally considered as a positive indicator of agricultural activity's integration in global economic activity, or as an indicator of technification (MAPA 2003). Undoubtedly a second reading is, however, possible, such as the one adopted in this book. And it is rather less rosy. The use of inputs has increased IC costs, pushing agricultural farms to compensate for this growth with more production or greater production specialization in a vicious circle, relying ever more on markets. This explains why, despite the relative production extensification and recent abandonment of "marginal" lands as described above, biomass extracted for commercial purposes has increased in cultivated lands. Generally, intermediate expenditure growth has only reflected a significant rise in non-renewable energy inputs, as well as increasing prices and negative environmental impacts as highlighted in Chap. 3. Similarly, from an economic viewpoint, IC growth has contributed to making agricultural activity ever less viable, by notably affecting the dynamics of agricultural income and the deterioration of terms of trade between farmers and the economy as a whole.

These conditions can be observed in agricultural income dynamics, reflected in Graph 4.7, both in absolute terms as well as per active and employed individual. Agricultural income in current euros shows constant growth until the turn of the century. However, in absolute terms and in constant year 2000 euros, two periods can be distinguished. A first period of growth, from 1950 to 1963, largely reflects the recovery from the depression caused by the Civil War and autarchy, as observed in the economy as a whole (Maluquer 2016; Prados de la Escosura 2017). But as of that date, agricultural income declined heavily and steadily, by 43% over the whole study period. The decade between 1977 (the Moncloa Pacts) and the entry of Spain into the European Union was particularly significant: income fell by 31% in absolute terms. As previously observed, this period also corresponded to an acceleration in the deterioration of terms of trade.

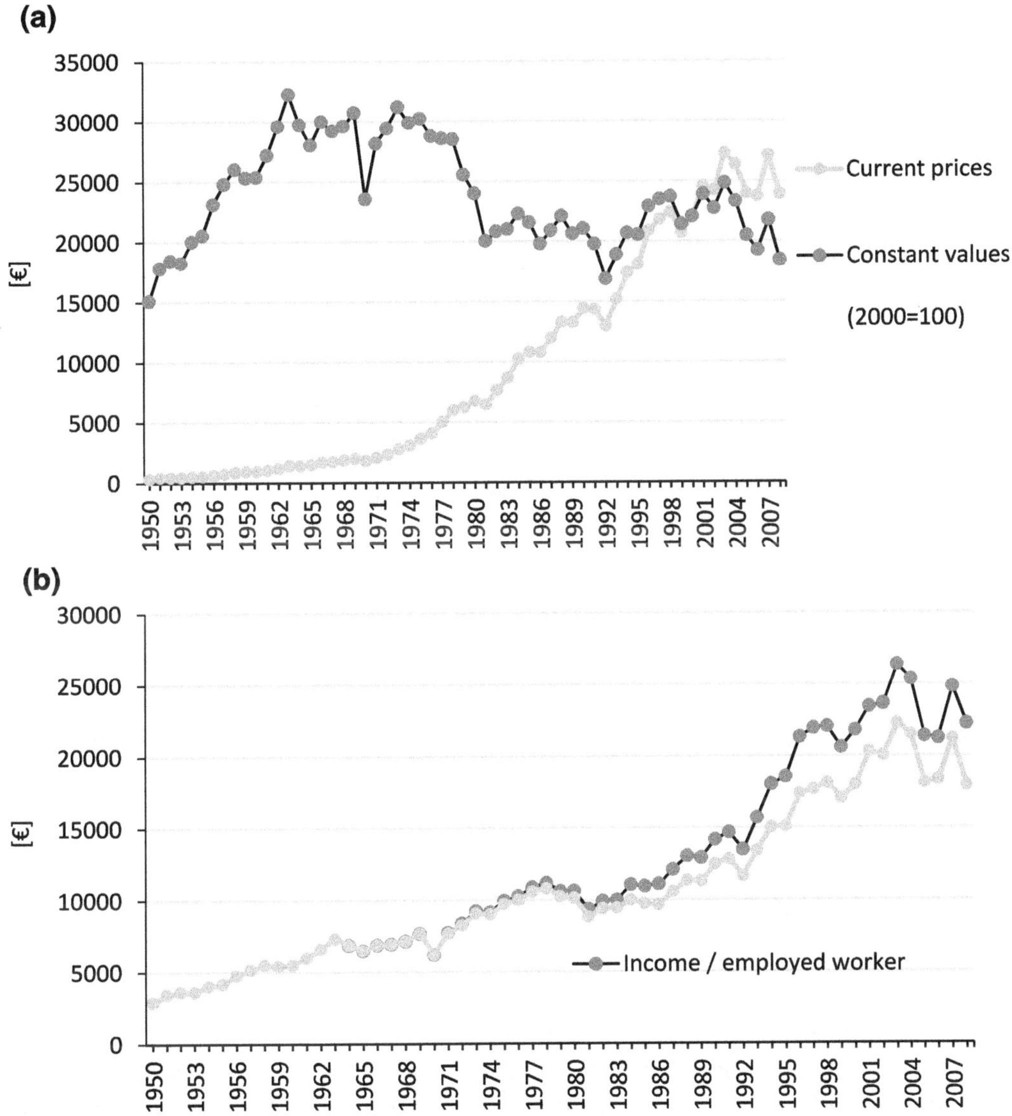

Graph 4.7 Income per active population member (**a**) and employed worker (**b**), in constant year 2000 euros. *Source* Agricultural Statistics Yearbooks (see justification in the text)

The drop has only been partly compensated by two other mechanisms. To start with, the fall was softened by the rising impact of operating subsidies that grew since Spain joined the EU to around 20% of the total income value (Graph 4.8). But the main compensation factor was the destruction of employment (partly linked, in turn, to the loss of farms) and the abandonment of agricultural activity that allowed not only to maintain income but to increase it per working population or employed worker until early this century. This increase, as indicated in the *White Paper on Agriculture* (MAPA 2003), has been above the EU average. Though the phenomenon could be read positively based on a strictly conventional analysis of the economy, it has led to negative consequences described earlier: land abandonment, rural depopulation,

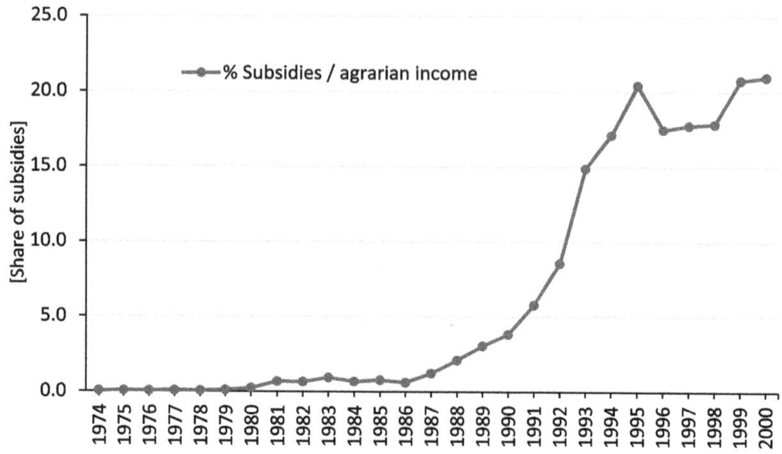

Graph 4.8 Share of subsidies in agricultural income. *Source* Agricultural Statistics Yearbooks (see justification in the text) and MAPA (2003) (It is impossible to build a consistent series of subsidies given the enormous changes in the accounting on the types and amount of included subsidies. We took (for indicative purposes) the longest uniform series of subsidies offered by the white paper on Spanish agriculture. (MAPA, 2003).

deagrarization, etc. Furthermore, in the context of the economic crisis, agriculture has done little to alleviate the effects of the unemployment crisis, as we will see in the following sections.

4.4 The Agricultural Population and Changing Living Standards

In this section, we examine the transformations in agricultural households' living standards. Agricultural households have shared, along with Spaniards and Europeans generally, considerable increases in living costs throughout the twentieth century, which have lasted to this day. As the metabolic profile of contemporary societies has increased, exosomatic consumption of energy and materials has become greater and its monetary cost has risen. Spain went from consuming 4.1 tons of materials per capita in 1950 to consuming 16.3 tons in the year 2000, a four-fold increase (Infante et al. 2015). Biomass represented 73% of materials in the fifties dropping to 19% in the year 2000. The growth rate has been slower than in central and western European countries, but in recent years a remarkable convergence has taken place, reaching an approximate EU-28 average of 15.6 t/cap in the year 2000.[5] This progressive increase in the reproduction costs of agricultural households is highly relevant to our research.

[5]Eurostat, consulted on 20 May 2018.

We dispose of sufficient information for the sixties onwards on the evolution of living costs, included in the *Household Budget Surveys* (*Encuestas de Presupuestos Familiares, EPF*) conducted and published by Spain's National Institute of Statistics (*INE*). As widely known, the surveys began to be designed in the fifties, following an unsuccessful attempt at the start of the previous decade. The first survey was published in 1958. By studying and homogenizing the data, we were able to reconstruct the evolution of Spanish households' average expenditure, both nationwide and for urban and rural areas. The panorama is that indeed, household expenditure has been on the rise ever since. Spanish households' average expenditure grew throughout the whole period, except in the years 1973/74/80/81 and 1990/1991–1996, during which a slight drop can be observed due to economic crises in those years. Growth took place despite a steady decline in average household size, which went from just over 4.5 members in 1958 to 2.6 in 2008 (Graph 4.9).

As shown in Graph 4.10, average household expenditure increased almost four-fold between 1958 and 2008 in constant 2008 euros. The graph shows two periods: the first period is one of accelerated growth and transition to mass consumption, strongly linked to the recovery from terrible living conditions under Francoism, and a second period of slower spending growth from the seventies onwards. This trend was interrupted by the oil crisis and the early nineties economic crisis, as illustrated in the graph (Graph 4.10).

Spending increases were widespread both in urban and rural areas, although household spending in rural areas was always significantly lower than the national average, reflecting somewhat lower living standards (Graph 4.10). The difference between the countryside and the city was not very notable in 1958 but increased sharply in 1964/65 and continued to do so in the following decades. It began to converge again in the mid-nineties. Either way, spending differences between urban and rural areas were always notable: they reached a peak of 41% in 1964–5 taking average national expenditure as a reference, and 9% in 2008, its lowest point (Graph 4.11).

Graph 4.9 Household composition, in number of household members. *Source* Compiled by the authors based on Household Budget Surveys of the National Institute of Statistics (*INE*)

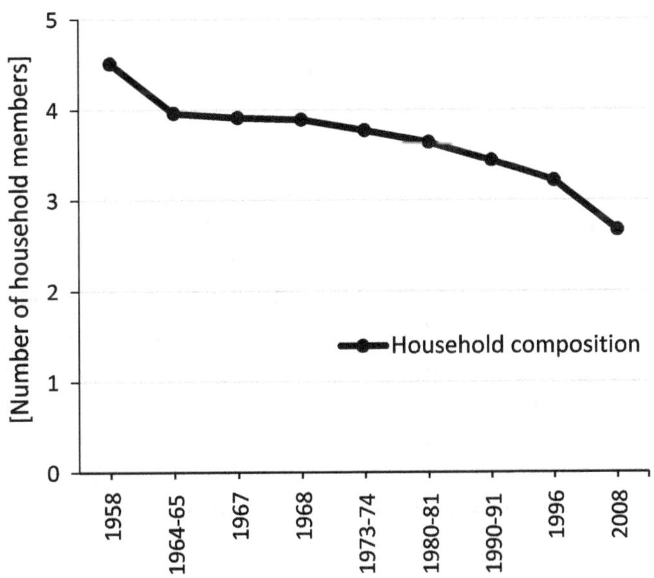

Graph 4.10 Evolution of average household expenditure according to territorial area, in thousands of the year 2008 euros. *Source* Compiled by the authors based on the household expenditure surveys of the National Institute of Statistics

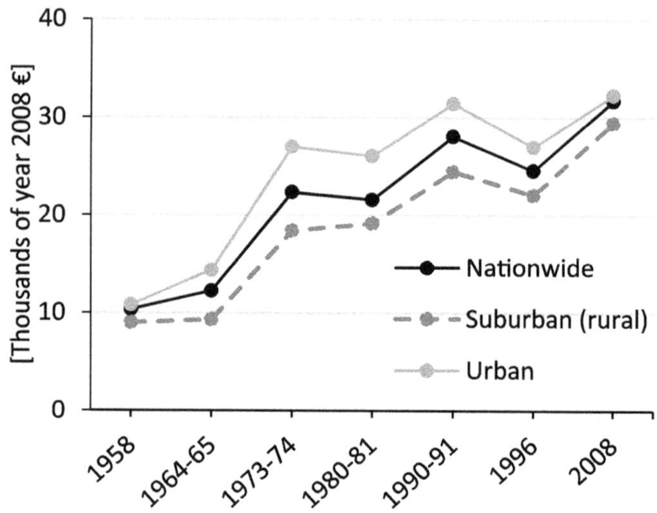

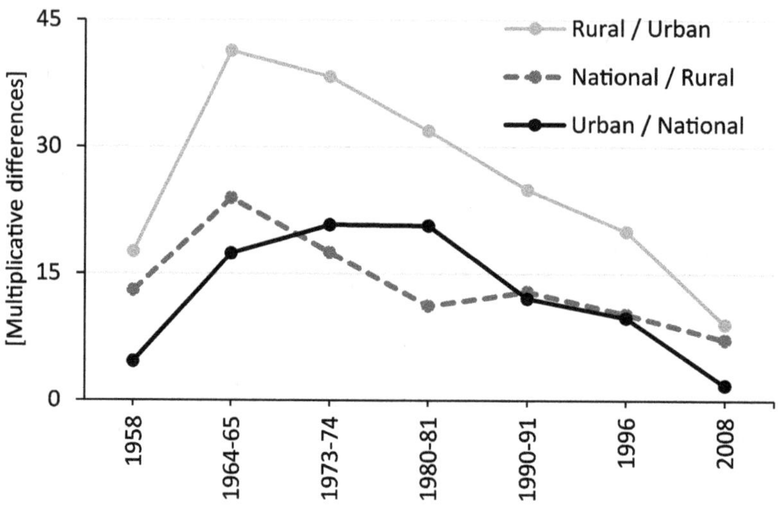

Graph 4.11 Differences in average urban, national and rural/suburban spending, in percentages. *Source* Compiled by the authors based on the household expenditure surveys of the National Institute of Statistics

Despite significant differences between the countryside and cities, the expenditure structure does not greatly differ. The categorization was not maintained over time in published Household Expenditure Surveys, as is the case for large amounts of statistical information. This obliged us to work on homogenizing the categories before comparing them. For the last years, the homogenization was conducted by aggregating categories. However, for 1958, the category "General Expenditure 1958" could not be broken down to make it comparable with the categories of the following years, so we decided not to modify it. Thus, this broad category is the sum of the categories "Other expenses", "Transport and communications", "Medical services and health conservation", and "Leisure, education and culture" (Graph 4.12).

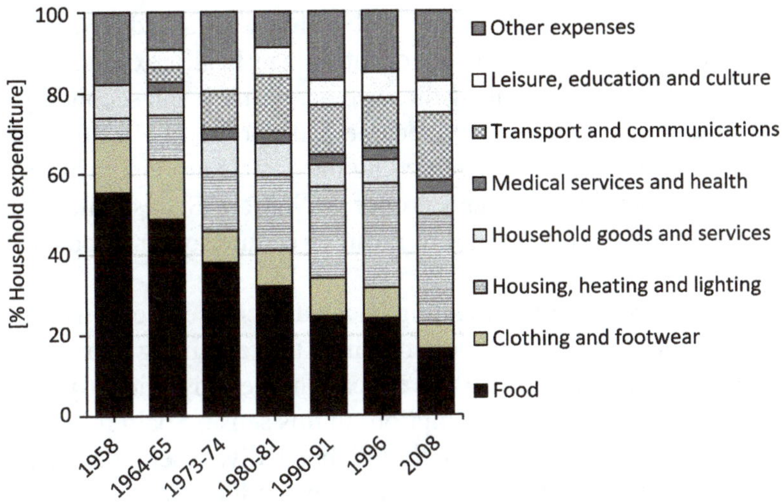

Graph 4.12 Percentage of average national household expenditure according to categories, percentage of the year 2008 euros. *Note* Other 1958 expenses include medical services, transport, and leisure. *Source* Compiled by the authors based on the *Household Budget Surveys* of the National Institute of Statistics *(INE)* The "1958 General Expenditure" category only exists for that year and as from 1964–65 it is divided into "Other expenses", "Leisure, Education and culture", "Transport and communications", "Medical services and health"

The expenditure structure and its evolution show that the amount of money dedicated to food grew between 1958 and 1973–74 and then decreased until 2008 in constant euros from that year. The amount invested in food by households went from €5725.11 in 1958 down to €5190.08 in 2008, a drop of 10.3%. However, the series shows an even greater percentage drop in food expenditure in relative terms, from 55.30% in 1958 to 16.37% in 2008.

Decreasing food expenditure has been common to rich Western countries and results from the combination of two phenomena: on the one hand, the growth of per capita income and household income, leading to bigger spending on other non-food goods; and on the other, constant food price reductions, despite remarkable diet shifts towards meat and dairy product intake (González de Molina et al. 2017; Collantes 2017a, b). This cheapening of shopping baskets has, as we shall see, a devastating impact on the agricultural sector, reducing the prices received by farmers and agricultural income. We will also see that this economic policy was led almost indistinctly by different governments, to lower wages and enlarge budget margins enabling households to purchase other types of goods and services.

In contrast, the "Housing, heating and lighting" item followed an opposite trend, from a meager 4.96% in 1958 to 27.26% in 2008. This item reflects the extent of household expenditure on housing and the weight of mortgage charges during the real estate boom. Expenditures related to "Transport and communications" and "Other expenses", also increased from 3.71 and 9.28%, respectively, in 1964–65, to around 17% for both in 2008, exceeding food expenditure. "Household goods and services" "Leisure, education and culture" and "Medical services and health conservation" expenses remained below 10% of household budgets, with no major changes in the series. The behavior of education and health expenditures does not seem a priori logical given that during these years, primary and secondary education became universal, access to university became widespread and health turned into an additional right. The trend can be justified by the State's gradual provision of these basic services, financed by taxes and proper to Welfare States. The public nature of these services undoubtedly left more room to acquire other goods and services (Graph 4.13).

In short, the series shows that spending on food decreased both in absolute terms and, above all, in relative terms. At the same time, the remaining household expenditure, except those dedicated to health and education, grew significantly, reflecting the constant growth of exosomatic consumption. In this sense, Spain followed a similar path to that of other wealthy countries, though belatedly, several decades later.

The Household Budget Surveys break down expenditure according to the household head's professional profile, enabling to uncover the behavior of three professional categories related to the agricultural sector. As shown in Graph 4.14, households belonging to "Farms with employees" were always above the national expenditure average, while households belonging to "Small owners with no employees" and "Day laborers" were below the average. The behavior of all three categories broadly followed the same trend as the national average. It was not possible to fully reconstruct the series due to a lack of data, therefore, we cannot ensure that this statement is true for the entire period. However, the series shows the relative "impoverishment" of small landowners with no employees who almost converged with day laborers in the mid-1990s. As we will see later (Graph 4.20), small farms paid the price of the impoverishment. Farms below 20 ha constitute the farm-size segment where most farms disappeared. Moreover, the moments of greatest decline coincided

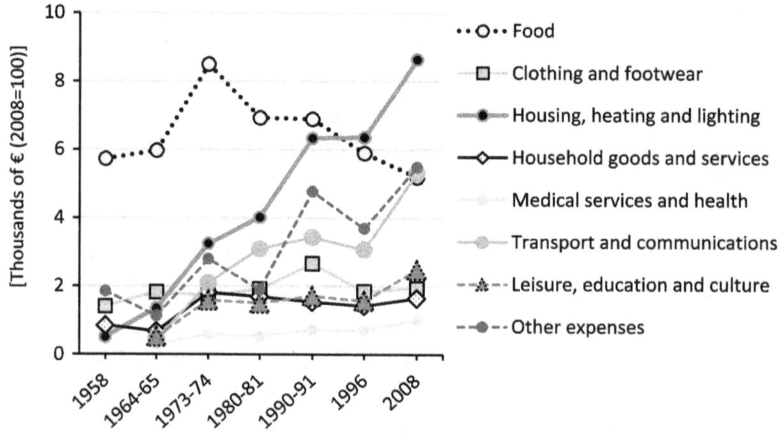

Graph 4.13 Household expenditure by category, in 2008 euros. Other expenses for 1958 include services, transport, and leisure. *Source* Compiled by the authors based on the Household Budget Surveys of the National Institute of Statistics (*INE*). The "1958 General Expenditure" category only exists for that year and is divided in 1964–65 into "Other expenses", "Leisure, Education and culture", "Transport and communications", "Medical services and health"

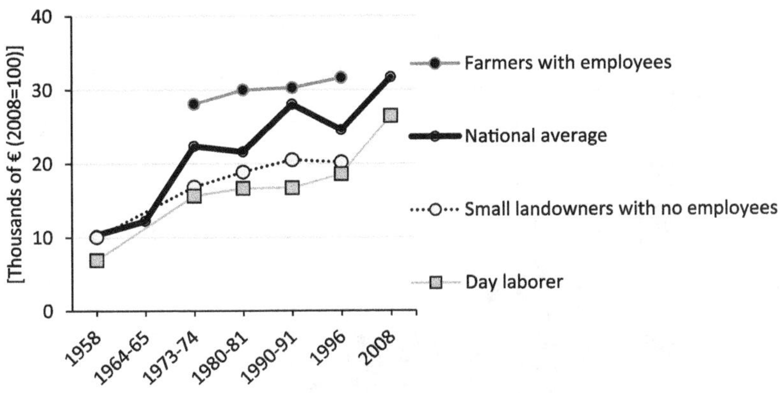

Graph 4.14 Household expenditure according to households' main breadwinner profession, in thousands of constant year 2008 euros per year and family. *Source* Compiled by the authors based on the Household Budget Surveys of the National Institute of Statistics (*INE*)

with that convergence process. In this sense, the CAP has not succeeded in stopping the tendency towards small farm destructions and the abandonment of activity by many owners.

A lack of data prevents us from analyzing the internal structure of household spending for these three household categories throughout the study period. However, the available data show that although there were no big differences in expenditure structure, there were variations in total expenditure size per household, as we saw in the previous section. The sources do offer information for the entire period on the "day laborers" category. It shows that day laborers spent more than average on food (Graphs 4.15 and 4.16).

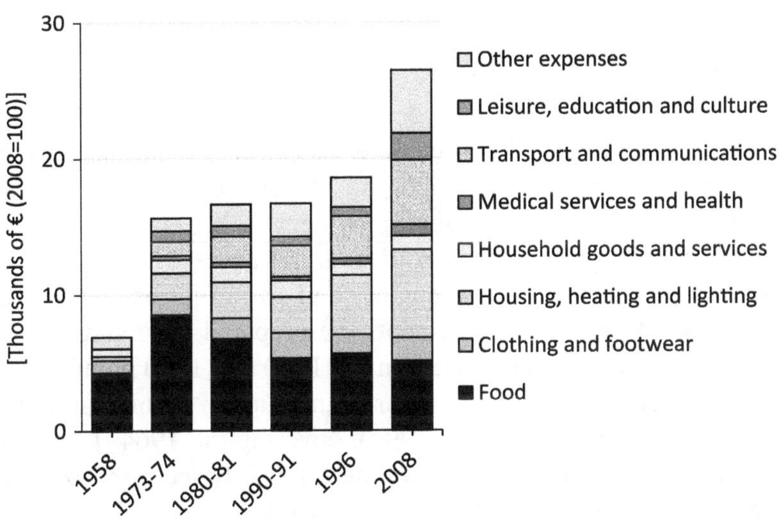

Graph 4.15 Household expenditure of day laborers according to expenditure categories, in 2008 euros. Other 1958 expenditures include medical services, transport and leisure

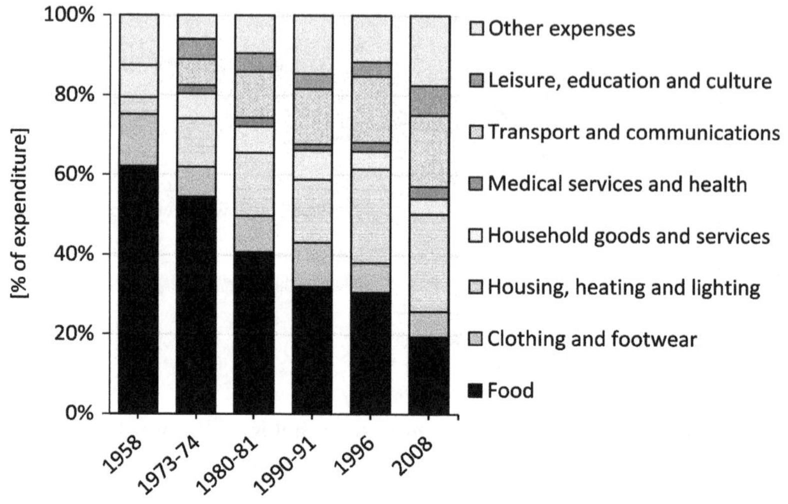

Graph 4.16 Structure of household expenditure of day laborers according to expenditure categories, in percentages of the year 2008 euros. Other 1958 expenditures include medical services, transport and recreation

To conclude, the expenditure of households who were dependent on agricultural income, except, unsurprisingly, of "Farms with employees", has consistently been below the national average since 1958. That is, access to goods and services by the majority of the agricultural population has been inferior to that of the rest of the country, especially that of day laborers and small landowners with no employees. This trend has been directly linked to agricultural activity's low remuneration, as studied above in the evolution of income.

4.5 The State of the Agricultural Population

What consequences have agricultural income reductions and household expenditure increases had on the agricultural population? In recent decades, one undeniable consequence has been the reduction of employment in the sector. Labor decline can be observed by analyzing the evolution of both the active agricultural population, those employed in the sector, and the number and size of farms, that is, the amount of resources constituting the fund element. We prepared the figures on the active population after 1950, (Table 4.6) based on the 1960 and 1970 population censuses and the series provided by Pilar García Perea and Ramón Gómez in their work for the Bank of Spain (García & Gómez, 1994), covering the 1964–1992 period. The report contains employment and unemployment series based on information found in the Active Population Survey (*EPA*), also published by the *INE*. The data from the *EPA* for the periods 1987–2004 and 2002–2008 helped to complete the entire study, allowing us to build a consistent series from 1960 to 2008. We based our series on the homogenization work carried out by García Perea and Gómez (1994) and filled in the previous and subsequent years using the same methodology. This meant reworking the series, separating male from female work, relying on the years for which information was available, and the interpolation of values for the remaining years. The interpolation was performed taking into account the percentage of the male active population with respect to total active population and the percentage of male agricultural active population of total agricultural active population, thus obtaining the absolute figures of the male active population as well as total male agricultural active population. The figures for active women were obtained by deducting male agricultural active population from total agricultural active population.

To draw up the series on the working population, we proceeded in the same way, using the Bank of Spain's standardized series and later filling the years with no data until completing the series. We then distinguished male and female workers, relying on the years for which there is sufficient information, the remaining values having been interpolated. However, active women/working women data obtained this way provided incoherent data for some years in which there were less active agricultural women than female farm workers, which is obviously impossible. We corrected these deficiencies, recalculating the series on active women adding unemployed women to the number of working women. The same strategy was followed to build the agricultural unemployment series (Table 4.6).

Poor working conditions and low wages during the first decades of Franco's era explain the speeding up of rural flight, at a time when industrialization resumed in the Basque Country and especially Catalonia. It was also influenced by the acceleration of industrial growth in Central Europe, attracting large quantities of emigrants from the south of the continent. It is no coincidence this coincided with the end of Spain's diplomatic isolation and integration into international markets. The rural population's shift to industry led to higher wages in the countryside and increased labor costs, favoring the introduction of tractors, harvesters and other machines thus initiating the mechanization of the main tasks in the field (Naredo 2004a, b). Mechanization explains, in turn, the quick decline of the active agricultural population during the

Table 4.6 Active agricultural population (in thousands of people) and its percentage over total active population and total population

Year	Men	Women	Active agricultural population	Active Population (*)	%	Total inhabitants	%
1950	4,827.6	409.5	5,237.2	10,793.1	48.5	28,172.2	18.6
1960	4,204.9	425.4	4,630.4	11,634.2	39.8	30,776.9	15.0
1970	3,404.4	401.8	3,806.2	12,362.7	30.8	34,041.5	11.2
1981	1,818.4	454.7	2,273.2	13,320.0	17.1	37,682.3	6.0
1991	1,110.3	432.5	1,542.8	15,602.2	9.9	38,872.2	4.0
2001	833.8	342.9	1,176.7	17,814.6	6.6	40,847.3	2.9
2008	734.8	291.2	1,026.1	23,065.5	4.4	45,668.9	2.2

Source INE (*) includes the official Spanish population and immigrants

sixties. In 1950, they represented more than 5.2 million active individuals, which accounted for almost half of the total active population and 18.6% of the Spanish population. In 2008, this figure came down to just over one million, barely exceeding 4% of the total active population and just over 2% of the population. The total number of the agricultural active population in 1950 had thus been reduced to one fifth, whereas the total active population had more than doubled, reducing agricultural activity as a source of employment. In just over fifty years, agriculture switched from providing work to almost half the active population to employing a tiny percentage, occupying an almost marginal place. Women's labor incorporation also took place in the agricultural sector, despite continued employment destruction. Women who used to represent only 8% of the active population accounted for 29.1% in 2000 and 28.4% in 2008. Women's agricultural work, which had always existed, became a little more visible.

Table 4.7 and Graph 4.17 show the joint evolution of the active agricultural population, both employed and unemployed, from 1950 to 2008. The number of active

Table 4.7 Evolution of active persons, employed, and unemployed persons in the agricultural sector, 1950–2008 (thousands of people)

Year	Active	1950 = 100	Employed	% Active	Unemployed	% Active
1950	5237.2	100	n.d.	–	52.8	1.1
1960	4630.4	82	4337.7*	100.0	45.2	0.9
1970	3806.2	73	3806.2	100.0	30.2	0.8
1980	2273.2	43	2258.5	99.3	109.0	4.8
1990	1542.8	29	1485.7	96.3	200.2	13.0
2000	1176.7	22	1012.1	86.0	216.3	18.4
2008	1026.1	19	828.2	80.7	137.7	13.4
2016	1016.8	19	774.5	76.2	242.2	23.8

Source Compiled by the authors based on sources referred to in the text
*The data corresponds to 1964. There were the same number of active individuals that year

Graph 4.17 Evolution of the active population, employed and unemployed persons in the agricultural sector (1950–2008) in millions of people

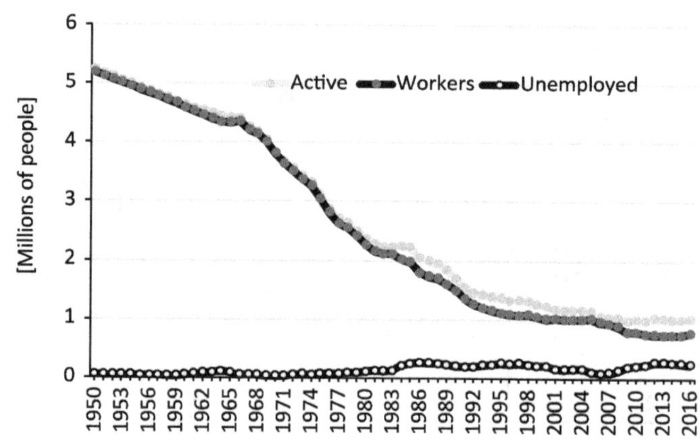

persons is almost matched by the number of persons working until the end of the eighties. From the early nineties onwards, the figures begin to diverge, with increasing numbers of unemployed persons. The unemployment figures gathered by the sources before 1980 doubtlessly refer to workers that were unemployed for several months a year but received a specific amount of daily wages.

As mentioned earlier, the way in which unemployment was recorded over the last decades does not adequately capture the seasonal and irregular nature of work, proper to some agricultural contexts such as that of Spain's, where labor demand is concentrated at specific moments of the year and leaving active persons out of work the rest of the year. It is common for agricultural employees to enter and leave unemployment lists once or several times during the year. The extent of this phenomenon has undoubtedly dropped in recent years due to the sector's drastic reductions in employment offers and because of unemployment benefits, which sometimes provide a more stable income than the job market itself.

According to the trend reflected in the graph and table, two distinct periods can be distinguished: a first period, from 1950 to 1980, is characterized by a narrow relationship between active and employed persons, with very low unemployment rates of around 1%, that are well under the rates considered as full employment rates; a second period from 1980 to 2008 is characterized by an increase in agricultural unemployment, that is moderate at first and then goes on to reach nearly 20% in year 2000. The Spanish economy's turnarounds and the rate of agricultural industrialization explain that the active and employed agricultural population dropped equally over the first period; therefore, agricultural unemployment (excluding seasonal unemployment) was practically non-existent. In a context in which industry and services' labor demand stimulated migrations from the countryside to cities, mechanization was the logical response to the rise in labor costs, especially for medium and large farms. Yet both processes did not necessarily evolve at the same speed. A feedback mechanism had to be established in which emigration made salaries more expensive, encouraged mechanization and decreased labor demand. Surplus workers chose to emigrate, making salaries even more expensive and encouraging mechanization. This mechanism kept agricultural unemployment down and continued to work as long as labor continued to be demanded by industry and services.

However, with the Political Transition and the establishment of the democratic regime, the effects of the 1973 oil crisis and subsequent industrial restructuring all led to increasing unemployment rates. According to the data of the *EPA*, unemployment rose above 10% in the early eighties and reached 25% in the mid-nineties, after a brief slow-down. Emigrating to cities in search of stable employment was no longer a solution. The destruction of employment in agriculture, resulting as we have seen from mechanization, accelerated and raised unemployment rates in the agricultural sector above 4%. That rate was exceeded in 1980 and would not stop increasing in subsequent years. The unemployment rate went from 5.6% in 1981, to 12% in 1990, to 17.6% in the year 2000 and exceeded 20% in 2010. Currently (2016) the agricultural unemployment rate is around 23.8%, after having exceeded 26% in 2014. The shortage of alternative jobs has aggravated unemployment, but this does not explain why it has continued to rise since the early seventies and especially since the eighties. The rate has been rising year after year (except during the three years between 2005 and 2007), reaching a peak in 2014. This means that unemployment in the sector is structural. As we will see in later sections, the reduction of labor costs has constituted the most widespread strategy to compensate continued reductions in farm income in real terms. In fact, labor costs accounted for 60.4% in 1964–5 and 31.9% in 2008, i.e., practically half. In other words, income increase per worker has come at the expense of employment. The intensity of this increase, as we will see below, jeopardizes the sector's generational replacement.

4.6 Changes in Farm Structures

The transformations described in the previous sections were correlated with farm structures. Their evolution can be reconstructed from two main sources: the "Agrarian Censuses" (1962, 1972, 1982, 1989, 1999, 2009) and the "Surveys on the Structure of Agricultural Holdings" (between 1987 and 2016). It is not possible, however, to build a consistent series using both sources. The structures survey only includes a very small share of the farms below one hectare, that are collected in the agricultural censuses. In addition, categories and definitions vary among the different censuses, especially regarding the composition of the utilized agricultural area (UAA). Although it is possible to solve this problem and homogenize census data (López Iglesias 2006), it is not possible with the latest agricultural census (the 2009 census), that uses the same definition as that of the structure surveys. Therefore, it is comparable with these surveys but not directly with the previous censuses. We thus dispose of two types of different series: one based on censuses, between 1962 and 1999, and another based on surveys as well as the latest agricultural census, between 1987 and 2016. Although they cannot be combined, they do at least show the trend followed regarding the amount and structures of farms.

Graphs 4.18 and 4.19, respectively, show the evolution in the number of farms and total area per farms. According to the censuses, the number of farms decreased between 1962 and 1999 by 39%, while the average area per farm increased by 55%. According to the structure surveys, the number of farms decreased by 48% and the average area grew by 71%, due to a 79% increase in the UAA per farm between 1987 and 2016. The figures thus show a considerable decline in the number of farms and a parallel increase in their average size. Regardless of comparison problems between the sources that prevent us from giving an exact figure, we can say that around two-thirds of farms disappeared during the agricultural industrialization process.

However, the rate of decline was not constant. Several periods can be distinguished. Between 1962 and 1972, the number of farms dropped quite sharply (-12.4%), and average surface areas also rose notably (16.9%), a phenomenon undoubtedly linked to the beginning of the industrialization process and the end of traditional organic agriculture (López Iglesias 2006). However, the rate slowed down during subsequent intercensal periods, differing considerably from the restructuring processes of neighboring countries (López Iglesias 2006). The number of farms

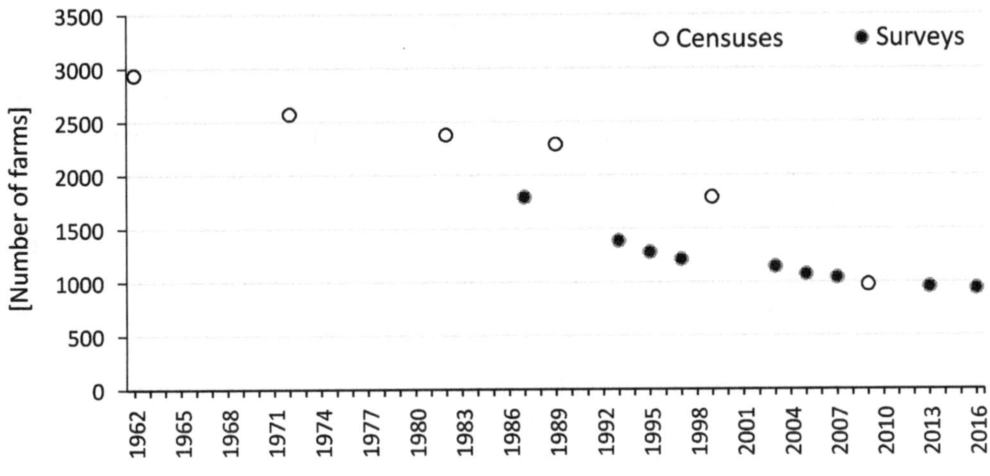

Graph 4.18 Number of agricultural holdings (in thousands) according to the agrarian censuses and the survey on agricultural holdings. *Source* Agrarian censuses and structure surveys

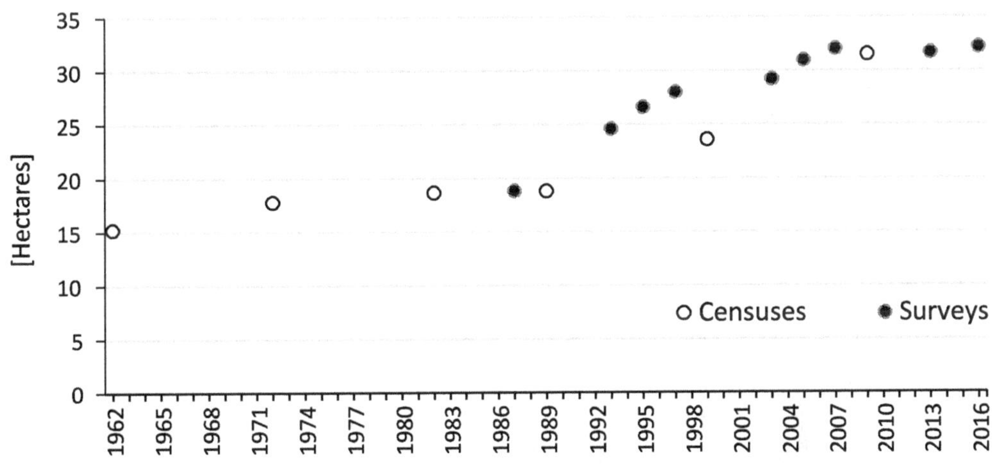

Graph 4.19 Total surface area per farm, in hectares. *Source* Agrarian censuses and structure surveys

decreased by 7.6% between 1972 and 1982 and by 3.8% between 1982 and 1989. It was after Spain joined the European Union that the restructuring process accelerated again and much more vigorously. Between 1989 and 1999, the number of farms decreased by 21.7%. The figures provided by the structure surveys are even more striking: between 1987 and 1997, the number of farms decreased by 32%. Regardless of the differences between the two sources, Spain's entry into the EU clearly had a substantial impact, accelerating the disappearance of farms and the concentration of land. Average surface areas grew by 48.9% between 1987 and 1997 and the UAA per farm grew by 53.3%. Since the turn of the century, farm has continued to disappear, but at a much slower pace: between 1997 and 2007 their numbers decreased by 14.2%, while between 2007 and 2016 they dropped by 9,9% in the midst of the economic-financial crisis. However, a new phenomenon took place in this later period: for the first time since Spain joined the EU, farm destruction has not been accompanied by an increase in farm surface area. This is a particularly relevant fact, since it implies that over 3 million hectares disappeared from farm surveys between 2007 and 2016 due to cessation of activity and, most likely, land abandonment.[6]

As mentioned, the decline in the number of farms came with an increase in the average size of remaining farms. Graph 4.20 shows the evolution of the number of farms according to strata from the structure surveys. Findings by López Iglesias (2006) regarding the previous period are confirmed by agrarian censuses for the 1987–2016 period. Although the number of farms decreased in all strata between 1987 and 2016 (with the exception of the upper stratum of farms equal to or above 100 hectares), smaller farms experienced the most difficulties in adapting to the industrial agricultural model. Farms under 10 hectares decreased by more than 50%; farms between 10 to 20 hectares dropped by 40% and farms between 20 and 30 hectares by 26%. Larger farms adapted much better to the process of industrialization. Since 2007, farms over 30 hectares have not even experienced any drop in their number.

For many economists, the disappearance of countless farms has been the result of a process of "classical structural adjustment" (Arnalte and López Iglesias 2002; Arnalte 2006b), a logical result of the sector's own industrialization. Its consequences are valued positively insofar as it has allowed to strengthen a smaller number of economically profitable farms. This process would also be environmentally positive

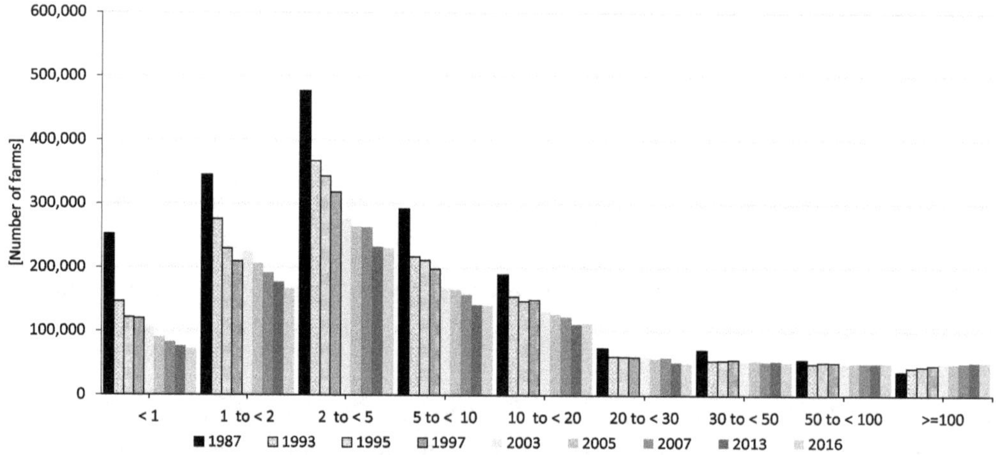

Graph 4.20 Evolution of the number of farms per size strata in hectares. *Source* Agrarian censuses and structure surveys

[6]Land abandonment also took place in previous years, although it is impossible to specify its rate or size due to the impossibility of comparing censuses and surveys, or even censuses from 1999 and 2009). The data from the censuses show relative stability of surface area surveyed between 1962 and 1982 (which even increased between 1962 and 1972) and a drop of more than 2 million hectares between 1982 and 1999. The drop in registered surface area in this period did not prevent average farm size from increasing considerably anyway.

since, according to the same argument, larger farms could provide environmental services at a lower cost. In contrast, other authors present a darker picture, considering that the disappearance of farms has negative social and environmental consequences, given that farms' purposes go beyond purely economic interests (Arnalte 2006a). Nonetheless, these authors do not question either whether adjustments are inevitable. It is, of course, debatable whether the adjustment is economically positive. In no way is it positive from the point of view of farmers themselves. Nor from the perspective of a functioning agricultural market model that differs from the prevailing model, in which income from the agricultural sector is continually transferred to other sectors of activity. This transfer is encouraged by the idea that agricultural policy should aim at ensuring cheap food in order to lower living costs.

Farm destruction and activity abandonment also explain that leasing spread as a mechanism for land mobility (López Iglesias 2006) and the rise of commercial companies. As we will see below, demographic aspects (average age of farm owners) as well as production aspects (work on the farm) must also be taken into account to understand these dynamics and future prospects. The data in Table 4.8 shows the link between farm destruction, which accelerated between 1989 and 1999 as we saw, and the rise of land leasing. The structure surveys show the same relationship, although in this case (Table 4.9), growth is especially visible from 1997 onwards.

López Iglesias (2006) pointed out that the spread of lease over the 1989–1999 period was not only linked to the resizing of farms, but to other related processes such as the rising number of agricultural holdings managed by trading companies. This trend is confirmed for the 1993–2016 period (Table 4.10). Although the type of farm that grew the most was the production cooperative (by 204%), its relative weight is still very small (2.4% of total UAA). Trading companies expanded considerably (by 51% between 1993 and 2016), going from 7.1% of the total at the beginning of the period to 11.5% of UAA in 2016. During that phase, the farms' restructuring process largely consisted of companies penetrating in agricultural production activities.

Farms owned and managed by physical persons decreased for all age brackets since 1987 (Graph 4.21), but the drop was especially significant and fast in holders aged under 35 (−65%), while that of holders aged over 65 decreased much less (−21.4%). This phenomenon shows that the land's production management model was beginning to change: it followed traditional patterns in higher age segments and

Table 4.8 Evolution of the Utilized Agricultural Area (UAA) by land tenure regimes according to the Agrarian Censuses (thousands of ha)

	1982	1989	1999
Property	16,836	17,929	17,632
Lease	4,826	4,901	7,073
Sharecropping	1,285	1,175	787
Others	725	735	824
Total	23,672	24,741	26,317

Source López Iglesias (2006)

Table 4.9 Evolution of the Utilized Agricultural Area (UAA) by tenure regimes according to the Structure Surveys (thousands of ha)

	1987	1993	1995	1997	2003	2005	2007	2013	2016
Property	17,256	17,961	18,248	18,530	17,457	17,067	17,213	13,749	13,712
Lease	5,472	5,163	5,455	5,667	6,761	6,937	6,797	7,670	7,573
Others	1,990	1,590	1,528	1,433	958	851	882	1,882	1,945
Total	24,719	24,714	25,230	25,630	25,175	24,855	24,893	23,300	23,230

Source Structure surveys

Table 4.10 Evolution of the Utilized Agricultural Area (UAA) according to legal personality (thousands of ha)

	1993	1995	2003	2005	2007	2013	2016
Physical person	18,849	18,679	17,444	17,114	16,980	16,212	16,106
Commercial society	1,766	2,031	2,239	2,322	2,451	2,591	2,662
Public entity	2,112	2,201	2,419	2,416	2,299	1,641	1,474
Production cooperative	183	169	211	230	260	580	557
Other legal status	1,804	2,149	2,863	2,773	2,903	2,277	2,432

Source Structure surveys

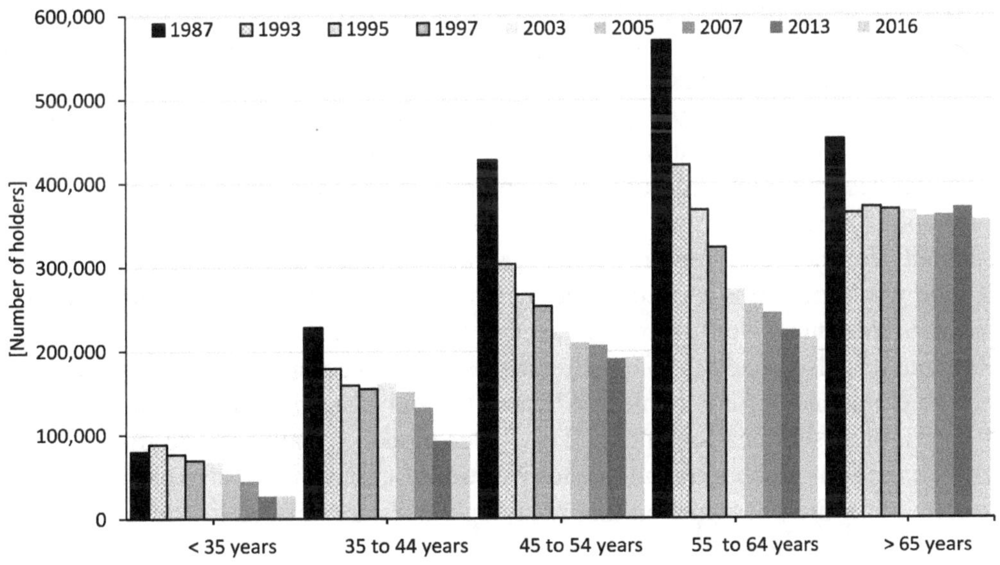

Graph 4.21 Age of farm holders, number of holders. *Source* Farm structure surveys

vice versa. As a result, demographic and production factors explain the changes in land tenure regimes and farm legal personalities, amid the disappearance of a large part of farms due to the cessation of activity.

In this respect, the aging of farmers is taking on worrying proportions and calling into question both the agricultural character of households and the survival of farms. Indeed, the classification of farmers according to age groups shows a very significant trend towards aging, clearly reflecting a lack of generational change. The model has shifted from a production model mainly supported by owners aged 45–64 years (56% in 1987) to a model in which up to 40% of the owners are over 65 years. This leads us to question the future of countless farms, and all the more so if one takes into account the ever declining weight of owners aged under 44 years, and especially those under 35 years.

The evolution of farm work (measured in annual working units, AWUs) also points in the same direction. The number of AWUs on farms decreased by 50.7% between 1987 and 2016, over three different periods: between 1987 and 1997, AWUs fell by 32.4%; between 1997 and 2007, the rate of decline was much slower (11.9%); and between 2007 and 2016 (17.2%), it accelerated again. This trend is consistent with figures on the evolution of the active agricultural population and income evolution described above. The data in Graph 4.20 indicate that the reduction concentrated on household work (−44%). Conversely, fix paid work grew by 67% and casual employment decreased by 11%. The growing trend in fix paid work is consistent with the changes regarding the legal personality of holdings described above and, especially with the spread of commercial companies.

The data shown in graph 4.22 generally shows that not only did total work on farms decrease, since many farms shut down, but the amount of work performed within surviving farms also dropped. This confirms that farmers' most common strategy to compensate drops in agricultural income has been work reduction. The graph also shows that employment was destroyed mainly within the household context and to a much lesser extent in casual employment. This phenomenon faithfully reflects households' progressive deagrarization, as they seek employment for their members, especially for their children, in other economic activities (Collantes 2007). Finally, fix paid work grow this related to the increase in the average size of farms, which are now more intensive than in the past and therefore require more constant use of labor; but it is also linked to the proliferation of trading companies that manage them and the emergence of companies dedicated to land management and the provision of agricultural services for others. This latter model, usually of a trading society nature, is ever more widespread due to cessations of activity, but also because of the part-time farming phenomenon that many operators are obliged to engage in given their reduced income.

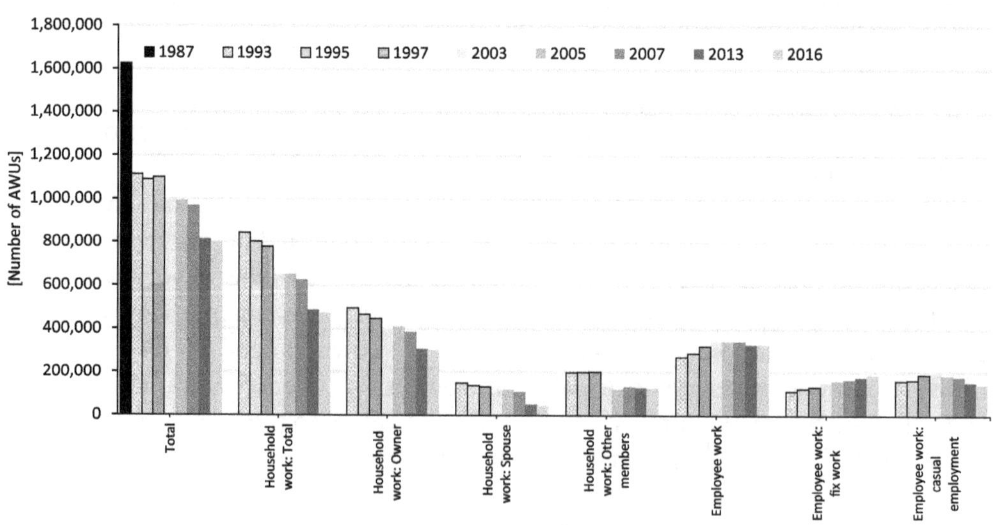

Graph 4.22 Number of farm Annual Working Units (AWUs.) *Source* Farm structure surveys

4.7 Breakdown of Agricultural Income and Coverage of Household Expenditure

Table 4.11 shows income broken down into remuneration of employees, net operating surplus and business income, which is equal to the net surplus minus the amounts paid by farms for rent, interest and indirect taxes. A similar evolution is found all around, except for business income that grew less, due to farms having to face bigger financial payments than in the past. Current values were used. However, agricultural income has been falling in constant values relatively sharply since the early sixties and this drop has become more significant in recent years. Almost half of agricultural income's real value has been lost since then.

Household expenditure, however, has followed an opposite trend and has risen steadily. This has been possible because of the reduction in number of farm workers, leading to growing income per person. Nonetheless, we need to know to what extent the growth of income per employed person has succeeded in covering average household expenditure. Table 4.12 compares the income per employed person, total income per agricultural worker and entrepreneurial income (the income perceived by farm owners) with the evolution of average Spanish expenditure from 1964 to 2008 and rural expenditure, referred to as "suburban" by the Household Budget Surveys. We only took into account the years for which we dispose of the surveys. Agricultural income, remuneration per agricultural worker and entrepreneurial income show income per farm owner or per employed person, while average expenditure is per household. They are not, therefore, strictly comparable. However, and with due caution, both variables allow us to apprehend the capacity of agricultural income to cover Spanish household average expenditure.

The data in Table 4.12 suggest there were two different periods. A first period, from the mid-sixties to the mid-nineties, in which income received by farmers and agricultural workers was clearly insufficient to cover household expenditure, whether in terms of national average or rural average. This largely explains differences in living standards between rural and urban spheres that were particularly significant during those years. A second period started in the mid-nineties, in which income from agricultural activity grew above household spending, but only thanks to the destruction of employment and the cessation of activity of many farms, as we will see below. It was not that agricultural incomes improved: they continued to deteriorate in constant terms but were distributed among fewer farmers and agricultural workers.

In short, in the early sixties, agricultural activity allowed to cover the household expenditure of the owner of an average farm. However, the steady drop in income and increases in average household spending significantly harmed farmers' living standards until the mid-1990s. Most farmers were able to cope by increasing production and reducing costs. The technologies associated with the industrialization of agriculture—fertilizers, phytosanitary products, improved and hybrid seeds, irrigation and mechanization—made it possible to increase productivity, even in very small farms. To the extent that increases in land productivity made it very difficult to save on inputs, the strategy of labor cost reduction became widespread. Indeed, the

Table 4.11 Agricultural Income breakdown in millions of current euros

Year	Agricultural income	Index figures	Remunerate workers	Index figures	Net op. surplus	Index figures	Owner income	Index figures
1964–5	1,484	100	226	100	1,259	100	1,224	100
1973–4	2,913	196	523	231	2,390	190	2,271	185
1980–1	6,600	445	1,300	575	5,300	421	4,800	392
1986	10,774	726	1,848	818	8,926	709	7,543	616
1990–1	14,377	969	2,271	1,005	12,105	961	9,967	814
1996	20,880	1,407	2,557	1,131	18,323	1,455	16,411	1,341
2008	23,899	1,610	3,630	1,606	20,269	1,610	17,445	1,425

Source Agricultural Statistics Yearbooks

Table 4.12 Evolution of agricultural income, income per agricultural worker, farm owner income per worker (after deduction of hired labor) and comparison with average household and average rural (suburban) household expenditure in current euros

Year	Income/employed person	Income/agricultural worker	Farm owner/household worker income	Average household expenditure	Suburban household expenditure
1964–5	342.6	207.0	377.8	475.4	361.5
1973–4	877.4	559.6	952.0	1,624.8	1,340.2
1980–1	2,996.9	2,104.4	3,029.4	4,958.6	4,403.6
1986	6,041.8	3,300.4	6,166.1	(*) 9,213.6	(*) 8,182.4
1990–1	10,156.8	4,849.6	10,522.5	15,188.2	13,233.2
1996	29,501.4	6,978.0	23,187.1	17,419.5	15,637.2
2008	26,901.9	8,578.7	37,497.1	31,711.0	29,426.8

Source Compiled by the author based on the Agricultural Statistics Yearbooks, Agricultural Holding Surveys, and Household Budgets Surveys. The estimates for employees and workers were calculated by the authors (see Table 4.7). (*) This value has been estimated based on before and after values

weight of wages on agricultural income declined steadily throughout the period, representing 60% in 1964–5 and 31.9% in 2008. However, income did not even become high enough to cover average rural household expenditure, largely because of the increase in intermediate costs and the rapid growth of household spending.

Possibilities of raising productivity through greater use of inputs or replacing labor with machines, however, were diminishing, especially in inland Spain. The conversion of rainfed lands into irrigated lands became increasingly difficult and only viable in traditional areas thanks to irrigation modernization programs. The use of chemical fertilizers reached a peak in the late nineties and the use of energy in mechanical traction would only recover early-nineties levels in 2008. Technologies that had led the industrialization of Spanish agriculture were of decreasing utility, especially for farms with low yields that could barely achieve higher incomes by incorporating inputs aimed at greater production. The most widespread strategy continued to be that of increasing labor productivity and farm size, but this strategy was beyond the reach of many farmers, leading to cessations of activity as well as farmland and pasture abandonment. This, in turn, explains employment destruction and the disappearance of countless farms in recent years, as described in the previous section. This job destruction has been taking place in a different context since the nineties: it is no longer possible to relocate agricultural labor surplus in industry, construction or services. Therefore, agricultural unemployment has begun to grow strong.

Employment and farm destruction are thus insufficient to reverse falling income, that has structural causes as we have seen. "Classical structural adjustment" has only alleviated the decline; but we cannot truly speak of a structural adjustment as it could be never-ending if no changes are made to current institutional arrangements. Current arrangements maintain an unbalanced relationship between prices received and prices paid as well as constantly declining income. In other words, decent income is only being achieved by suppressing jobs and farms, i.e., through the sector's self-destruction.

Moreover, agricultural income has not been distributed evenly. Farm structures have been and continue to be deeply unequal both in terms of size and economic viability. Countless small farms have been unable to sustain the same consumption standards as the rest of the population. They have implemented intensification and specialization strategies to achieve them. However, not all farms have been capable of applying these strategies and have ended up abandoning their activity. Data from farm structure surveys is explicit. Between 1993 and 2007, the Gross Margin per farm (approximately equivalent to GVA), measured in numbers of ESUs,[7] increased from 7 to 20.7 (in current prices). With this margin, the possibilities of covering average household expenditure increased (from 64 to 85%). Farms whose Gross Margin reached amounts equal to average household expenditure also increased (from 14% of farms in 1993 to 27% in 2007). However, half the farms with less than 6 ESUs

[7]ESU: European Size Units. Corresponds to 1200 €.

disappeared. Farms above 16 ESU (€19,200) doubled and those under 40 ESUs more than tripled. As a result, the increase in gross operating margin came at the expense of the disappearance of almost a quarter of farms and the average size increase of the remaining farms. A systematic relationship between farm viability and degree of coverage cannot be established, given that other factors linked to income from other gainful non-agricultural activities exist. But these data show that a major share of farms continued to be at risk of disappearing, despite this percentage having fallen from 86% in 1993 to 81% in 2007.

Between both years, Gross Margin per farm increased by 81% in constant euros. Farmers achieved this increase in different ways. On the one hand, by increasing the physical size of their farms, as we saw in Graph 4.19, and on the other, by orienting production towards higher gross margin products. Graphs 4.23 and 4.24 show the technical-economic orientation of farms in 1987 and in 2007. They show the four product orientations offering highest gross margins, but do not show all production orientations. Four others have also been included: they are below the average but have notable territorial impact. As can be observed, intensive livestock activity in 1987, based on monogastric animals, pigs and poultry, quadrupled farms' average gross margin. Intensive livestock thus represented the highest gross income-earning activity. This activity was followed by various crops and horticulture, cereals and milk cattle. At the other extreme, we find crops such as olive groves, viticulture, fruit, and citrus. This is certainly due to the territorial weight of low-yield farms. In 2007, this orientation had hardly changed, although livestock farms dedicated to the breeding of monogastric animals (granivores) had raised their gross margin,

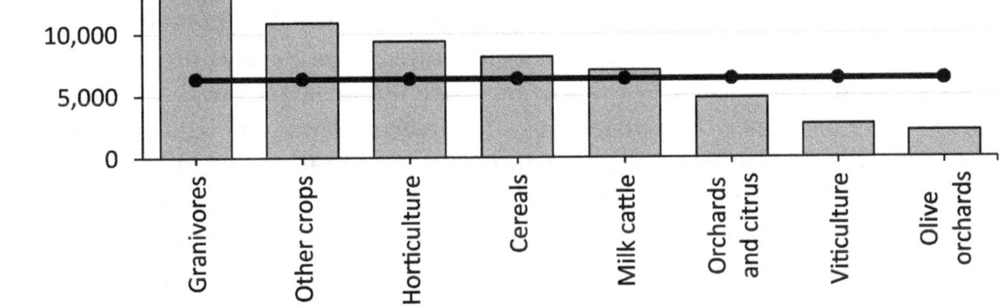

Graph 4.23 Gross margin (GM) per farm according to productive orientation, year 1987, euros. *Source* Farm structure surveys

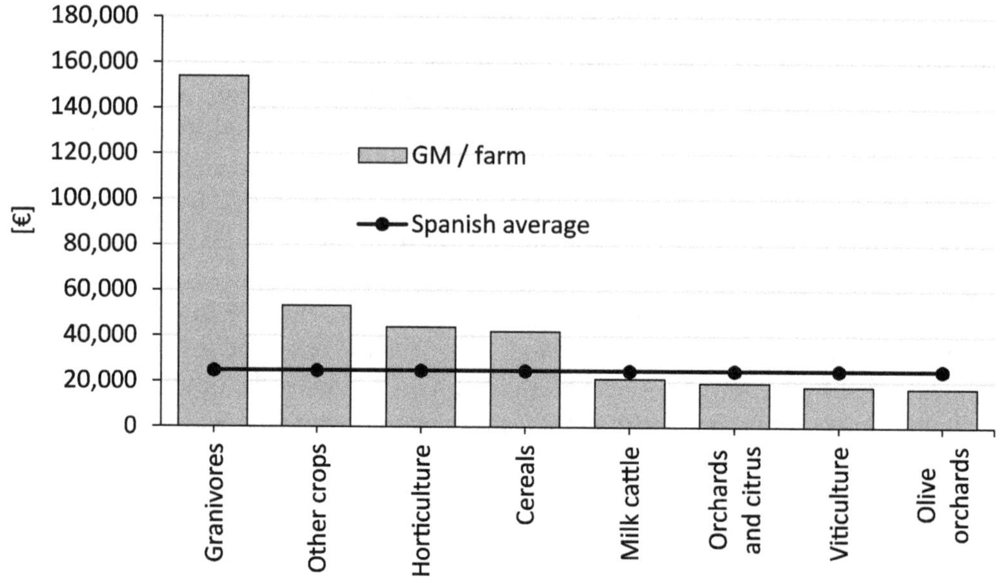

Graph 4.24 Gross margin (GM) per farm according to productive orientation, year 2007, euros. *Source* Structure surveys

reaching 6 times more income than the average. Horticultural farms come next, due to the weight of farms under plastic, various crops, and dairy farms. Farms presenting lowest gross margins continued to be viticulture, fruit, citrus, and olives, accompanied by farms dedicated to cereal production. These farms have survived only thanks to subsidies that are now almost completely decoupled from production. They are very dependent on the European Economic Community budget and successive CAP reforms. Agricultural holdings of great territorial impact, located in Spain's interior, are still seriously at risk of disappearing if the current regulatory framework does not change.

In 1987, granivores, various crops and horticultural production generated income above average rural household expenditure. In 2007, production margins had risen, matched by that of milk cattle. In short, the graphs show that gains in European size units (ESU) were realized through a clear orientation towards production with higher gross income, especially towards intensive livestock and crops under plastic. It also shows that, once the possibilities of increasing labor and land productivity diminished, the possibilities of tackling drops in prices received by farmers consisted of livestock and horticultural specialization. Both orientations have led to growing external inputs in energy terms, that is, they have increased the scope of Spanish agriculture's intensification and specialization. In the case of pork and poultry production, high gross margins are possible thanks to massive imports of very cheap feed based on corn and soybeans. We will come back later to this issue and the socio-environmental impacts generated in the countries of origin. Overcoming agriculture's profitability crisis has thus brought about ongoing intensification and specialization of horticultural production and the sector's *livestocking*. Both processes have accentuated environmental sustainability problems analyzed in this book. Neither process has led to solutions of social equality applicable to the whole of Spain's rural society.

Inland extensive farms have been less likely to adopt more intensive production techniques such as greenhouses, for which access to irrigation and high initial investment are essential. Intensive farms have followed a similar path, although in recent years they have been relocated inland based on vertical integration systems of large livestock or meat distribution companies. Production alternatives to olives, fruit trees, almond trees, and other woody crops or vineyards have been limited. However, certified organic production has represented a solution adopted by extensive farms. The latter have achieved income supplements thanks to agri-environmental measures allowing them to increase the number of ESUs. Extensive cattle ranching, woody crops and rainfed cereals are good illustrations. These types of farms accounted for 83% of areas classified as organic or reconverted as of December 31, 2008. At the end of 2011, over half of extensive livestock (53.7%) became organically managed in Andalusia (González de Molina 2012).

Organic agriculture has also been contributing to maintaining agricultural activity in areas of lowly competitive conventional agriculture. Regions such as the Pedroches Valley, the Sierra de Huelva, the Sierra de Segura, the North of the province of Granada, etc. contain most of Andalusia's organic surface areas. In these areas, subsidies account for a major percentage of operating surpluses (Guzmán Casado and Alonso Mielgo 2009). They offset the low productivity of rainfed treenuts, olives and cereal crops, and may even "explain" that in some cases production is not harvested. According to a study on the economic performance of organic farms conducted in 2007 in Andalusia (Soler et al. 2009), total subsidies (those from the first pillar plus agri-environmental subsidies) accounted for 43% of the value of final olive production, 71% of the production of nuts and no less than twice the value of extensive crops production. In the latter case, income was practically equal to the value of subsidies and almost 78% in the case of nuts. Agri-environmental subsidies played an even more decisive role in organic livestock. Subsidies amounted to 69.4% of the 2005 production value. Without subsidies, income would have been reduced by 41%.

Organic agriculture and livestock already account for more than 2.02 Mha in Spain, which is the first country in Europe per certified area, representing 8.7% of the agricultural area (MAPAMA 2017a). It has halted the intensification process, though it has not necessarily stopped the use of external inputs nor completely reversed falling income trends. The process sometimes referred to as *conventionalization* threatens to nullify the advantages of this production model, bringing it ever closer to the conventional model (Ramos García et al. 2018). However, organic production does represent an opportunity to stop the deterioration of agroecosystems of inland Spain. Agri-environmental measures could be conceptualized as payments for environmental services. This "new" form of agroecosystem management is partly based on the rationale developed so far and has paradoxically emerged as a result of the non-feasibility of the described model. We will come back to this issue in more detail in the epilogue.

4.8 Conclusions

We can conclude that the processes of industrialization and subsequent globalization have significantly decreased the weight of the agricultural population fund element at the expense of increasing the size of the other social fund element, technical the means of production. As we saw in Chap. 3, technical means of production have acquired an excessive weight that threatens not only the environmental health of agroecosystems but also the viability of agricultural activity as we know it today. The rate of destruction of agricultural employment has reached such worrying levels, associated with rural depopulation and the "deagrarization" of rural households, that generational change and the viability of agricultural activity as such are at risk. This deterioration has intensified under the "classical structural adjustment" undergone by Spanish agriculture. The causes should be sought among the institutional mechanisms that have allowed agricultural sector income to be continually transferred to other sectors. These mechanisms make adjustments inevitable and threaten to prolong them until the sector eventually collapses, i.e., until most agricultural holdings disappear due to lack of profitability. Structural adjustment and its consequences do not, therefore, appear to be conjunctural. The agricultural sector has been adapting this way since the 1960s at least.

It is worth asking whether the destruction of jobs and farms leading to the ongoing depopulation of our countryside may not, in fact, represent an obstacle to the necessary transition towards sustainable agriculture. To answer this question, we must challenge the prevailing evaluation narratives, and consider their relevance in our analysis. The size of the agricultural population, closely linked to the amount of work required in agriculture, has almost always been associated with development levels and economic growth. The decrease in the agricultural population is viewed, therefore, in a positive way by more conventional economists. Although it is not possible to establish this fund element's optimal size without referring to its space-time context, we cannot simply consider its evolution throughout the twentieth century in Spain as positive and put an end to the discussion. The continuity of the agricultural sector's current institutional arrangement jeopardizes the future of sustainable agriculture. It also weakens the supply of ecosystem services that are essential for all economic activity to be sustainable, both in rural and urban areas.

References

Acosta F, Cruz S, González de Molina M (2009) Socialismo y democracia en el campo (1880–1930): los orígenes de la Federación Nacional de Trabajadores de la Tierra. Madrid. Ministerio Medio Ambiente y Medio Rural y Marino

Arnalte Alegre E (2006a) Economía política del proceso de ajuste estructural en la agricultura de los países desarrollados. In Arnalte Alegre E (ed) Políticas agrarias y ajuste estructural en la agricultura española. Madrid, Ministerio de Agricultura, Pesca y Alimentación, pp. 17–54

Arnalte Alegre E (ed) (2006b) Políticas agrarias y ajuste estructural en la agricultura española. Madrid, Ministerio de Agricultura, Pesca y Alimentación

Arnalte Alegre E, López Iglesias E (2002) Análisis del ajuste estructural clásico: trabajo preparatorio del libro blanco de la agricultura y el desarrollo rural del ministerio de agricultura, pesca y alimentación. Madrid. Ministerio de Agricultura, Pesca y Alimentación

Calatayud S, Martínez Carrión JM (1999) El cambio técnico en los sistemas de captación e impulsión de aguas subterráneas para riego en la España Mediterránea. In: Garrabou R, Naredo JM (eds) El agua en los sistemas agrarios. Una perspectiva histórica. Madrid. Argentaria/Visor, pp 15–39

Carreras de Odriozola A, Tafunell Sambola X (2005) Estadísticas históricas de España, siglos XIX–XX. Madrid. Fundación BBVA

Carrión P (1932 [1975]) Los latifundios en España. Su importancia, origen, consecuencias y solución. Barcelona. Editorial Ariel

Collantes Gutiérrez F (2007) La desagrarización de la sociedad rural española, 1950–1991. Historia Agraria 42:251–276

Collantes Gutiérrez F (2017a) Because they just don't want to: dairy consumers, food quality, and Spain's nutritional transition in the 1950s and early 1960s. Agric Hist 91(4):536–553

Collantes, Gutiérrez F (2017b) Nutritional transitions and the food system: expensive milk, selective lactophiles and diet change in Spain, 1950–65. Hist Agrar 73:119–147

Gallego D (1986) Transformaciones técnicas de la agricultura española en el primer tercio del siglo XX. In: Garrabou R, Barciela López C, Jiménez Blanco JI (eds) Historia agraria de la España contemporánea, vol 3: el fin de la agricultura tradicional (1900–1960). Barcelona. Editorial Crítica, pp 170–229

Gallego D (2001) Historia de un desarrollo pausado: integración mercantil y transformaciones productivas de la agricultura española. In: Pujol et al (eds) El pozo de todos los males. Sobre el atraso de la agricultura española contemporánea. Barcelona. Crítica, pp 147–214

García P, Gómez R (1994) Elaboración de series históricas de empleo a partir de la Encuesta de Población Activa (1964–1992). Documentos de Trabajo del Banco de España 9:1–63

GEHR (Grupo de Estudios de Historia Rural) (1983). Notas sobre la producción agraria española, 1891–1931. In: Revista de Historia Económica 1(2):185–252

González de Molina M (2001) The limits of agricultural growth in nineteenth century: a case-study from mediterranean world. Environ Hist 7(4):473–499

González de Molina M (2012) Luces y sombras del crecimiento de la producción ecológica en Andalucía durante el último quinquenio (2007–2011) Cuad Interdiscip Desarro Sosten 9:153–192

González de Molina M, Guzmán Casado GI (2006) Tras los pasos de la insustentabilidad. Agricultura y Medio ambiente en perspectiva histórica (siglos XVIII–XX). Barcelona. Editorial Icaria

González de Molina M, Soto D, Aguilera E, Infante-Amate J (2014) Crecimiento agrario en España y cambios en la oferta alimentaria, 1900–1933. Hist Soc 80:157–183

González de Molina M, Soto D, Infante-Amate J, Aguilera, E, Vila Traver J, Guzmán GI (2017) Decoupling food from land: the evolution of spanish agriculture from 1960 to 2010. Sustainability 9(12):23–48

Guzmán Casado G, Alonso Mielgo A (2009) Evaluación de la medida agroambiental Agricultura Ecológica en el periodo 2004–2006. In: González de Molina M (ed) El desarrollo de la agricultura ecológica en Andalucía (2004–2007): crónica de una experiencia agroecológica. Barcelona. Editorial Icaria, pp 67–80

Infante-Amate J (2011) Los temporeros del olivar: una aproximación al estudio de las migraciones estacionales en el sur de España (siglos XVIII–XX). In: Revista de Demografía Histórica no 29, vol 2, pp 87–118

Infante-Amate J (2014) ¿Quién levantó los olivos? Historia de la especialización olivarera en el sur de España (s. XVIII–XX). Madrid. Ministerio de Agricultura, Alimentación y Medio Ambiente

Infante-Amate J, Soto D, Aguilera E, García Ruiz R, Guzmán G, Cid A, González de Molina M (2015), The spanish transition to industrial metabolism long-term material flow analysis (1860–2010) J Ind Ecol, 19(5):866–876. Available in: https://doi.org/10.1111/jiec.12261

INE (Instituto Nacional de Estadística) (1902) Censo de población de España 1900. Madrid

INE (Instituto Nacional de Estadística) (1913) Censo de población de España 1910. Madrid

INE (Instituto Nacional de Estadística) (1922) Censo de población de España 1920. Madrid

INE (Instituto Nacional de Estadística) (1932) Censo de población de España 190. Madrid

INE (Instituto Nacional de Estadística) (1943) Censo de población de España 1940. Madrid

INE (Instituto Nacional de Estadística) (1946) Anuario de estadística de 1944-1945. Madrid

INE (Instituto Nacional de Estadística) (1952) Censo de población de España 1950. Madrid

INE (Instituto Nacional de Estadística) (1958) Encuesta Presupuestos Familiares de 1958. Madrid

INE (Instituto Nacional de Estadística) (1974) Censo Agrario 1972. Madrid

INE (Instituto Nacional de Estadística) (1986) Censo Agrario 1982. Madrid

INE (Instituto Nacional de Estadística) (1992) Censo Agrario 1989. Madrid

INE (Instituto Nacional de Estadística) (1999) Censo Agrario 1999. Madrid

INE (Instituto Nacional de Estadística) (2008) Estadística e indicadores del agua. Madrid

INE (Instituto Nacional de Estadística) (2009) Censo Agrario 2009. Madrid

INE (Instituto Nacional de Estadística): Encuesta de Estructura de las Explotaciones Agrícolas. Madrid

INE (Instituto Nacional de Estadística): Encuesta de Población Activa. Madrid

INE (Instituto Nacional de Estadística): Series históricas de datos censales de España, (1900-2001), [web page]. Madrid. Available in: http://www.ine.es/censo2001/historia.htm

López Iglesias E (2006) El proceso de ajuste estructural en la agricultura española: caracterización general de las tendencias en las dos últimas décadas. Arnalte Alegre E (ed) Políticas agrarias y ajuste estructural en la agricultura española. Madrid. Ministerio de Agricultura, Pesca y Alimentación, pp 55–92

Maluquer de Motes J (2009a) Del caos al cosmos: una nueva serie enlazada del Producto Interior Bruto de España entre 1850 y 2000. Rev Econ Apl XVII(49):5–46

Maluquer de Motes J (2009b) Viajar a través del cosmos: la medida de la creación de la riqueza y la serie histórica del PIB. Rev Econ Apl XVII(51):25–54

Maluquer de Motes J (2013) La inflación en España: un índice de precios de consumo, (1830–2012). Madrid. Servicio de Publicaciones del Banco de España

Maluquer de Motes J (2016) España en la economía mundial: series largas para la economía española (1850–2015). Madrid. Instituto de estudios económicos

MAPA (Ministerio de Agricultura, Pesca y Alimentación) (2003) Encuesta sobre superficies y rendimientos de cultivos del año 2002. Memoria. Madrid

MAPAMA (Ministerio de Agricultura y Pesca, Alimentación y Medio Ambiente) (2017a) Agricultura Ecológica. Estadísticas 2016. Madrid. Available in: http://publicacionesoficiales.boe.es/

Marco I (2017) Dialogues on nature, class and gender: revisiting socio-ecological reproduction from past organic advanced to industrial agricultures (Sentmenat, Catalonia, 1860–1999). Barcelona. Universitat de Barcelona, Tesis doctoral inédita

Naredo JM (2004a) Reflexiones metodológicas en torno al debate sobre El pozo y el atraso de la agricultura española. Hist Agrar 33:151–164

Naredo JM (2004) La evolución de la agricultura en España (1940–2000). Granada. Universidad de Granada

Naredo JM, Sumpsi JM (1984) Caracterización de los modelos disciplinarios de trabajo agrario en las zonas de gran propiedad. Agricultura y Sociedad, no 33:45–86

Prados de la Escosura L (2003) El progreso económico de España (1850–2000). Bilbao. Fundación BBVA

Prados de la Escosura L (2009) Del caos al cosmos: la serie del PIB de Maluquer de Motes. Revista de Economía Aplicada XVII(51):5–23

Prados de la Escosura L (2017) Spanish EconomicGrowth, 1850–2015; Suiza. Springer Nature Prees

Ramos García M, Guzmán GI, González de Molina M (2018) Dynamics of organic agriculture in Andalusia: moving towards conventionalisation? Agroecol Sustain Food Syst 42(3):328–359

Simpson J (1997) La agricultura española (1765–1965): la larga siesta. Alianza, Madrid

Soler Montiel M, Pérez Neira D, Molero Cortés J (2009) Cuentas económicas de la agricultura y ganadería ecológicas en Andalucía 2005. In: González de Molina M (ed) El desarrollo de la agricultura ecológica en Andalucía (2004–2007): crónica de una experiencia agroecológica. Barcelona. Editorial Icaria, pp 135–148

Villa I (2017) Transformaciones en el Metabolismo Agrario y su impacto socio-ecológico: Montefrío, 1750–1920. Universidad Pablo de Olavide de Sevilla, Tesis doctoral inédita, Sevilla

5

Industrialization Process of Agriculture and Environment

In this chapter, we focus on the changes to the functioning and structure of Spain's agroecosystem during the industrialization process in the twentieth century. Specifically, we aim at showing how changes in the quantity and quality of energy and material flows contributed to deteriorating the quality of the land fund element, that includes soil, biodiversity, water, etc., and that supports the provision of agroecosystemic services, among which biomass production. The degradation of the agroecosystem's biophysical structure is reflected in the progressive decline of energy returns in the form of biomass. We will also show how Spanish agricultural production had direct environmental impacts in remote regions of the planet due to the outsourcing of the land cost of food through massive imports of biomass from these regions. The methodology used for calculations is not described in this book but has been published in the following texts: García Ruiz et al. (2012), Soto et al. (2016), Guzmán and González de Molina (2017), Guzmán et al. (2017), Aguilera et al. (2018), Guzmán et al. (2018).

5.1 Functioning of the Agroecosystem

Chapter 3 showed how the intensification of Spanish agriculture was based on the increasing use of external inputs. The external energy invested in the agroecosystem over the 1900–1960 and 1960–2008 periods multiplied, respectively, by a factor of 3.9 and 5.5 (Graph 5.1 and Annex 3).

First, the increase of external inputs allowed to overcome limiting factors (e.g., nutrients, water) to some extent and ensure greater protection against heterotrophic organisms that translated into a higher NPP (Graph 5.1). However, this growth has been limited (10% from 1900 to 1960, and only 18% between 1960 and 2008), implying, as we shall see, a negative return on external energy invested to achieve this increase.

Graph 5.1 Energy flows in Spain's agroecosystem in 1900, 1960 and 2008, in petajoules

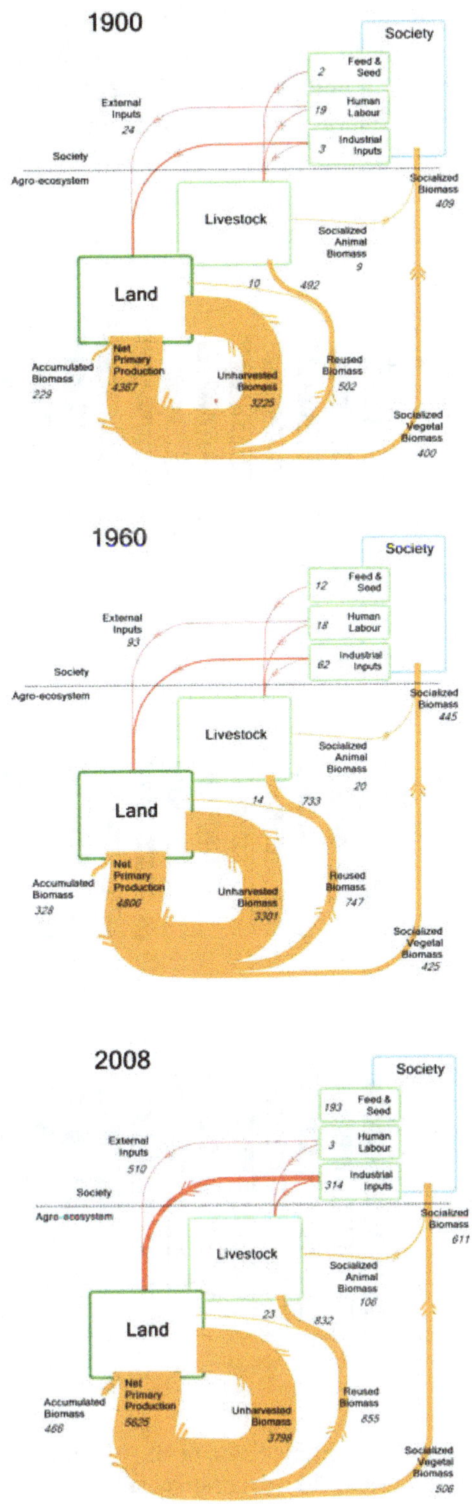

Second, ever-increasing external inputs led to modifications in the NPP's social use pattern (Graph 5.2). These changes are important because as we explained in Chaps. 1 and 2, the fundamental mechanism underlying agroecosystems is their use of energy in the form of biomass that is recirculated and stored within them.

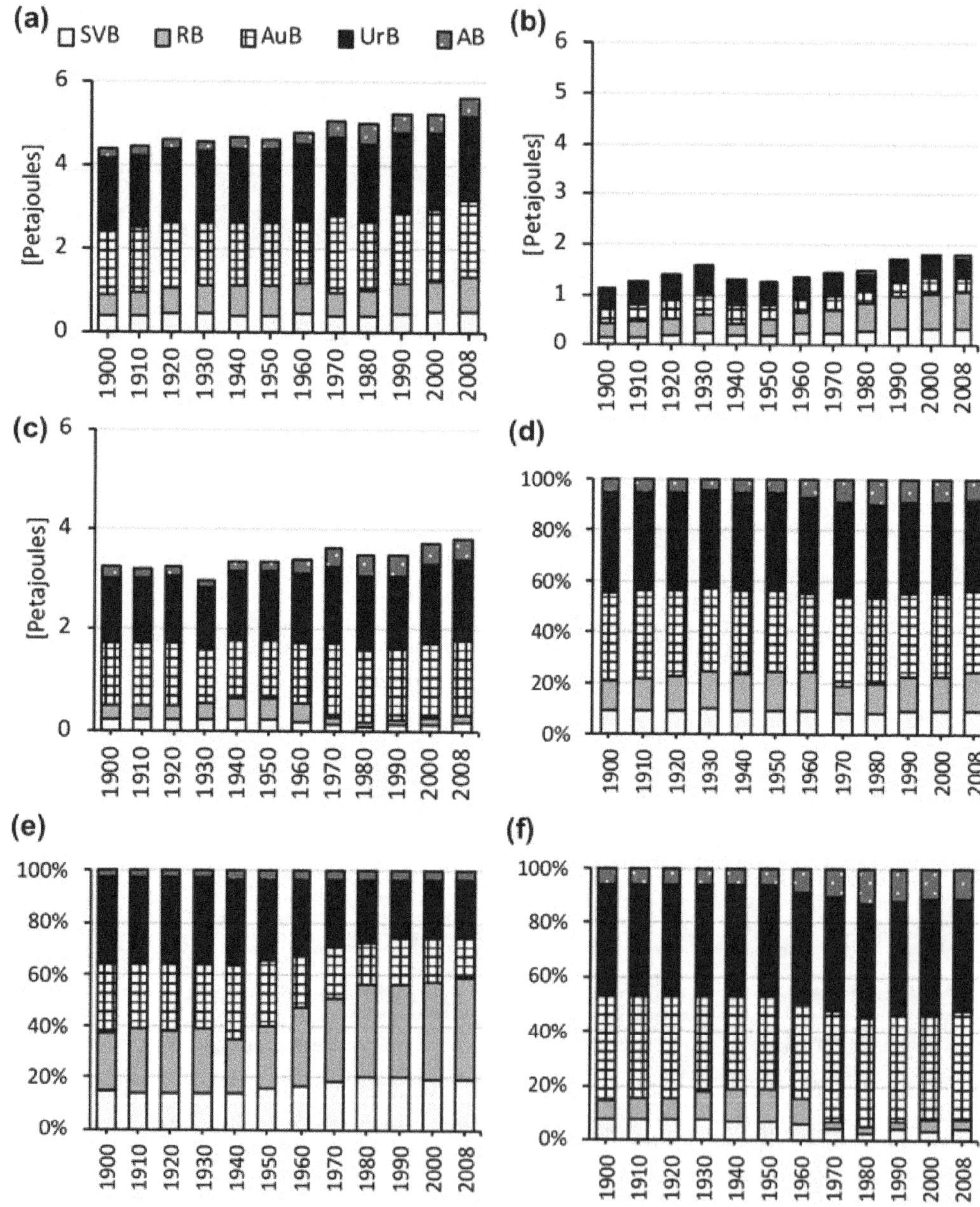

Graph 5.2 The evolution of NPP according to its uses in absolute terms: Spanish agroecosystem (**a**), croplands (**b**), silvopastoral areas (**c**); and in relative terms: Spanish agroecosystem (**d**), croplands (**e**), silvopastoral areas (**f**), petajoules. **Note** *SVB* Socialized Vegetal Biomass; *RB* Reused Biomass; *AuB* Aerial unharvested Biomass; *UrB* Unharvested root Biomass; *AB* Accumulated Biomass

Therefore, a balance is achieved between the energy that is extracted (Socialized Vegetal Biomass (SVB) + Reused Biomass (RB), with respect to the non-extracted energy (Unharvested Biomass (UB) + Accumulated Biomass (AB)) that must be respected to ensure its long-term functioning.

In terms of energy, the share of biomass extracted and not extracted from croplands remained relatively constant between 1900 and 1950. Extracted biomass rose from 38 to 40% and, non-extracted biomass dropped from 62 to 60% over the period (except for 1940 when extraction fell to 35% due to the Civil War) (Graph 5.2b and e). In pasturelands, the tendency was similar. Between 1900 and 1950, extracted biomass increased from 15% in the first decades to 18% in the 1930–50 period (Graph 5.2c and f).

As from the 1950s, massive incorporations of external energy reinforced the process of substitution of internal flows, though differently in croplands and silvopastoral areas. This process deteriorated, as we will show later, the quality of the agroecosystem land fund element. Between 1950 and 2008, changes in RB were driven by big increases in livestock, mainly in monogastric animals (pigs and poultry), and the shift from extensive to intensive management (see Chap. 2 for more details). This profound change in the composition and management of livestock would not have been possible without massive imports of feed (mainly soybean and corn), that is difficult to produce in Spain for agroclimatic and economic reasons. As a result, pastures were partially abandoned. Meanwhile, increasing amounts of high-quality biomass (grains and fodder) from croplands were dedicated to livestock. In these lands, RB went from 25% of NPP in 1950, to 40% in 2008 in terms of energy. Meanwhile, in silvopastoral areas, it fell from 12 to 4% (Graph 5.2e and f). These land-use imbalances (intensification of agricultural land use and abandonment of pastures) also occurred in the interior's croplands. Cultivation areas were reduced during this period mainly because of the abandonment of non-irrigated lands that responded poorly to external inputs. Changes in the social use pattern of biomass were also visible mainly in the 80s and 90s through increasing practices of burning of straw and other crop residues that was no longer used to feed livestock.

Conversely, a smaller share of biomass was abandoned on croplands. UB went from 57% of NPP to 38% (Graph 5.2e), a clear sign that soil fertility had become reliant on mineral fertilizers, to the detriment of organic matter. The use of herbicides also explains this relative fall. In silvopastoral areas, however, UB and AB increased by 6 and 5 percentage points, respectively, as a result of abandonment (Graph 5.2f).

As a result of the processes mentioned above, in terms of energy, Socialized biomass (SB) grew by 5.5% between 1900 and 1950, SVB accounted for an increase of 4.4% and Socialized Animal Biomass (SAB) an increase of 54.1%. Subsequently, SB growth accelerated, increasing by 41.5% between 1950 and 2008, mainly boosted by a SAB increase of 636%. The SVB grew barely by 21% over the period, as a result of a 75% increase in the crops SVB and a 26% drop in woodland SVB. The latter decline was due to the replacement of firewood by fossil fuels in households. The increase in cultivated land SVB (75%) was higher than that recorded for those lands' NPP (43%) (Guzmán et al. 2017). The increase in the harvest share of cereal varieties introduced by the Green Revolution contributed to the disparity, significantly increasing grain yields without any major changes to total aerial biomass (Austin et al. 1980; Vita et al. 2007; Sánchez García et al. 2013). The changes in the plant biomass distribution pattern were driven by the loss of agricultural residues' functions of feeding livestock and replacing soil fertility, in turn, caused by external energy imports.

Third, external input imports made it possible to simplify rotations and replace legumes, which were no longer essential for incorporating nitrogen into the agroecosystem. From 1960 to 2000, the surface area devoted to legumes fell from 1.4 to 0.55 million hectares (Mha). Consequently, biological fixation N accounted for 28% of the entries in 1960, and only 11% in the year 2000 (Guzmán et al. 2018). To summarize, the intensification of agricultural land in Spain unfolded based on the substitution of internal energy loops (particularly through the relative decline of UB, and the marginalization of legumes), with external energy inputs, increasing the generated entropy and, as we will see later, degrading the land fund element and its components.

5.2 The Energy Efficiency of Agricultural Production

The total energy consumed to sustain the functioning of agroecosystems not only includes external flows but also the internal biomass flows that are reinvested in the agroecosystem (unharvested biomass plus reused biomass) (Guzmán and González de Molina 2015; Tello et al. 2016; Guzmán and González de Molina 2017). In terms of energy, the flows necessary for the functioning of Spain's agroecosystem rose from 3761 PJ in 1900 to 5163 PJ in 2008 (Graph 5.1). This represents a 37.3% increase in energy consumption. However, NPP only achieved an increase of 28.8% (Graph 5.2). This energy return was low despite the significantly higher amount of energy invested and also despite an increase in irrigated areas that rose from 0.8 Mha in 1900 to 3.3 Mha in 2008. Access to additional water flows is key to increasing plant production in semi-arid regions such as the Mediterranean. Therefore, it would be expected that the combined additional contributions of energy and water would have been synergistic and increased NPP further. However, this did not occur because the decoupling between internal energy and material flows led to the degradation of the land fund element, as we will see later, resulting in a negative return of the invested energy. We call this rate of return on energy investment NPP-EROI. The NPP-EROI (Net Primary Productivity-Energy Return on Investment) is part of the battery of agroecological indicators of energy efficiency (Agroecological EROIs) that we advanced in other studies to estimate the return of energy invested by society, in terms of biomass flows that support the agroecosystem's fund elements. In these texts, we defended that the indicator of energy efficiency NPP-EROI = NPP/(RB + UB + external inputs) systematically expresses the evolution of the quality of the agroecosystem's fund elements, to the extent that the agroecosystem's total biomass production is included in the numerator—not only the production share that is of interest to society—and all the energy consumed in the production process is present in the denominator. The degradation processes affecting the fund elements (e.g. salinization, soil erosion, genetic erosion, etc.) must be compensated by the incorporation of increasing amounts of energy to alleviate the agroecosystems' loss of productive capacity. Therefore, the negative evolution of this indicator indicates the agroecosystem's degradation (Guzmán and González de Molina 2015; Guzmán and González de Molina 2017; Guzmán et al. 2017; Guzmán et al. 2018). In the case of Spain, it remained stable at 1.16 between 1900 and 1960, when it began to fall gradually reaching 1.09 in 2008 (Guzmán et al. 2017; Guzmán and González de Molina 2017).

Another interesting agroecological EROI is the so-called Biodiversity-EROI (= UB/(RB + UB + External inputs) that defines the return on energy invested in the agroecosystem in the form of available phytomass for wild heterotrophic species (UB). The relationships between energy flows and biodiversity have been put forward by ecologists, based on concrete studies that show that systems with higher amounts of energy entering the food web can support longer food chains and, therefore, increased biodiversity (Thompson et al. 2012). The EROI of Spanish agriculture decreased more slowly in the first half of the century and fell more sharply in the second half, especially as of 1970 (Guzmán et al. 2017). As we explained earlier, this decline was encouraged by the increase in biomass extraction for human and animal food, and by harmful practices such as the burning of crop residues and the use of herbicides on cultivated land. If we calculate the Biodiversity-EROI exclusively for croplands, the drop is much bigger (Graph 5.3). Therefore, the heterotrophic species associated with agricultural areas were the first to be affected by the relative scarcity of available phytomass. This effect has not been compensated by the abandonment of pastures and woodlands. In summary, the agroecosystem's dissociation between intensive production areas and abandoned and/or protected areas (e.g., 40.5% of total forest areas of Spain are protected, according to MAGRAMA 2014) has not achieved a significant increase in trophic energy available for transfer from plants to other levels in the food webs.

Graph 5.3 Biodiversity-EROI in the Spanish agroecosystem and in cultivated lands

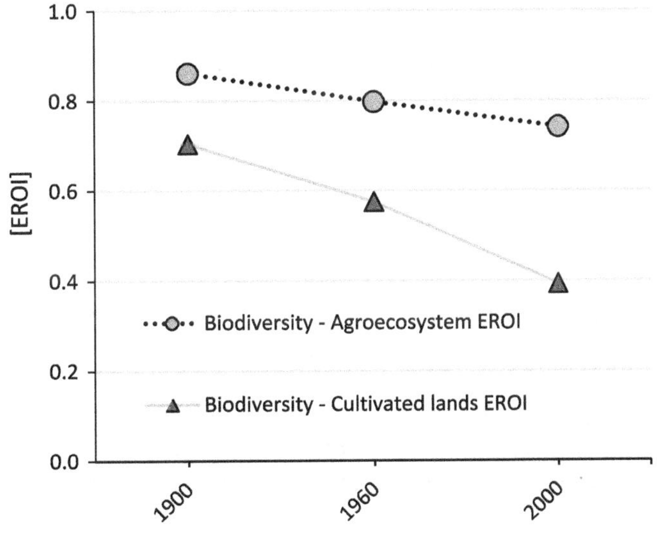

According to the supporters of "land sparing" (see Phalan et al. 2011 as an example), the intensification of agriculture based on the incorporation of external inputs to create areas free from human intervention is an appropriate way of maintaining biodiversity. Our results put this stance into question. Land sparing was implemented in the second half of the twentieth century in Spain, and there seems to have been no benefits for wildlife, as we will see later.

It is worth noting that the sharp fall in cultivated land UB also brought about a drastic reduction in organic matter soil inputs, negatively affecting the quality of this fund element. To evaluate these effects, we chose to use material flows, specifically carbon, as exposed in the next section.

Lastly, we will describe the evolution of the relationship between socialized biomass and external inputs, called "net efficiency" or "external final EROI" (EFEROI = SB/external inputs). This ratio constitutes the most widely used energy efficiency indicator in agricultural studies; it has the advantage of being comparable with other studies, although it also has the serious drawback of considering the agroecosystem as a black box because it does not take internal flows into account (Guzmán and González de Molina 2015; Tello et al. 2016; Guzmán and González de Molina 2017). Its use is, therefore, more economic than agroecological. The evolution of this indicator is even more baffling. It fell from 17.3 in 1900, to 4.8 in 1960 and 1.2 in 2008 (Guzmán et al. 2017). These figures are similar to those obtained by Carpintero and Naredo (2006) for Spanish agriculture, i.e., 6.10 in 1950–51 and 1.27 in 1999–2000.

To summarize, the energy efficiency of Spanish agriculture followed a downward trend, with serious economic and environmental consequences: dependence on inputs from outside the sector increased to the detriment of internal biomass flows that feed the components of the land fund element.

5.3 State of the Components of the Land Fund Element

The changes described in the system's functioning modify the state of the land fund element's components, as explained below.

5.3.1 Soil

(a) Replacement of edaphic macronutrients (N, P, and K) closing the nutrient cycle at the agroecosystem scale

Throughout the twentieth century, changes in the amount of nutrients in the land's soil, resulting from the balance between inputs and outputs, followed three trends based on an intricate set of socio-economic and political factors.

Until 1950, the balances of N and P (inputs minus outputs) were fairly balanced (from +0.23 kg N ha^{-1} $year^{-1}$ and −0.28 kg P ha^{-1} $year^{-1}$ in 1900, up to 3.9 kg N ha^{-1} $year^{-1}$ and 2.4 kg P ha^{-1} $year^{-1}$ in 1950) in cultivated fields (Graph 5.4). In the case of K, it was negative during the first 20 years of the twentieth century (−0.26 kg K ha^{-1} $year^{-1}$ on average) and was slightly positive until 1950 (1.1 kg K ha^{-1} $year^{-1}$) (Graph 5.4). These balanced or slightly positive balances in the mid-twentieth century became markedly positive from the 1960s, especially in the case of nitrogen, with average annual increases of 0.79, 0.20 and 0.38 kg of N, P and K ha^{-1}until 2000 (Graph 5.4). In 2000, the surplus entering annually in the fields was 40.3 kg N ha^{-1}, 12.5 kg P ha^{-1} and 16.7 kg K ha^{-1}. This value in the case of N is lower than that provided by Leip et al. (2011) who found a positive annual balance for N of 50 kg N ha^{-1} $year^{-1}$ for Spain during the 2001–2003 period, slightly lower than the European average (EU27). However, the values were similar to those quantified for the set of herbaceous crops (39.7 kg N $ha^{-1}year^{-1}$) and woody crops (41.1 kg N ha^{-1} $year^{-1}$) for Spain in 2011 (MAGRAMA 2013).

The patterns of change in the nutrient balance were mainly due to changes in the amount and input entry routes of N, P and K. During the first third of the twentieth century, the annual rate of N, P, and K inputs increased slightly (Graph 5.5). In the case of N, between 57.3 and 63.4% of annual inputs during this period corresponded to natural (precipitation and fixation of atmospheric N) and recycled (crop residues) inputs on the production site itself, while the contribution of natural and recycled inputs of potassium was low (20.0–25.5%) or very low in the case of phosphorus (10.3–17.7%). Manure constituted between 41.6 and 56.0% and 65.2–69.6% of the entries of P and K, respectively. During this period, annual inputs from synthetic fertilizers in the case of N and especially of P, as well as rising production levels led indirectly to input increases in the form of crop residues and manure (Grap 5.5).

During the 1940s, the annual input of nutrients decreased compared to the previous decade (Graph 5.5), mainly due to a decrease in inputs from crop residues, due, in turn, to lower production levels associated with a decrease in N and P inputs from synthesis fertilizers. From 1960 to 2000, annual inputs of N, P, and K, respectively, multiplied 2.3, 2.0, and 2.25 times (Graph 5.5). This increase has been mainly due to the increase of inputs through (i) synthetic chemical fertilizers, which in 2000 were 4.1, 2.0, and 6.0 times higher than in 1960 for N, P, and K, respectively; (ii) organic fertilizers, which increased between 1.3 and 1.9 over the same period; and (iii) crop residues that in 2000 were 3.8, 2.5 and 3.5 times higher than those of 1960 for N, P, and K, respectively. Worthy of note, the annual entries of N by biological fixation in 2000 were only 7% higher than those of 1960, revealing that the agricultural model changed from one based on inputs by biological fixation and manure to a model relying on synthetic fertilizers.

The notable increase in N, P, and K inputs in Spanish lands, which were 4.2, 5.8, and 3.4 times higher for N, P, and K in 2008 than in 1900, did not occur simultaneously to outputs due to aerial biomass production. During the study period, the amount of N, P and K in the produced biomass multiplied between 2.8 and 2.9 times, but at different rates over the three clearly identified periods (Graph 5.6).

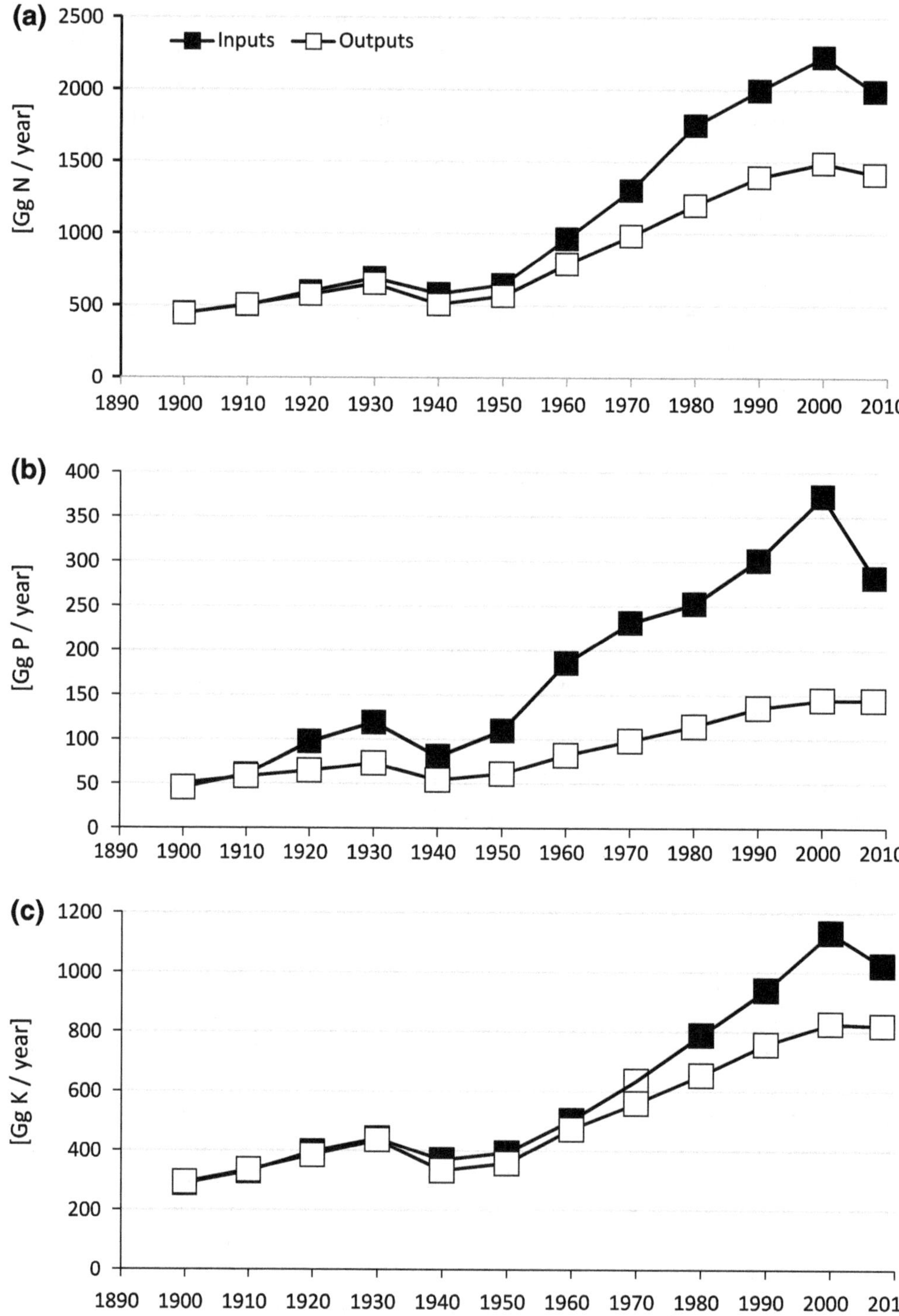

Graph 5.4 Inputs and outputs of nitrogen (**a**), phosphorus (**b**) and potassium (**c**) in Spanish cultivated lands from 1900 to 2008, in gigagrams of N, P and K. The balance is equal to the difference between the inputs and outputs

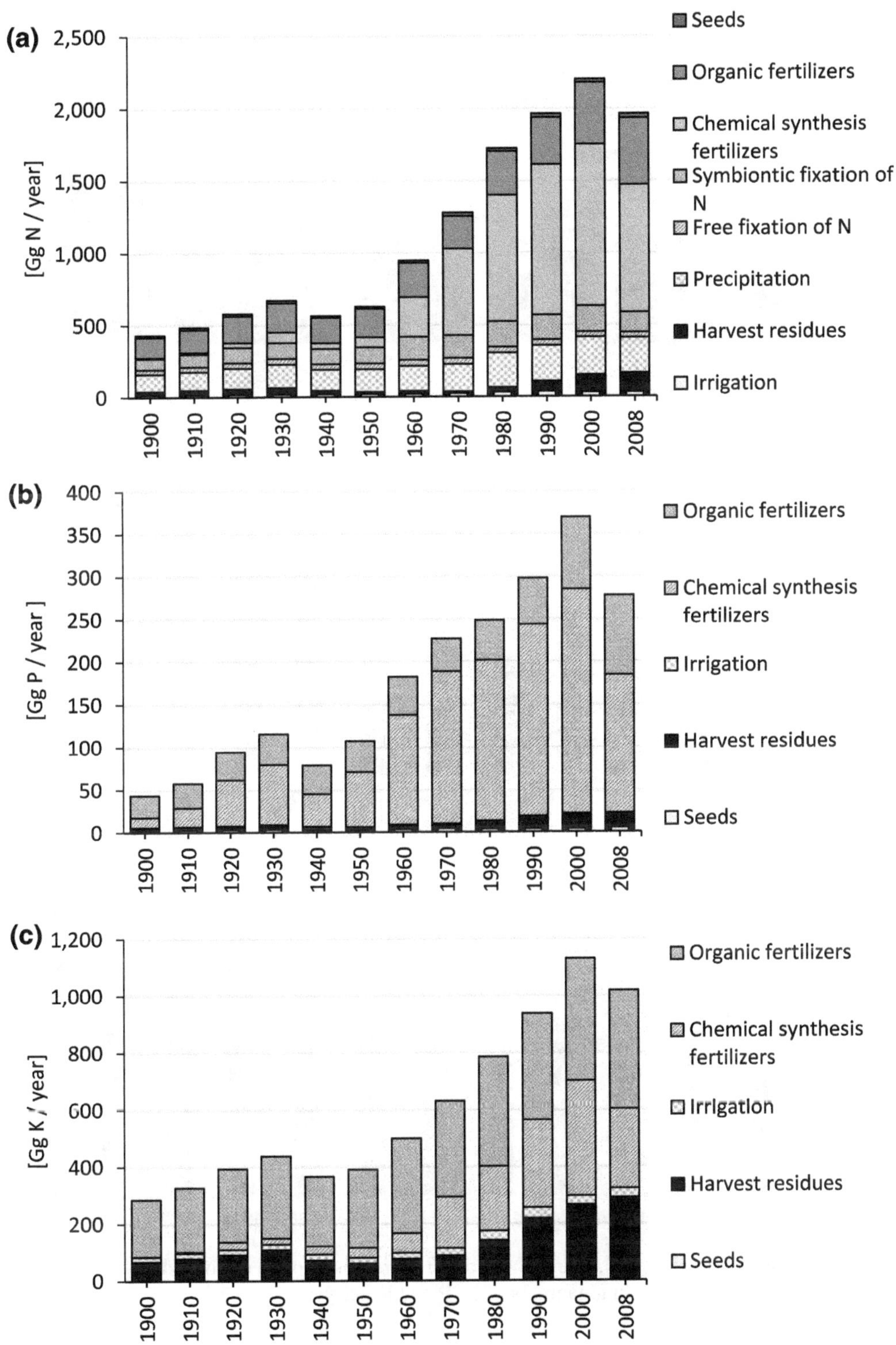

Graph 5.5 Annual inputs of N (**a**), P (**b**) and K (**c**) through different entry routes in cultivated lands over the 1900 to 2008 period in Spain, in gigagrams of N, P and K per year

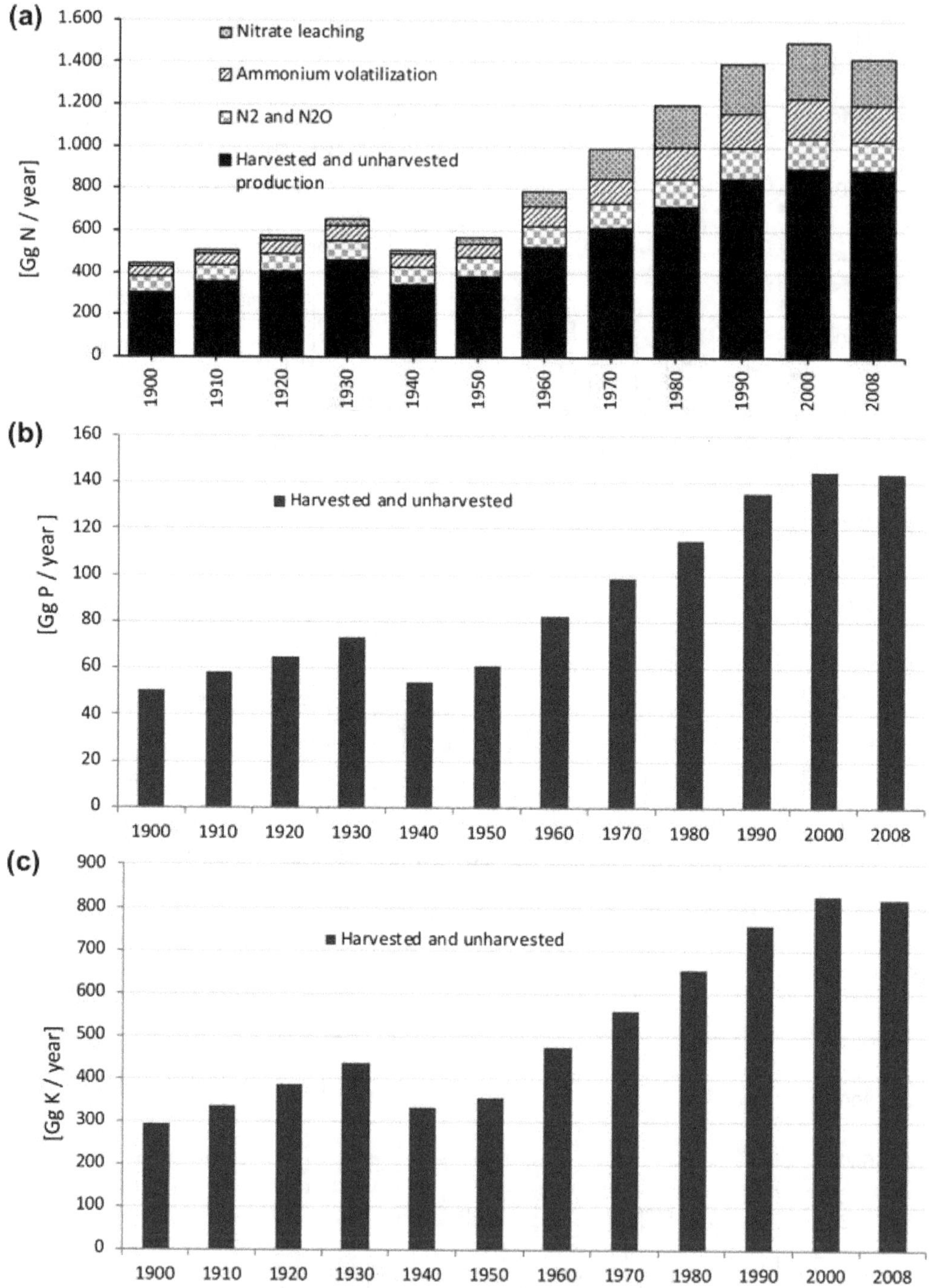

Graph 5.6 Annual outputs of N, P and K via different routes into the cultivated lands during the 1900 to 2008 period in Spain, in gigagrams of N, P and K per year

At the end of the first third of the century, the outputs of N, P, and K were 45.2–49.2% higher than those recorded for 1900, and between 69.2% and 70.9% corresponded, for the case of N, to outputs in the form of aerial biomass, the main destination being animal feed (Graph 5.6). Indeed, between 61.4 and 63.7%, 54.7–57.7% and 75.7–77.5% of N, P and K produced in the cultivated lands were aimed at stock feed. During this period, 13.6, 11.0, and 23.1% of N, P, and K were harvest residues that remained in the field.

As in the case of inputs, there was a turning point in the 1940s when N, P and K outputs, dropped compared to the end of the first third of the century, mainly due to declining production (Graph 5.6). Nutrient outputs increased markedly from the 1960s and in 2000, N, P, and K were 89.0, 74.9, and 75.2% higher than those estimated in 1960. In 1900, the total annual outputs of N, P, and K were 26.8 kg N ha^{-1}, 3.05 kg P ha^{-1} and 17.7 kg K ha^{-1}, while they were 81.9 kg N ha^{-1}, 8.3 kg P ha^{-1} and 47.5 kg K ha^{-1} in 2008. The productivity of N, P, and K in the form of aerial biomass rose slightly during the first third of the century, decreased during the 40s, and increased steadily after the 60s (Graph 5.7). In 2008, productivity had multiplied twofold compared to 1960 and multiplied 2.68–2.77 times compared to 1900, for N, P, and K.

The share of recycled nutrients—i.e., harvest and manure residues exclusively from livestock feed produced in crop fields—in nutrient inputs, tended to decrease throughout the study period (Graph 5.8). In the cases of N and P, this share decreased at a fairly similar rate until the end of the 1950s, after which its decline accelerated and in 2008, the recycling of N and P compared to total inputs was approximately half that of 1900. In the case of K, there was a sharp drop from the 1970s, mainly due to the increase in K inputs in the form of synthetic fertilizers.

During the first third of the twentieth century, annual net transfer of N, P, and K from pastures to cultivated lands, in the form of manure that was collected while the livestock was stabled, was more or less constant (Graph 5.9). It increased during the 1940s and part of the 1950s (Graph 5.9). From the 60s, this pattern was reversed and

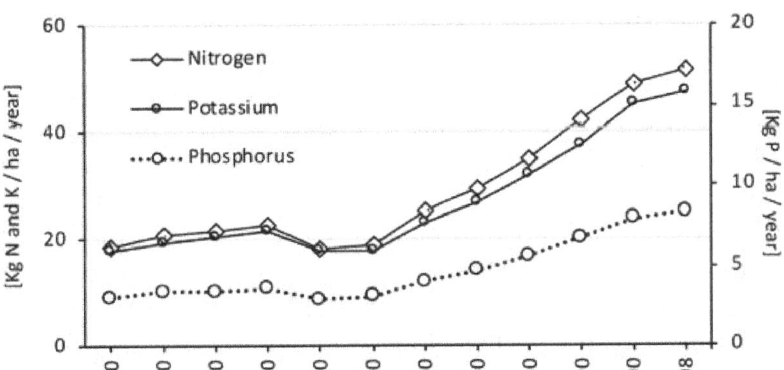

Graph 5.7 Productivity in terms of N, P and K in the aerial biomass produced in Spanish cultivated lands over the 1900 to 2008 period, in kilograms of N, P and K per hectare and per year

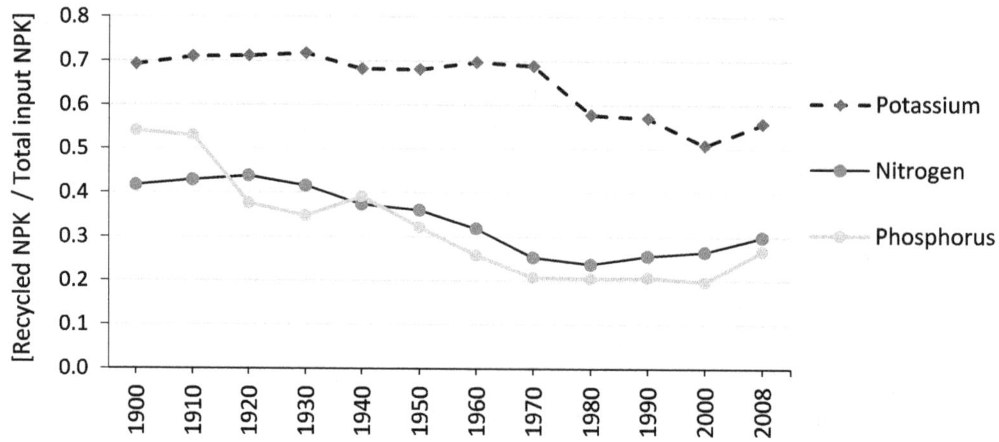

Graph 5.8 Share, out of 1, of the recycling of N, P and K of the total input in Spanish cultivated lands over the 1900 to 2008 period

in 2008 there was a net transfer of N, P, and K from cultivated lands to pastures in the form of excretions originating in food from the land—imported or not.

The gap in quantitative terms between total inputs and aerial biomass outputs led to drops in the efficiency of use of N, P, and K and rising losses of N to the atmosphere and other ecosystems with potential negative effects. The efficiency of use of N, P, and K (NPK contained in the aerial NPP divided by the total number of NPK inputs) shows a clear pattern related to the notable increase in nutrient inputs via synthetic fertilizers and manure produced from imported food since the last third of the twentieth century. The efficiencies of N use at the beginning of the twentieth century were above 66% and dropped to 40% in 2000 (Graph 5.10). In the case of P, the efficiency fell from 110% in 1900 to 38% in 2000, and in the case of K, that has always been higher than N and P, it ranged between 90 and 100% during the first half of the twentieth century, declining as of 1960 to reach 73% in 2000. For all three macronutrients, use efficiencies increased in 2008, mainly due to a decrease in inputs through synthetic fertilizers, despite the fact that productivity did not fall.

The decrease in N use efficiency in Spanish agricultural land took place at the same time as the absolute losses of N and those relative to the N produced in the aerial biomass, especially as of the 1960s. In 2000, the annual amount of lost N (30.3 kg N ha^{-1} $year^{-1}$ on average) was 3.7 times higher than in 1900 (8.3 kg N ha^{-1} $year^{-1}$), although the largest annual increases were observed since the 1960s (Graph 5.11). In 1900, 0.41 kg N was lost for each kg of N produced, and this value increased by 59% in 2000 (0.66 kg N per kg of N produced) (Graph 5.11).

These losses of N are linked to environmental impacts that have been described in detail (Erisman et al. 2013). The N cycle in agroecosystems has indeed many escape routes because nitrate is a very mobile nutrient, and non-retained nitrogen in ecosystems, mainly in the form of organic nitrogen, can be transferred to the atmosphere, in the form of N_2O, N_2, and NH_3 or to other aquatic and terrestrial ecosystems in the form of nitrate, where it contributes to a large amount of adverse

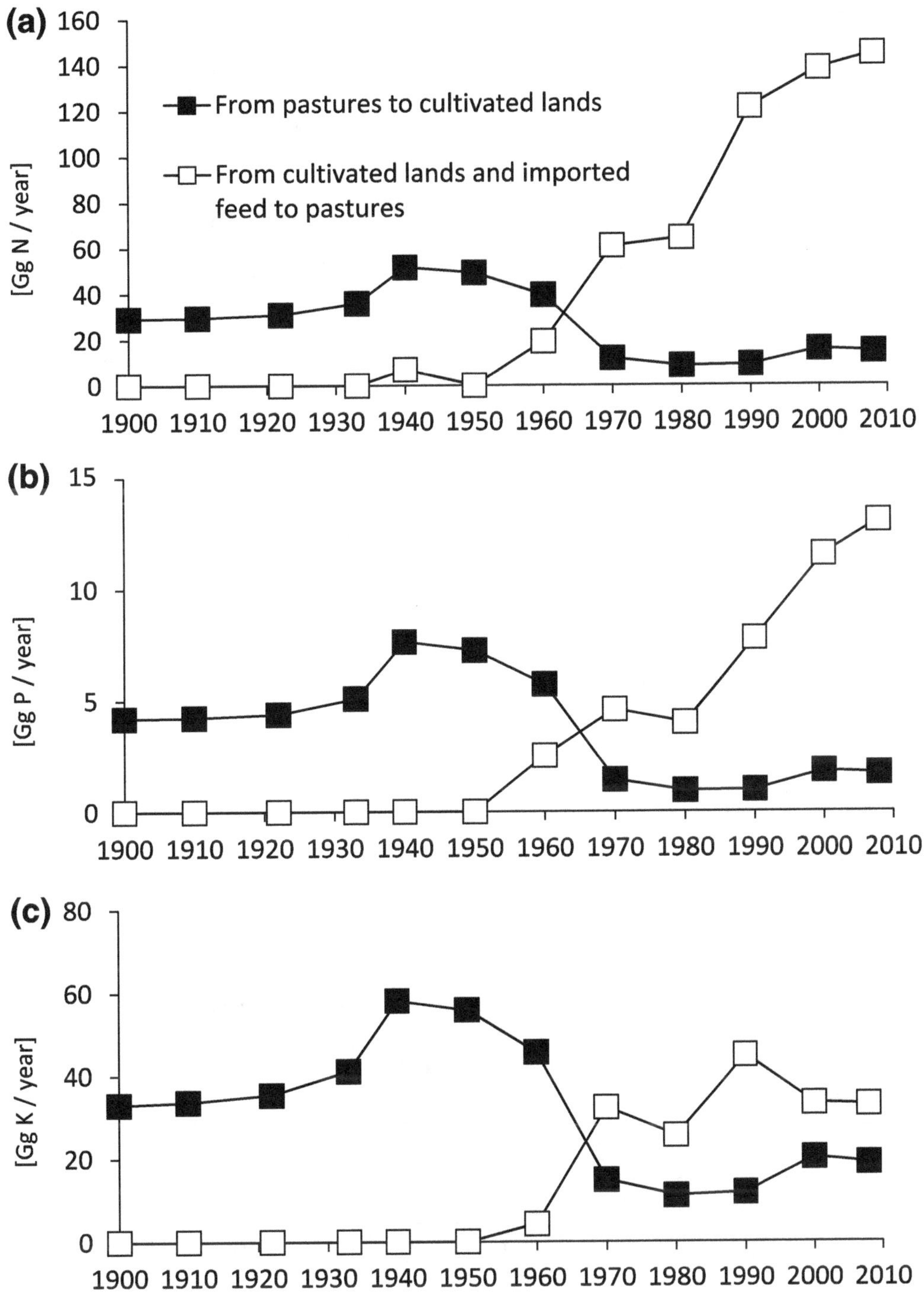

Graph 5.9 Annual transfer of N(**a**), P(**b**) and K(**c**) from the pastures to cultivated lands and from the cultivated lands and imported feed through excretions over the 1900–2008 period in Spain, in gigagrams of N, P and K per year

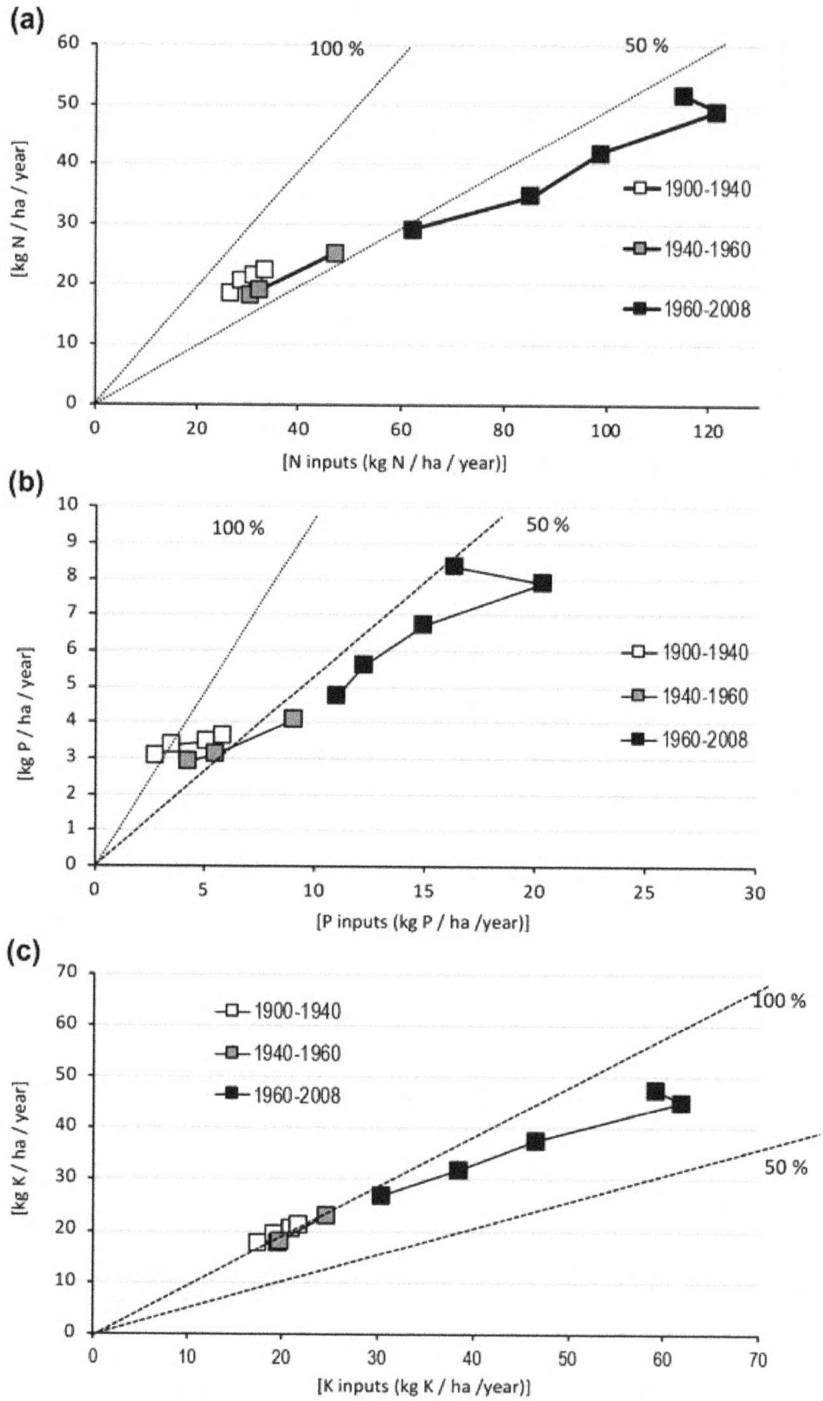

Graph 5.10 Relationship between the productivity in terms of N(**a**), P(**b**), and K(**c**) and the annual inputs of N, P, and K in Spanish cultivated lands. The dashed lines reflect the 100% and 50% use efficiencies of the three macronutrients

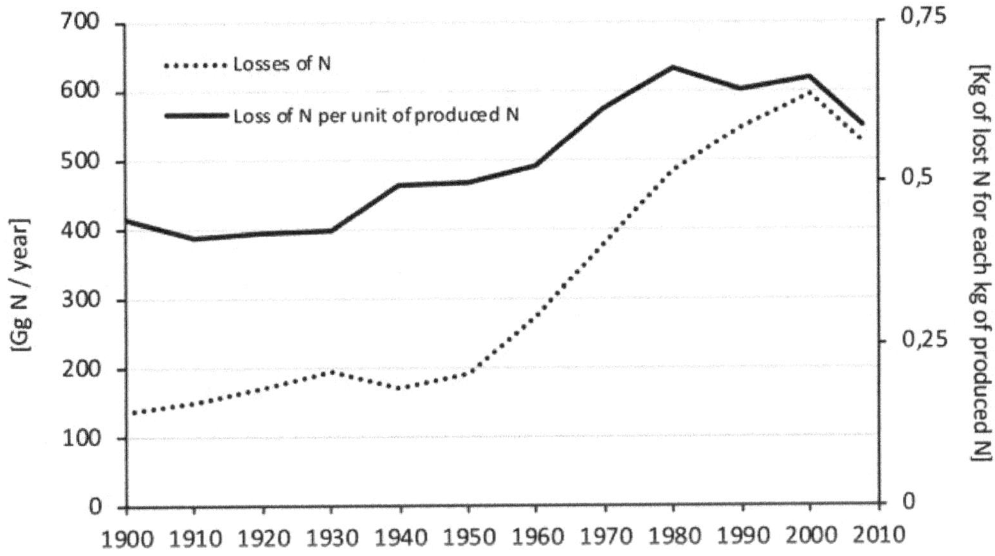

Graph 5.11 Annual losses of N (Gg N, in the form of N_2 and N_2O, volatilization of ammonium and leaching of nitrate) and losses relative to produced N (kg N lost kg N produced^{-1}) of crops

effects. Nitrate lost in agroecosystems can become a pollutant in surface waterways, encouraging eutrophication processes that result in a loss of biodiversity and loss in water quality, and in aquifers. Relatively high levels of nitrate (>25 mg NO_3^- l^{-1}) in drinking water have been linked to the incidence of colon cancer (Grizzetti et al. 2011). Moreover, nitrate can be lost in agroecosystems in the form of N_2O, which is the third greenhouse gas and contributes to the destruction of the ozone layer (Ravishankara et al. 2009) therefore, to climate change (Galloway et al. 2003; Galloway et al. 2008). Finally, ammonium can be transformed into ammonia gas that, in high concentrations, can be toxic to plants (Nordin et al. 2011) and is considered an atmospheric pollutant (Sutton and Fowler 2002).

5.3.2 Replacement of Organic Carbon

In both ecosystems and agroecosystems, the main driver of soil organic carbon (SOC) is the input of biomass (Aguilera et al. 2013). The fundamental difference is that in agroecosystems, the size of this entry is conditioned by the cultivation method both directly and indirectly. It is conditioned directly because a number of management practices (burning waste, organic fertilization, application of herbicides, etc.) intentionally modify the size of this entry; and it is conditioned indirectly because the NPP is affected by farming methods, to the extent that they affect the state of the fund elements and/or modify the availability of limiting factors. On the other hand, management practices also affect SOC because of impacts on the mineralization of organic matter, thus altering the system's output. For example, irrigation stimulates microbial activity and therefore mineralization and tillage can generate a provisional increase in mineralization, while it can also move carbon to deeper layers, where it is more protected. Therefore, the SOC balance is the result of distinct and sometimes opposite processes.

In the case of Spain, it is worth asking to what extent and when UB entries to the soil have been compensated by external biomass imports. Graph 5.12 shows that in cultivated lands, carbon inputs in the soil reached their maximum levels in the first third of the twentieth century, after which they shrunk and reached their minimum levels in the 1980s. A progressive recovery is currently underway related to a certain increase in UB and the greater availability of manure due to continually expanding livestock. The UB increase results from bigger restrictions to stubble burning and to an increase in residue production that has accompanied the expansion of irrigation and the drop in grazing on agricultural land over the last decade. The drop in carbon entry during the twentieth century helps to explain why half of Spain's agricultural land currently has an organic carbon content of less than 1% (Rodríguez-Martín et al. 2009).

Graph 5.13 shows the stocks of equilibrium SOC in each period, after calculating the balance between inputs and outputs. As shown, there are significant differences in SOC stocks per hectare between the different land uses (Graph 5.13b). Therefore, changes in the total stocks of C (Graph 5.13a) were partly due to changes in land use, and partly due to the evolution of equilibrium C stocks for each type of land use. The highest levels of equilibrium SOC were reached at the beginning of the twentieth century, due, on the one hand, to relatively high levels of C inputs and, on the other, to relatively low mineralization rates because of lower average temperatures, to still reduced irrigated land surface areas, and to a good vegetation cover in woody crops (Aguilera et al. 2018). The equilibrium SOC began to fall in 1920, first because of expanding cultivated areas to the detriment of pastures, and from the 50s onwards, due to ever more widespread burning practices, herbicides, and the reduction of harvest indexes due to varietal changes (Graph 5.13b).

As of 1990, stocks of equilibrium SOC of agricultural lands recovered but did not attain the levels of the early twentieth century. The slight increase is due to the increase in C entries, described above. Despite this, the levels were below their potential with respect to current levels of productivity and are not sufficient to maintain the

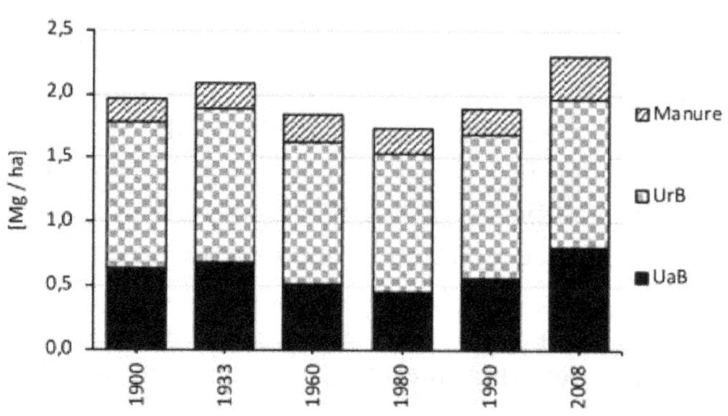

Graph 5.12 Evolution of annual soil carbon entries in agricultural land in the twentieth century, Mg/ha/year. Note: Unharvested aerial Biomass (UaB); Unharvested root Biomass (UrB)

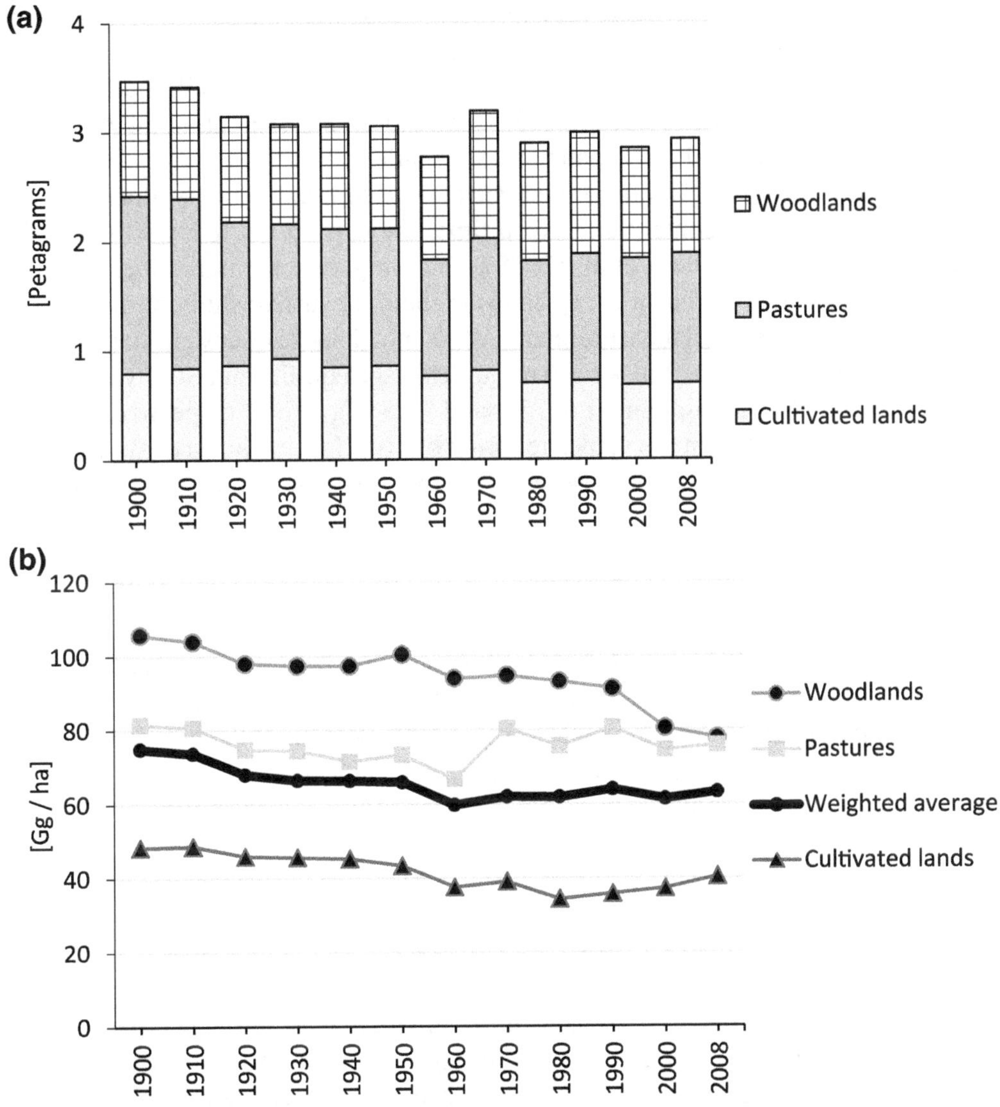

Graph 5.13 Evolution of stocks of equilibrium organic soil carbon for different land uses and for the agroecosystem as a whole, 1900–2008, petagrams, total (**a**) and per hectare (**b**)

SOC in a context of increased mineralization because of rising average temperatures, associated with climate change, of extensive irrigated areas, and scarce plant cover in woody crops. The effect of climate change in recent decades has also been visible in pastures and woodlands: while it stabilizes equilibrium levels in a context of increased inputs in pastures, it leads to lower equilibrium SOC levels in a context of stable inputs in woodlands (Graph 5.13b).

In terms of the stock of equilibrium SOC with respect to the total biomass recycled in cultivated lands (UB + RB + imported feed consumed by livestock in cultivated lands), the evolution is markedly negative. This indicator falls from 844 g C MJ^{-1}in 1950 to 451 g C MJ^{-1} in the year 2000. In other words, it is necessary to recycle 87%

more biomass to obtain the same stock of SOC in agricultural lands. The current biomass management strategy is clearly inadequate in relation to climate change mitigation. Monogastric livestock, whose manure, often handled in liquid form, has a low C: N ratio contributes to this low efficiency, in addition to greater mineralization due to climatic change and the expansion of irrigation.

The equilibrium SOC data presented correspond to hypothetical values and not true estimates of actual content at a given time. We carried out these estimates for agricultural lands in Aguilera et al. (2018), obtaining very similar values to those estimated by Rodríguez-Martín et al. (2016) in their exhaustive study of the SOC in Spanish soils based on field measurements. We do not dispose of any similar estimate for pastures and woodlands, but we can use the average equilibrium SOC values over the period studied as a guideline, i.e.,75.8 and 94.5 Mg C ha^{-1}respectively. These values are situated between the values reported by Rodríguez-Martín et al. (2016) of 64.1 and 69.3 respectively, and those reported by Doblas-Miranda et al. (2013), of 103.0 and 101.6, respectively.

The relatively low values in agricultural lands indicate that these soils are at a degradation threshold (Romanyà et al. 2007; Rodríguez-Martín et al. 2016). The increase in NPP resulting from intensification, together with massive feed imports, could have theoretically led to a substantial increase in the return of organic carbon to the soil, thus enabling to face increased mineralization brought about by climate change and the spread of irrigation. However, the breakdown of the balance between the uses of biomass and the preferential use of feed to nourish poultry and pigs prevents this from happening.

5.3.3 Biodiversity

Non-agricultural biodiversity is another fund element to have been seriously harmed by Spanish agriculture's metabolic transformation. According to the "country profile" developed for Spain by the "Convention on Biological Diversity", Spanish biodiversity has "dropped significantly" in recent decades, and 40–60% of the evaluated species have been included in some threatened category (CBD 2017). There are multiple causes. According to the Evaluation of Spain's Ecosystems of the Millennium report, the greatest threats to biodiversity include the expansion of intensive agriculture, urbanization and habitat fragmentation caused by the increase in linear transport infrastructures (EEME 2011, pp. 52). The Observatory on Sustainability also establishes a strong relationship between the intensification of agriculture and the loss of biodiversity in Spain (OS, 2016).

The impact of intensive agriculture on non-agricultural biodiversity is of a strong and complex nature. First, contamination by fertilizers and pesticides has been found to be a major problem, and it goes against the European Union's clearly downward trend (OS 2016). In this book, we described the evolution of N flows from agriculture that is dissipated in the atmosphere and waters, affecting terrestrial and aquatic biodiversity. We also presented the increasing size, in terms of energy, of the flows of pesticides that have been poured into the agroecosystem (see Chap. 4).

Second, several authors have emphasized how the industrialization of traditional agriculture has damaged biological diversity due to the loss of landscape heterogeneity at multiple spatial and temporal levels (Benton et al. 2003; Firbank et al. 2008; Lindborg and Eriksson 2004; Perfecto and Vandermeer 2010; Schuch et al. 2012; Vos and Meekes 1999). In other case studies, we showed how the local generation of energy and material flows that sustain organic-based agriculture is reflected in the landscape, leading to complex land matrices (Guzmán and González de Molina 2009; Garzón et al. 2011) that favor biodiversity. Conversely, simplified landscapes proper to industrialized agriculture harm biodiversity (Marull et al. 2015).

Third, we showed how agricultural industrialization has generated a major imbalance in the uses of phytomass in cultivated lands, which further weakens the conservation of non-agricultural biodiversity. If we consider threatened bird species as an indicator of biodiversity, 17.5% of them are associated with pseudo-stem cereals and 5% with high diversity agricultural areas (orchards, irrigated tree crops, etc.) (EEME 2011, pp. 80). MAPAMA (2013) thus shows that between 1998 and 2012, the trend for birds associated with the agricultural medium, expressed as % change in populations, was 4.8% in tree crops and 25% in cereal crops. The cereal crop group does not use fertilizers and pesticides particularly intensively. However, the fiasco of unharvested biomass reached a peak in the case of cereals. Modern varieties have little straw, are mostly low in size and of short cycle, the use of herbicides and the burning of stubbles mean these crops leave hardly any useful biomass for heterotrophic species, which affects the size of the populations they can sustain and the trophic chains they are part of. This is the case of populations of birds of prey, such as the Lesser Kestrel. The decline of these birds is linked to the fact that they need to invest much greater efforts to catch their prey (arthropods and small vertebrates) in the cereal fields since they have been modernized (Ministry of the Environment 2004). For other birds linked to cereals, UB also offers a refuge for them to reproduce themselves. On this subject, the reader can consult the actions recommended in the "Life Project" for the conservation of steppe birds issued by the Ministry of Environment of Andalusia (2003).

It is very difficult to discern the isolated effects of each of these processes on declining biodiversity. Interactions and synergies are likely to exist between them. In recent years, the conversion of farms from Industrialized Agriculture (also called conventional agriculture) to Organic Farms (managed without agrochemicals and using organic fertilization making them similar to traditional farms) provide us with keys to build a deeper understanding of the interactions. These studies have proliferated in recent years and various available meta-analyses show that converting to organic farming improves biodiversity in cultivated areas (Bengtsson et al. 2005; Hole et al. 2005; Norton et al. 2009; Tuck et al. 2014; Gomiero 2015). Only some of the studies collected in the meta-analyses assess the specific causes of this rise in biodiversity. Some studies show that the main causes consist of a greater complexity of the landscape and ecological connectivity and the reductions in the pressure of biocides. However, in recent years, several authors have found that the increase of forage resources (availability of phytomass) for heterotrophic species is one of the drivers of this increase in biodiversity. For example, Rundlöf et al. (2008) state that the two main favoring factors of bumblebees in organic farms, compared to conventional ones, are the prohibition of agrochemicals and the provision of additional fodder resources. Döring and Kromp (2003), in their literature review on carabid beetles in conventional versus organic agriculture, found that in most cases, the species was richer in organic farms, because ecologically managed fields provided larger food supplies for herbivorous fauna. Wickramasinghe et al. (2004) found that organic management was beneficial for bats both through the provision of more structured habitats, as well as greater food resources (insect prey). Gabriel et al. (2013) showed that differences in the biodiversity of different species (bumblebees, bees, butterflies, epigeous arthropods) between organic and conventional fields could be due to the fact that UB grass was more abundant in organic farms and, consequently, so was arthropod biodiversity.

5.4 A Diet Rich in Food of Animal Origin: The Outsourcing of Its Land Costs

To finish, we must point out that the negative effects on agroecosystems' fund elements are not limited to Spain's territory: they have also been partly outsourced abroad. While the effects of bad management affected local areas in traditional agriculture, with rise of international trade, these effects have become globalized.

To study this impacts Infante-Amate et al. (2018) estimated land embodied in Spain's biomass trade, i.e., the land required to produce the biomass exported and imported by Spain. Graph 5.14 shows the evolution and composition of land embodied in both imports and exports of biomass. Both of them depict strong growth in the period analyzed, mostly concentrated from the 1960s onwards. In fact, between 1900 and 1933, the land traded remained stable, while between ca. 1940 and 1960 both imports and exports declined. Land embodied in imports multiplied 8-fold over the whole period analyzed, from 1.3 million hectares (Mha) in 1900 to 1.9 Mha in 1960 and 11.0 Mha in 2008. Similarly, land embodied in exports increased 7-fold, from 0.7 to 1.0 Mha and 4.5 Mha, respectively. When distinguishing the final uses of traded land, feed appears as the main product traded, mainly as of 1960 in the case of imports. Its share in land embodied in imports rose from 0.6 Mha in 1960 to 5.6 Mha in 2008. This result is in relation to nutritional transition already explained in other parts of this book. At the beginning of the twentieth century, fibers and industrial products accounted for 41% of total land embodied in imports, while they dropped to 11.9% by 2008.

Net flows depict, therefore, a clear pattern of historical external dependence, which was especially evidenced from the 1960s. In all (twelve) benchmark years analyzed, Spain was a net importer of land, although the magnitude varied over the course of the period studied. In 1900, net imports accounted for 0.7 Mha or 372 m^2/inhab, while the

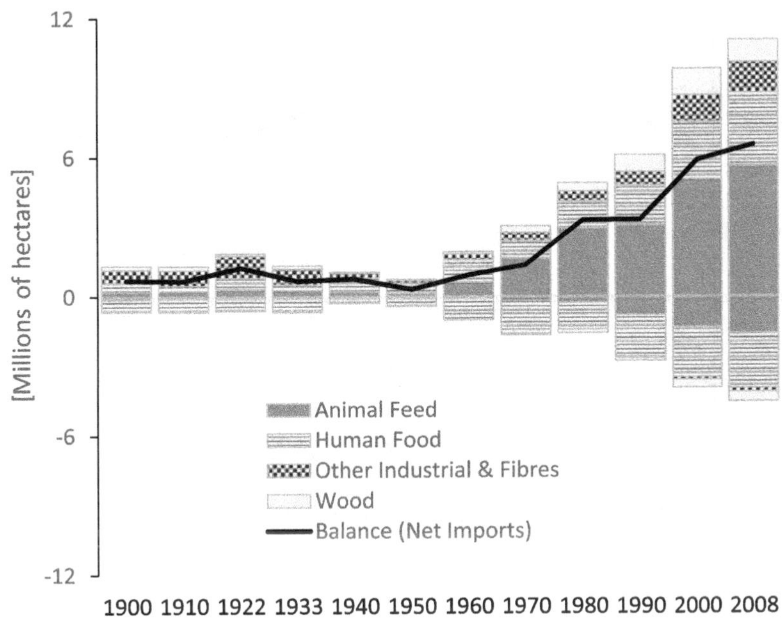

Graph 5.14 Land embodied in biomass imports (positive values) and exports (negative values), in millions of hectares. *Source* Infante-Amate et al. (2018)

figure was 9 and 4 times higher in 2008, respectively (9.1 Mha or 1954 m^2/inhab). The minimum value was found in 1950, due to the particular historical context related to the Spanish Civil War and the Autocratic policies developed under Franco's dictatorship, as discussed in Chap. 2.

In the case of cropland products, Infante-Amate et al. (2018) also provide evidence on embodied land in consumption activities, not only in trade. Knowing the actual cropland occupied in Spain (domestic cropland) and having estimated cropland embodied both in imports and exports, authors were able to quantify land requirements related to cropland-based biomass consumption. Cropland does not reveal any abrupt changes over the last century in Spain, moving from 16.5 Mha 1900 to 17.3 Mha in 2008, and peaking at 20.9 Mha in 1970. However, actual cropland use, considering the effect of land embodied in trade, shows more significant changes, mainly from 1960. In the early stages of the century, cropland use was basically explained by domestic cropland, due to low international trade levels. Actual cropland requirements were mainly met with domestic resources. Actual cropland use accounted for 17.1 Mha in 1900, a similar figure to domestic cropland. As trade gained importance, the gap between domestic cropland and actual cropland requirements increased (Graph 5.15). Today, actual cropland use (22.8 Mha) is 1.3 times higher than the cropland area occupied within the country (17.3 Mha). In the 1960s cropland embodied in imports amounted 1.9 Mha while cropland embodied in exports was 1.0 Mha. Net imports barely represented 4% of actual cropland demand and cropland embodied in imports equalled only 11.6% of domestic cropland, i.e., cropland occupied in third countries was only one tenth of land occupied for cropland

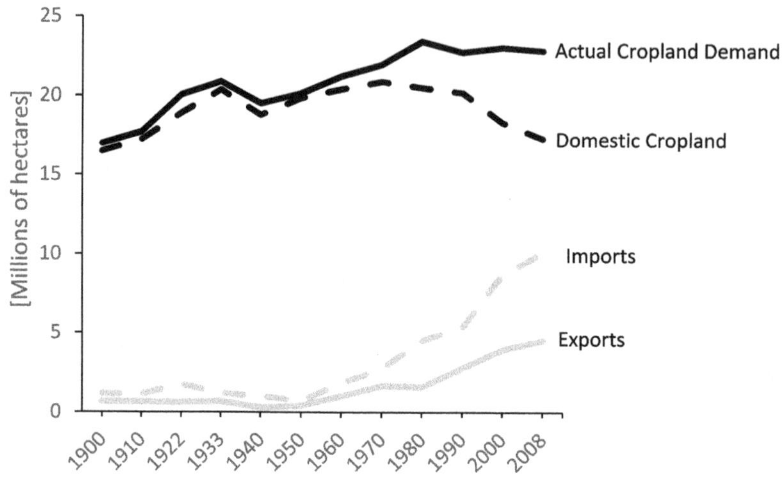

Graph 5.15 Actual cropland demand, domestic cropland, imports and exports of land, in millions of hectares. *Source* Infante-Amate et al. (2018)

within the country. Only five decades later, in 2008, net imports of cropland grew 7-fold, amounting to 37.7% of actual requirements, and total cropland imports made up 64.0% of domestic cropland. Given that a significant part of domestic cropland is devoted to exports (26.2% in 2008), today, Spanish inhabitants occupy a similar amount of cropland domestically as they do overseas to meet their cropland needs.

The expansion of the farming frontier represents one of the most pressing problems at a global level (Foley et al. 2005; Rockström et al. 2009), associated with deforestation (Kaplan et al. 2005; DeFries et al. 2010), which in turn leads to serious environment problems such as increased climate change (Cramer et al. 2004), losses of biodiversity (Pereira et al. 2012), and the alteration of the nitrogen cycle (Austin et al. 2006; Billen et al. 2014). In this chapter, we have explained the impacts of land intensification in Spain. Nevertheless, many other problems associated with the expansion of farmland and its intensification have taken place outside of Spain, and especially in highly sensitive areas from an environmental perspective. Spain's main imports of biomass are grain and oilseeds for animal feed (see also Lassaletta et al. 2014b; Soto et al. 2016), and the majority of these imports come from countries in which export agriculture generates serious social and environmental problems. According to Kastner et al. (2014), the main two exporters of agricultural land to Spain in 2009 were Brazil and Argentina, with just over one million hectares exported in each case. Many studies have indicated that agro-exports in these two countries are responsible for processes of deforestation and agrarian intensification (e.g., Kastensen et al. 2013; Lassletta et al. 2014) that, as well as serious environmental problems, generate serious social problems that often lead to violent conflict, displacements and the breaking down of traditional communities (Hecht and Corckburn 2010; van Solinge 2010; Mayer et al. 2015). Recently, exports from Eastern Europe have grown in importance. Currently, Ukraine, Romania, and Bulgaria allocated almost 1.5 Mha to exporting agrarian products to Spain. Agro-exports in these regions are also a growing cause for concern in terms of the socio-political and environmental problems (Visser and Spoor 2011) associated with them. In short, imports of biomass not only involve hidden land flows but also other impacts such as the loss of biodiversity, pollution, precarious labor, etc.

References

Aguilera E, Lassaletta L, Gattinger A, Gimeno S (2013) Managing soil carbon for climate change mitigation and adaptation in Mediterranean cropping systems: a meta-analysis. Agr Ecosyst Environ 168:25–36

Aguilera E, Guzmán GI, Álvaro-Fuentes J, Infante-Amate J, García-Ruiz R, Carranza-Gallego G, Soto D, González de Molina M (2018) A historical perspective on soil organic carbon in Mediterranean cropland (Spain, 1900–2008). Sci Total Environ 621:634–648

Austin RB, Bingham J, Blackwell RD, Evans LT, Ford MA, Morgan CL, Taylor M (1980) Genetic improvements in winter wheat yields since 1900 and associated physiological changes. J Agric Sci 94:675–689

Austin AT, Pineiro G, Gonzalez-Polo M (2006) More is less: agricultural impacts on the N cycle in Argentina. Biogeochemistry 79:45–60

Bengtsson J, Ahnstrom J, Weibull A (2005) The effects of organic agriculture on biodiversity and abundance: a meta-analysis. J Appl Ecol 42:261–269

Benton TG, Vickery JA, Wilson JD (2003) Farmland biodiversity: is habitat heterogeneity the key? Trends Ecol Evol 18(4):82–188

Billen G, Lassaletta L, Garnier J (2014) A biogeochemical view of the global agro-food system: Nitrogen flows associated with protein production, consumption and trade. Glob Food Secur 3(3–4):209–219

Carpintero O, Naredo JM (2006) Sobre la evolución de los balances energéticos de la agricultura española, 1950–2000. Hist Agrarno 40:531–554

CBD (Convention on Biological Diversity) (2017) Spain-Country Profile. Available in: https://www.cbd.int/countries/profile/default.shtml?country=es#facts

Cramer W, Bondeau A, Schaphoff S, Lucht W, Smith B, Sitch S (2004) Tropical forests and the global carbon cycle: impacts of atmospheric carbon dioxide, climate change and rate of deforestation. Philosophical Transactions of the Royal Society B: Biological Sciences 359(1443):331–343

de Vita P, Nicosia OLD, Nigro F, Platani C, Riefolo C, di Fonzo N, Cattivelli L (2007) Breeding progress in morpho-physiological, agronomical and qualitative traits of durum wheat cultivars released in Italy during the 20th century. Eur J Agron 26:39–53

Defries RS, Rudel T, Uriarte M, Hansen M (2010) Deforestation driven by urban population growth and agricultural trade in the twenty-first century. Nat Geosci 3(3):178–181

Doblas-Miranda E, Rovira P, Brotons L, Martinez-Vilalta J, Retana J, Pla M, Vayreda J (2013) Soil carbon stocks and their variability across the forests, shrublands and grasslands of peninsular Spain. Biogeosciences 10:8353–8361

Döring TF, Kromp B (2003) Which carabid species benefit from organic agriculture?-a review of comparative studies in winter cereals from Germany and Switzerland. Agr Ecosyst Environ 98:153–161

EEME (Evaluación de los Ecosistemas del Milenio de España) (2011) La Evaluación de los Ecosistemas del Milenio de España: síntesis de resultados. Madrid. Ministerio de Medio Ambiente, y Medio Rural y Marino. Madrid

Erisman JW, Galloway JN, Seitzinger S, Bleeke RA, Dise NB, Petrescu AMB, Leach AM, de Vries W (2013) Consequences of human modification of the global nitrogen cycle. Philos Trans R Soc 368(1621):1–9

Firbank LG, Petit S, Smart S, Blain A, Fuller RJ (2008) Assessing the impacts of agricultural intensification on biodiversity: a British perspective. Philos Trans R Soc 363:777–787

Foley JA, Defries R, Asner GP, Barford C, Bonan G, Carpenter SR, Chapin FS, Coe MT, Daily GC, Gibbs HK, Helkowski JH, Holloway T, Howard EA, Kucharik CJ, Monfreda C, Patz JA, Prentice IC, Ramankutty N, Snyder PK (2005) Global consequences of land use. Science 309(5734):570–574

Gabriel D, Sait SM, Kunin WE, Benton TG (2013) Food production versus biodiversity: comparing organic and conventional agriculture. J Appl Ecol 50:355–364

Galloway JN, Aber JD, Erisman JW, Seitzinger SP, Howarth RW, Cowling EB, Cosby BJ (2003) The nitrogen cascade. Bioscience 53:341–356

Galloway JN, Townsend AR, Erisman JW, Bekunda M, Cai Z, Freney JR, Martinelli LA, Seitzinger SP, Sutton MA (2008) Transformation of the nitrogen cycle: recent trends, questions, and potential solutions. Science 320:889–892

García-Ruiz R, González de Molina M, Soto Fernández D, Guzmán Casado G, Infante-Amate J (2012) Guidelines for constructing nitrogen phosphorous and potassium balance in historical agricultural systems. J Sustain Agric 36(6):650–682. https://doi.org/10.1080/10440046.2011.648309

Garzón Casado B, Iniesta-Arandia I, Martín-López B; García-Llorente M, Montes C (2011) Entendiendo las relaciones naturaleza y sociedad de dos cuencas hidrográficas del sureste semiárido andaluz desde la historia socio-ecológica. VII Congreso Ibérico sobre Gestión y Planificación del Agua 'Ríos Ibéricos +10': mirando al futuro tras 10 años de DMA. (Talavera de la Reina, España, 16–19 de febrero de 2011)

Gomiero T (2015) Effects of agricultural activities on biodiversity and ecosystems: organic versus conventional farming. In: Robinson GM, Carson DA (ed) Handbook on the Globalisation of Agriculture. Northampton, MA, USA. Edward Elgar Publisher, pp 77–105

Grizzetti B, Bouraqui F, Billen G, Van Grinsven H, Cardoso AC, Thieu V, Garnier J, Curtis C, Howarth R, Johnes P (2011) Nitrogen as a threat to European water quality. In: Sutton MA, Howard CM, Erisman JW et al (eds) The European nitrogen assessment. Cambridge University Press, Cambridge, pp 379–404

Guzmán Casado GI, González de Molina M (2009) Preindustrial agriculture versus organic agriculture: the land cost of sustainability. Land Use Policy 26:502–510

Guzmán GI, González de Molina M (2015) Energy efficiency in agrarian systems from an agroecological perspective. Agroecol Sustain Food Syst 39:924–952

Guzmán GI, González de Molina M (2017) Energy in agroecosystems: a tool for assessing sustainability. CRC Press, Boca Raton (FL)

Guzmán GI, González de Molina M, Soto D, Infante-Amate J, Aguilera E (2017) Spanish agriculture from 1900 to 2008: a long-term perspective on agroecosystem energy from an agroecological approach. RegNal Environ Chang 149:335–348

Guzmán GI, Aguilera E, García-Ruiz R, Torremocha E, Soto D, Infante-Amate J, González de Molina M (2018) The agrarian metabolism as a tool for assessing agrarian sustainability, and its application to Spanish Agriculture (1960–2008). Ecol Soc 23(1), art. 2

Hecht SB, Cockburn A (2010) The fate of the forest: developers, destroyers, and defenders of the Amazon. University of Chicago Press, Chicago, Illinois

Hole DG, Perkins AJ, Wilson JD, Alexander IH, Grice PV, Evans AD (2005) Does organic farming benefit biodiversity? Biol Cons 122:113–130

Infante-Amate J, Aguilera E, Palmeri F, Guzmán G, Soto D, García-Ruiz R, González de Molina M (2018) Land embodied in Spain's biomass trade and consumption (1900–2008): historical changes, drivers and impacts. Land Use Policy 78:493–502

Karstensen J, Peters GP, Andrew RM (2013) Attribution of CO2 emissions from Brazilian deforestation to consumers between 1990 and 2010. Environ Res Lett 8(2):024005

Kastner T, ERB KH, Haberl H (2014) Rapid growth in agricultural trade: effects on global area efficiency and the role of management. Environ Res Lett 9(3). Available in: https://doi.org/10.1088/1748-9326/9/3/034015

Lassaletta L, Billen G, Romero E, Garnier J, Aguilera E (2014) How changes in diet and trade patterns have shaped the N cycle at the national scale Spain (1961–2009). Reg Environ Change 14(2):785–797

Leip A, Britz W, Weiss F, de Vries W (2011) Farm, land, and soil nitrogen budgets for agriculture in Europe calculated with CAPRI. Environ Pollut 159:3243–3253

Marull J, Tello E, Fullana N, Murray I, Jover G, Font C, Coll F, Domene E, Leoni V, DECOLLI T (2015) Long-term bio-cultural heritage: exploring the intermediate disturbance hypothesis in agroecological landscapes, (Mallorca, c. 1850–2012). Biodivers Conserv 24(13):3217–3251

Lindborg R, Eriksson O (2004) Historical landscape connectivity affects present plant species diversity. Ecology 85(7):1840–1845

MAGRAMA (Ministerio de Agricultura, Alimentación y Medio Ambiente) (2013) Anuario de Estadística Agraria 2013. Madrid

MAGRAMA (Ministerio de Agricultura, Alimentación y Medio Ambiente) (2014) Diagnóstico del sector forestal español. Análisis y Prospectiva, Serie Agrinfo/Medioambiente, nº 8. Madrid

Mayer A, Schaffartzik A, Haas W, Rojas-Sepúlveda A (2015) Patterns of global biomass trade. Implications for food sovereignty and socio-environmental conflicts. Environ Justice Organ, Liabilities Trade, 20

Nordin A, Sheppard LJ, Strengbom J, Bobbink R, Gunnarsson U, Hicks WK, Sutton MA (2011) New science on the effects of nitrogen deposition and concentrations on Natura 2000 sites (theme 3): background document. In Hicks WK; Whitfield CP, Bealey WJ, Sutton MA (eds) Nitrogen deposition and Natura 2000: science and practice in determining environmental impacts. COST Office—European Cooperation in Science and Technology, pp 115–129

Norton L, Johnson P, Joys A, Stuart R, Chamberlain D, Feber R, Firbank L, Manley W, Wolfe M, Hart B, Mathews F, Macdonald D, Fuller RJ (2009) Consequences of organic and non-organic farming practices for field, farm and landscape complexity. Agr Ecosyst Environ 129:221–227

OS (Observatorio de la Sostenibilidad) (2016) Sostenibilidad en España 2016. Cumplimiento de los Objetivos de Desarrollo Sostenible de Naciones Unidas[web page]. Available in: www.observatoriosostenibilidad.com

Pereira HM, Navarro LM, Martins IS (2012) Global Biodiversity Change: The Bad, The Good, And The Unknown. Annu Rev Environ Resour 37:25–50

Perfecto I, Vandermeer J (2010) The agroecological matrix as alternative to the land-sparing/agriculture intensification model. PNAS 107:5786–5791

Phalan B, Onial M, Balmford A, Green RE (2011) Reconciling food production and biodiversity conservation Land sharing and land sparing compared. Science 333:1289–1291

Ravishankara AR, Daniel JS, Portmann RW (2009) Nitrous oxide (N2O): the dominant ozone-depleting substance emitted in the 21st century. Science 326:123–125

Rockström J, Steffen W, Noone K, Persson Å, Chapin FS, Lambin EF, … FOLEY JA (2009) A safe operating space for humanity Nature 461(7263):472–475

Rodríguez-Martín JA, López Arias M, Grau Corbi JM (2009) Metales pesados, materia orgánica y otros parámetros de los suelos agrícolas y pastos de España. Madrid. INIA-MAGRAMA, MCI

Rodríguez-Martín JA, Álvaro-Fuentes J, Gonzalo J, Gil C, Ramos-Miras JJ, Grau Corbí JM, Boluda R (2016) Assessment of the soil organic carbon stock in Spain. Geoderma 264(part A):117–125

Romanyà J, Rovira P, Vallejo R (2007) Análisis del carbono en los suelos agrícolas de España Aspectos relevantes en relación a la reconversión a la agricultura ecológica en el ámbito mediterráneo. Ecosistemas 16(1):50–57

Rundlöf M, Nilsson H, Smith HG (2008) Interacting effects of farming practice and landscape context on bumble bees. Biol Cons 141:417–426

Sánchez-García M, Royo C, Aparicio N, Martín-Sánchez JA, Álvaro F (2013) Genetic improvement of bread wheat yield and associated traits in Spain during the 20th century. J Agric Sci 151:105–118

Schuch S, Bock J, Krause B, Wesche K, Schaefer M (2012) Long-term population trends in three grassland insect groups: a comparative analysis of 1951 and 2009. J Appl Entomol 136(5):321–331

Soto Fernández D, Infante-Amate J, Guzmán GI, Cid A, Aguilera E, García-Ruiz R, González de Molina M (2016) The social metabolism of biomass in Spain, 1900-2008: From food to feed-oriented changes in the agro-ecosystems. Ecol Econ 128:130–138

Sutton MA, Fowler D (2002) Introduction: fluxes and impacts of atmospheric ammonia to national, landscape and farm scales. Environ Pollut 119:7–8

Tello E, Galán E, Sacristán V, Cunfer G, Guzmán GI, González de Molina M, Krausmann F, Gingrich S, Padró S, Marco I, Moreno-Delgado D (2016) Opening the black box of energy throughputs in farm systems: a decomposition analysis between the energy returns to external inputs, internal biomass reuses and total inputs consumed, (the Vallès County, Catalonia, c.1860 and 1999). Ecol Econ 121:160–174

Thompson RM, Brose U, Dunne JA, Hall RO Jr, Hladyz S, Kitching RL, Martinez ND, Rantala H, Romanuk TN, Stouffer DB, Tylianakis JM (2012) Food webs: reconciling the structure and function of biodiversity. Trends Ecol Evol 27(12):689–697

Tuck SL, Winqvist C, Mota F, Ahnström J, Turnbull LA, Bengtsson J (2014) Land-use intensity and the effects of organic farming on biodiversity: a hierarchical meta-analysis. J Appl Ecol 51:746–755

van Solinge TB (2010) Deforestation crimes and conflicts in the Amazon. Crit Criminol 18(4):263–277

Visser O, Spoor M (2011) Land grabbing in post-Soviet Eurasia: the world's largest land reserves at stake. J Peasant Stud 38(2): 299–324

Vos W, Meekes H (1999) Trends in European cultural landscape development: perspectives for a sustainable future. Landsc Urban Plan 46:3–14

Wickramasinghe LP, Harris S, Jones G, Jennings NV (2004) Abundance and species richness of nocturnal insects on organic and conventional farms: effects of agricultural intensification on bat foraging. Conserv Biol 18(5):1283–1292

6

Agrarian Metabolism: An Approach to Spanish Agriculture

In the previous chapters, we studied the evolution of each fund element. We will now try to examine them in an integrated fashion to understand the metabolic dynamic and its drivers. First, however, it is necessary to know the biomass demands of the Spanish economy and society as well as how the agrarian sector fulfilled them. This requires an analysis of domestic consumption and of the role played by foreign trade in its evolution. In a country with a long export tradition like Spain, this is essential. The objective is to characterize the structure, functioning, and dynamics of Agrarian Metabolism (AM), taking into account its place in the metabolism of the Spanish economy as a whole, based on its main indicators and its behavior in relation to domestic consumption. Next, we will try to analyze the biggest drivers of the agricultural sector's metabolic activity, both on the supply side and demand side. The analysis reveals at least four differentiated periods in the biophysical evolution of Spanish agriculture.

6.1 The Agrarian Sector in the Metabolism of the Spanish Economy

What happened in Spanish agriculture cannot be disassociated from the economy as a whole and, therefore, from the final uses of the animal and vegetable biomass that society has demanded. In two recent studies, we analyzed the metabolism of materials in the Spanish economy between 1860 and 2008 (Infante-Amate et al. 2015) and examined the role played by biomass throughout the twentieth century (Soto et al. 2016). As has happened with developed countries, Spanish industrialization was accompanied by an accelerated increase in the consumption of materials, both in absolute terms and per capita, thanks mainly to the growing extraction of abiotic materials. On a global scale, it was at the end of the 1950s when the extraction of abiotic exceeded the extraction of biomass, according to the data composed with the EW-MFA methodology (Kraussman et al. 2009). In Spain, the growth of the consumption of materials was equally important but also experienced a certain lag in time. The great transformation of the country did not take place until the 1960s, both in terms of extraction and consumption of resources (Infante-Amate 2015). The per capita direct consumption of materials in Spain was almost 20% lower than the world average in 1960. In 1970, it was similar and in the year 2000, the consumption of Spanish materials already doubled the global average (Krausmann et al. 2009; Infante-Amate et al. 2015). The transition was not, as happened in England or the USA (Schandl and Schulz 2002; Gierlinger and Krausmann 2012), an unhurried process but a swift one. Its patterns match well with countries called latecomers, which went through industrialization in a much faster way than firstcomers did, also in biophysical terms (Krausmann et al. 2009). In this sense, the first phase of the metabolic transition was a "weak transition", characterized by significant qualitative changes, though without any notable transformation in quantitative terms.

From the point of view of the Domestic Extraction (DE) of materials, it went from 58.9 million tons (Mt) from 1860 to 422.6 Mt in 2010, multiplying by 7.2 (Graph 6.1). The DE was multiplied by 1.4 between 1860 and 1950, while between 1950 and 2008 it did so by 5, mainly due to the growing weight of abiotic materials. The biomass remained relatively stable, with extreme values of 49.5 Mt (1900) and 68.5 Mt (2008), while the inorganic materials grew significantly: in 1860, the DE of these materials barely amounted to 1.2 Mt, in 1950 to 23.7 and in 2000, the year in which the historical maximum was reached, it was 425.7 Mt. From the middle of the XIX century and until the first decades of the XX century most of the extraction of abiotic products was due to metallic materials, mainly to iron ore that was exported to industrialized countries, especially to England. Still in 1910, the iron exported to that country accounted for almost 50% of the national extraction. Until Franco's autarky, when there was a blackout of international trade in the country, Spain was, therefore, a net exporter of resources, mainly minerals destined for the industrialized countries. Since 1920, energy products replaced metallic minerals in extractive importance. Its extraction continued to increase until 1990 when the peak was reached with 38.6 Mt. Since then it has continuously decreased to 8.3 Mt in 2010, a figure lower than that of 1940. From the sixties, the prominence has corresponded to the quarry products that have grown continuously until now accounting for almost 90% of the total DE of the country. They passed between 1860 and 2010 from 0.06 Mt to 342.1 Mt. These materials reflect the strong growth experienced in the last decades of the twentieth century, and until the economic-financial crisis, by the Spanish real estate sector, and that these goods were used in construction and to a lesser extent in the country's own industrialization (Graph 6.1).

In any case, until the beginning of the sixties the extraction of abiotic materials did not surpass the biotic ones. Until that time, the metabolism of the Spanish economy remained essentially "organic", although the relative share of biomass in the overall metabolism of the Spanish economy declined steadily throughout the twentieth century. In 1860, 97.9% of the extraction was biomass, while in 2008 it only represents 16%, a pattern shared with the developed countries. Paradoxically, the biomass DE has not stopped growing, as we have seen in Chap. 2, stimulated by a growing and

Graph 6.1 DE in millions of tons (**a**), DE in percentage (**b**), Imports (positive), exports (negative) and total PTB (**c**) and PTB per inhabitant of biotic, abiotic and total, in tons per inhabitant and year (**d**). *Source* Infante-Amate et al. 2015

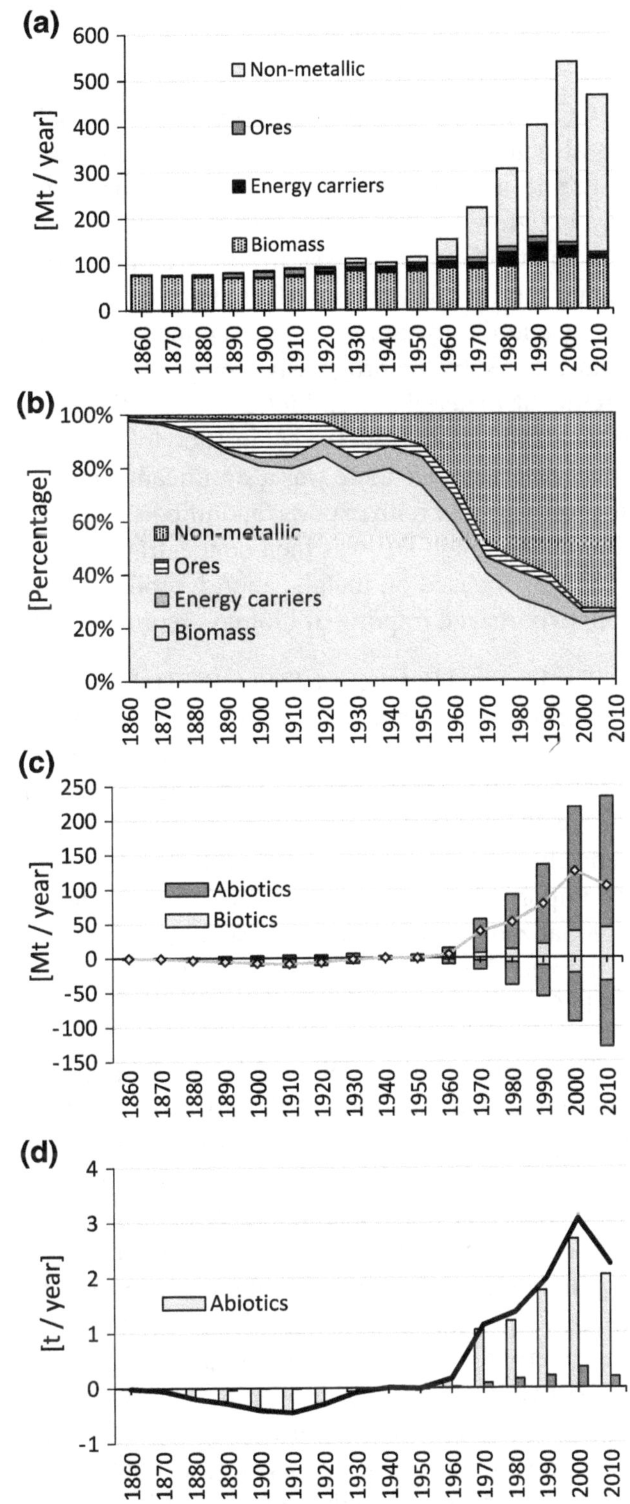

specialized demand. This evolution is coincident with the evolution of the DE on a planetary scale (Krausmann et al. 2009).

On the other hand, the integration of the Spanish economy in international markets has also gone through two periods that should be retained. A first period comprises from mid-nineteenth century to mid-twentieth, when trade flows were little significant but growing (with stoppages due to historical junctures such as the First World War and the Great Depression) to suffer a kind of widespread commercial blackout during the First Francoism. This was due to the autarkic policy that cut the incipient economic dynamism of the country. The second begins in the sixties and is characterized by the progressive integration of the Spanish economy in international circuits of materials, both biotic and abiotic, to a large extent depending on them. Total exports went from 0.5 Mt in 1860 to 132.6 Mt in 2010, and imports from 0.4 to 245.9 Mt. Here also there was a significant change in the composition of trade flows: in 1860 a 20% of both exports, and imports were biotic and the remaining 80% were abiotic. Although DE was then concentrated in agricultural products, the country's trade was focused on metals, energy products, and other minerals. In recent years, both exports and imports of biomass represent only between 8–14% of total sales.

The Physical Balance of Trade (PTB) shows the changing behavior of the Spanish economy. As we have said, Spain was a net exporter of materials until 1950 and, since then, it has been a net importer in increasing magnitudes. During the second half of the nineteenth century and much of the first half of the twentieth century, we saw that most of the national production of iron was sold to other countries. The trend changed in the second half of the twentieth century. Exports grew substantially, but imports did so at a much higher rate, both biotic and abiotic. The transition to a globalized economy became, rather quickly: in 1950 the PTB was 0.2 Mt, while in the year 2000 it was 126.5 Mt, going from 0.01 to 3.1 t/inhab/year. (Graph 6.3, Table 6.1).

The changes in the DE and in the PTB explain the behavior of the domestic consumption of materials of the Spanish society (Graph 6.2), which like all of the developed West has experienced a very significant increase, especially in the second half of the twentieth century. This pattern corresponds to the so-called Great Acceleration of the consumption of energy and materials (Constanza et al. 2007). Between 1860 and 1950 the DC grew moderately, going from 58.7 to 85.5 Mt. The population increase made that there was even a fall in consumption by inhabitant: the DMC per capita went from 3.8 to 3.1 t/inhab/year. During this period, most of the materials mobilized were logically of biotic origin. But during the second half of the twentieth century, consumption multiplied, putting pressure on the DE and causing resources from other countries to flow into Spain, that is, the PTB became more and more positive. In this way, the DC went from 85.5 Mt in 1950 to 619.8 Mt in 2000 and from 3.1 t/inhab/year to 15.2. Due to the economic crisis, consumption fell to 11.6 t/inhab/year in 2008 (Table 6.1, Graph 6.3).

Table 6.1 Indicators of the metabolic profile in Spain

		1860	1950	2000
Population	Millions	15.6	28.0	40.7
GPD/capita	000$ 1990	1.2	2.3	171.5
DEpc	t/capita	3.8	3.0	12.1
Biotic		3.7	2.2	1.7
Abiotic		0.1	0.8	10.5
PTB	t/capita	−0.0	0.0	3.1
Biotic		−0.0	0.0	0.4
Abiotic		−0.0	0.0	2.7
DMC	t/capita	3.8	3.1	15.2
Biotic		3.7	2.2	2.1
Abiotic		0.1	0.9	13.1
Biomass/total DMC	%	98.1	72.1	13.6

Source Infante-Amate et al. 2015

6.2 Foreign Trade and Domestic Consumption of Biomass

What role has the agricultural sector played in the dynamics that we have just seen? As we have seen, the growth of consumption has been based on abiotic materials, relegating biomass to a secondary place. The consumption of biomass per capita has decreased in line with this, but this has not meant a reduction of the biomass consumed in absolute terms, as we have seen, but quite the opposite. The demand for biomass has also grown throughout the twentieth century for various reasons that we will have occasion to analyze with particular intensity in its second part. Next, we will focus on the evolution and composition of the biomass DC and to what extent the supply came from the DE or foreign trade.

Against the traditional belief, Spain has not been an agro-exporting country if viewed from the biophysical point of view since it has received more biotic products than it has exported. To a large extent, this is due to the fact that the main exports of Spain were composed of fruit and vegetable products, with a high water content and therefore with a much lower dry weight than abiotic materials or imported biomass. Even so, the percentage of biomass traded in international markets was quite small compared to the total biomass ED. Despite this, the importance of foreign trade in biomass has increased significantly. Total imports went from 0.8 to 31.9 Mt between 1900 and 2008. Exports from 0.5 to 12.7 Mt. However, this growth has not been continuous throughout the century. Until the 1960s the weight of foreign trade was low, even contracted after 1933, but from 1970 there was an accelerated growth that has not yet stopped. There has also been a significant change in its composition: until the 1960s most of the biomass exports were concentrated in the category of human food and tended to diversify as of that date. In 2008, the main export categories were wood and firewood (especially wood), followed by human food and animal feed.

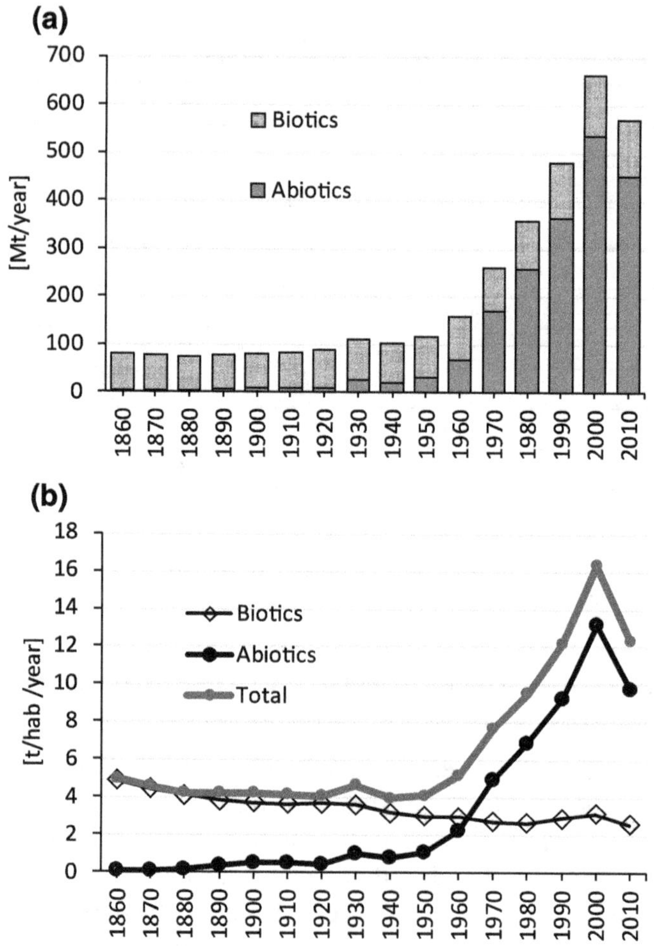

Graph 6.2 Total consumption, millions of tons (**a**) and consumption per inhabitant, in tons per inhabitant and year (**b**)

With regard to imports until 1933, about half were wood, while the main category between 1940 and 1960 was that of human food. Since 1970, imports of animal feed increased to almost half of the total imported biomass (42%) (Table 6.2).

The PTB shows that effectively and contrary to what the monetary values say, Spain has been a net importer of biomass throughout the period of our study. Only in some years between 1900 and 1970 has it exported more than it has imported into the food category with human destiny. Likewise, the weight of the PTB has been very insignificant until the 1960s, in such a way that the DC evolved in parallel with the DE between 1900 and 1960 (Graph 6.1b). The percentage of PTB on the DC oscillated between 0.9 and 2.4% during those years (with an extreme value of 0.1% in 1950). However, since 1970, the role of foreign trade in biomass DC has had an increasing importance, from 6.2% in that date to 22.2% in 2008. The greatest commercial integration of Spain in the last 40 years explains that the DC of biomass has grown at a higher rate (74%) than the DE (38%) between 1900 and 2008, from

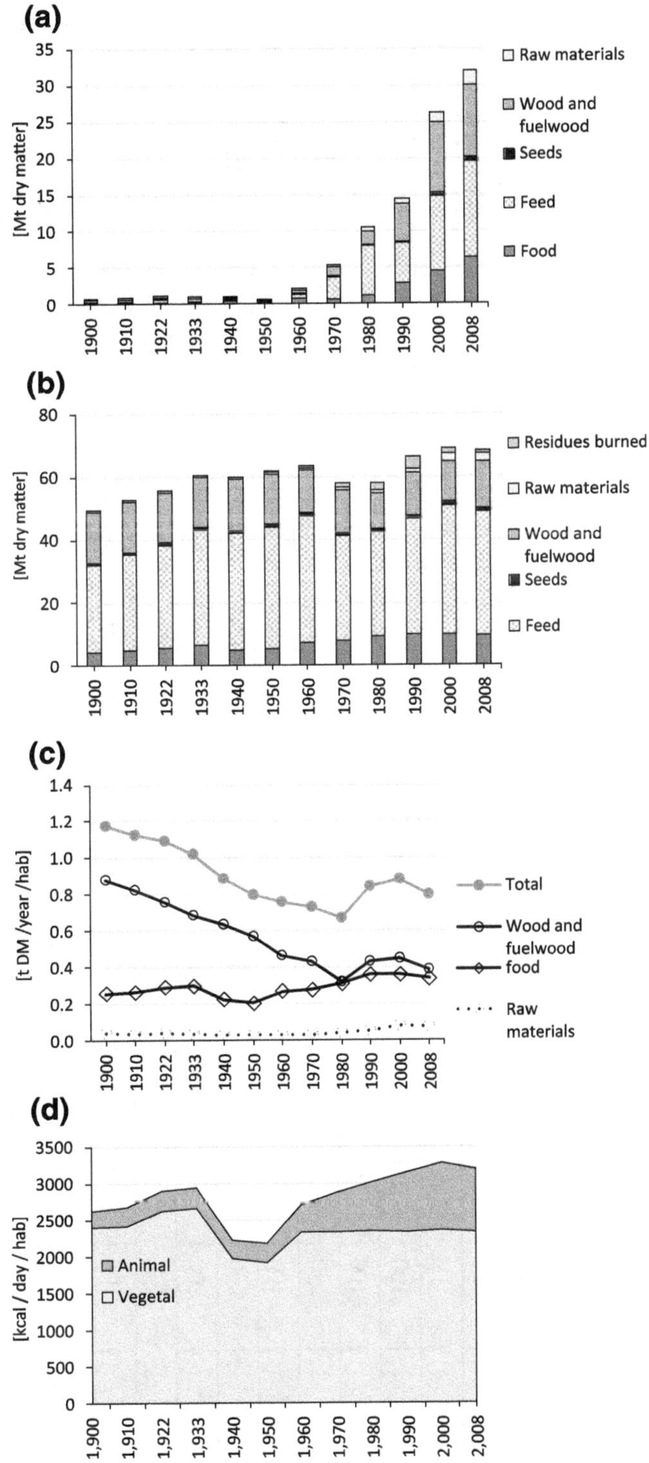

Graph 6.3 Biomass trade and consumption. (**a**) PTB, Mt of dry matter (**b**) DMC, Mt of dry matter (**c**) Consumption per capita, Mt of dry matter (**d**) Net food balance, kilocalories per inhabitant per year. *Source* Soto et al. (2016)

Table 6.2 Physical trade balance, Kt dry matter

	1900	1910	1922	1933	1940	1950	1960	1970	1980	1990	2000	2008
Imports												
Food	100	100	200	100	500	200	700	600	1100	2800	4400	6300
Feed	100	200	300	100	200	100	600	3000	6900	5600	10,400	13,300
Seeds	0	0	0	0	100	0	100	0	100	100	400	600
Wood and fuel wood	400	400	300	500	100	100	500	1300	1900	5200	9600	9800
Raw materials	200	200	300	300	100	100	300	300	500	600	1400	2000
Total	800	800	1200	1000	900	500	2200	5300	10,500	14,300	26,200	31,900
Exports												
Food	100	200	200	300	100	400	500	900	1100	1900	3700	4000
Feed	0	0	100	0	0	0	100	200	300	1400	1400	1900
Seeds	0	0	0	0	0	0	0	0	0	100	100	100
Wood and fuel Wood	100	100	100	100	0	0	100	300	1100	1800	4400	6200
Raw materials	100	100	100	0	0	0	0	100	300	300	400	500
Total	300	400	500	400	200	500	600	1500	2800	5400	9900	12,700
Physical trade balance												
Total	500	500	700	600	700	0	1600	3800	7600	8900	16,300	19,300

Source Soto et al. (2016)

50.0 to 86.7 Mt of dry matter. In other words, biomass consumption in Spain depended on the DE until the late 1960s. From that moment, it began to depend increasingly on imports. The Spanish consumption of biomass today represents a considerable percentage, 27.6%, of the NPP, seven points more than in 1900, but a part is actually extracted in other countries, given that the DE is a 21.8% of NPP produced by Spanish agroecosystems. The difference between the DE and the DC, which did not exist in 1900, stood at 18.3 Mt of dry matter in 2008, the highest of the period. This gap means that international trade contributes a fifth of domestic consumption, as we have seen (Graph 6.3).

6.3 The Main Indicators of Agrarian Metabolism

How did the agricultural sector respond to the consumption increase described in the section above? Despite international trade's increasing role, the bulk of the required biomass continued to be supplied by Spanish agroecosystems. To meet growing demand, it was necessary to expand Spanish agricultural metabolism by almost 50% (see Table 6.3). This was achieved not only by increasing net primary productivity but also by raising the share appropriated by society as seen in Chap. 2, as well as net biomass imports. Maximum relative extraction took place in the 1950s, coinciding with the end of traditional agriculture when production difficulties favored by Francoism encouraged maximum appropriation of biomass, in a context of falling yields. However, in absolute terms, the maximum volume of extracted biomass was reached in the year 2000, when almost seventy million tons of dry matter was obtained mostly from crops. Therefore, greater production efforts essentially concentrated in cultivated lands and in certain crops or livestock specialties. Indeed, DE growth was driven by cultivated land intensifications, increasing by 236% with respect to 1900. In contrast, the abandoning of pastures or their underusage, together with conservation and reforestation, with scarce biomass energy use, explain DE drops, respectively, by 46 and 17% in these lands since 1900. The NPP_{act} showed an opposite trend concerning the three major land uses from the perspective of human biomass appropriation: DE decreased by 17% per hectare in forests and by 81% in pastures, while it increased for crops. Such a remarkable growth in agricultural production can be explained not only by cultivated lands' productivity growth, but also by changes in biomass use patterns extracted from cropland, as seen in Chap. 2: productivity multiplied threefold for the main crops, but only grew by 40% relating to residues.

There has also been significant changes in the final use of DE. Biomass aimed at human food consumption increased from 9 to 14%, biomass aimed at raw materials went from 1 to 4% and biomass aimed at animal feed went from 56 to 57.5%, reaching two-thirds in the 1960s. Since that decade, around 40 million tons of dry matter per year have been used to feed livestock, despite the fact that animal traction is no longer used. This development can be explained by the Spanish agricultural sector's increasing orientations towards livestock and was especially visible in the case of cereals. The role of cereals as feed is heavier today than in 1900 when it

Table 6.3 Evolution of the indicators of Spanish agriculture metabolism, 1900–2008 in Mt of dry matter

	1900	1933	1950	1970	1990	2008
Imports (input)	0.03	0.3	0.4	2.8	4.3	4.7
Actual NPP	244.6	255.8	257.9	283.3	292.6	314.2
Unharvested biomass	183.4	184.3	184.2	203.5	202.9	222.1
Accumulated biomass	11.7	10.8	12	21.7	23.4	23.7
Domestic extraction (DE)	49.5	60.7	61.7	58.1	66.3	68.5
Reused biomass	28.4	37.6	39.9	35.8	41.3	41.9
Recycled biomass (Unharvested biomass + Reused biomass)	211.7	221.8	224.1	239.4	244.2	263.9
Socialized biomass (export)	21.5	23.7	22.4	23.7	27.7	30.6
Socialized vegetal Biomass	21.2	23.1	21.8	22.2	24.9	26.6
Socialized animal Biomass	0.4	0.6	0.6	1.4	2.8	4
Domestic Consumption (DC) (DE + Import-Export)	28.4	37.9	40.3	38.7	45.6	46.6
TMR (DE + I)	49.6	61	62.1	60.9	70.6	73.2
Metabolic profile (per capita)	1.5	1.5	1.4	1.1	1.2	1
TMR/per capita	1.1	1.3	1.3	1.3	1.5	1.6
1900 = 100						
Imports (input)	100	974	1168	8279	12.674	13.832
NPPreal	100	105	105	116	120	128
Unharvested biomass	100	100	100	111	111	121
Accumulated biomass	100	93	103	186	201	202
Domestic extraction (DE)	100	123	125	117	134	138
Reused biomass	100	132	140	126	146	148
Recycled biomass (Unharvested biomass + Reused biomass)	100	105	106	113	115	125
Socialized biomass (export)	100	110	104	110	129	142
Socialized vegetal Biomass	100	109	103	105	118	126
Socialized animal Biomass	100	159	160	406	800	1148
CD (DE + Import-Export)	100	133	142	136	161	164
TMR (DE + I)	100	123	125	123	142	148
Metabolic profile (per capita)	100	99	94	75	76	66
TMR/ha SAU	100	123	125	122	141	148

Source author's compilation

was still the basis of food. Biomass aimed at fuel followed an opposite trend: it represented 32% of total extracted biomass in 1900 and 21% in 2008, replaced by gas and electricity in households. As commented earlier, significant wastage was generated by the destruction of crop residues, which reached 3.6 Mt, i.e., 5.5% of DE in the 1990s. This squandering continues today.

Livestock has undergone a fundamental change. Livestock used to be organic-based and have close ties to the land, but it became industrial. Animals were mostly housed and landless. Therefore, livestock became much more dependent on feed supply and industrial inputs which mostly came from international trade. Total livestock increased sharply due to this transformation, from 54 million heads in 1900 to 838 million heads in 2008. Livestock composition, the destination of its products and services as well as its management and feeding thus significantly changed. In terms of liveweight, livestock size multiplied by 2.4, from 2.8 million tons at the beginning of the twentieth century (with peaks of around 3.4 million in the 1930s), to almost 7 million tons around 2008. These figures reflect a growing livestock specialization in poultry, pigs, and, to a lesser extent, cattle, mainly oriented towards the production of meat and dairy products. An illustration is the spectacular growth of Socialized Animal Biomass (SAB), which increased from 0.4 million tons of dry matter to more than 4 million, i.e., a multiplication factor of 11.5. This remarkable growth reflects the specialization in intensive livestock farming, which has generated big environmental impacts as described earlier, especially the increase of GHG emissions or the alteration of the nutrient flows caused by intensive livestock (Lassaletta et al. 2014).

In short, agroecosystems have undergone a profound change. They passed from agrosilvopastoral integration, where livestock and forestry agricultural activity was closely linked to the territory, to a growing segregation of land uses, causing links with the territory to break and the progressive substitution of internal flows by flows external to the sector, a significant portion of them coming from abroad (Infante-Amate et al. 2018). A large part of the pastures was abandoned or clearly underutilized and forested areas have grown either for commercial exploitation or for "conservation". Traditional uses of these areas have diminished substantially. We can say that Spanish agriculture has specialized in a group of crops (fruit and vegetable production, olive groves) and in intensive livestock. Other agricultural activities have been abandoned or are being underutilized.

The behavior of Accumulated Biomass (AB) reflects this development. Its contribution to the NPPreal has been the largest of all, from 11.7 million tons of dry matter in 1900 to 23.7 million in 2008, more than doubling and occupying an increasing percentage of the NPPact, from 4.8 to 7.5%. Accumulated biomass in the aerial part of forests has been mainly responsible for this increase, multiplying by almost 20. This was due on the one hand to a threefold increase in forested areas, and on the other, because the use of firewood in Spanish forests was disappearing as the household energy transition progressed. The implementation of public conservation policies and the declaration of protected natural spaces also played a part. The case of Spain seems to fit with the so-called "Forest Transition" where forest areas grow at the cost of farmland. Academic literature associates this phenomenon with so-called *land sparing*, where one part of the territory is used so intensively that other parts, especially those that cannot achieve very high yields, can be dedicated to forestry purposes. But it may also be due, as we shall see later, to soil imports from foreign countries.

Spanish society directly or indirectly appropriated a fifth (and currently almost a quarter) of all NPP_{act}. However, its interventions in the dynamics of agroecosystems have become more visible than at the beginning of our study period. Decisions on land uses have ended up directly affecting the rest of the non-appropriated biomass. Unharvested biomass (UB) grew by 21% in absolute terms throughout the period, but in relative terms, it was the type of biomass that contributed the least to the growth of total NPP, clearly below the average. Its evolution somewhat reflects the deterioration of the land fund element: in relative terms, its importance fell from 75% of the NPP_{act} in 1900 to 69.3% in 2008. This explains why Recycled Biomass (RcB) in agroecosystems grew generally less than DE, although the RcB increased. This drop was bigger in pastures and croplands.

The territorial imbalance described above and the breakdown of the internal loops of agroecosystems explain the importance that external inputs have acquired for the functioning of Agrarian Metabolism. As shown in Table 6.3, imports, i.e., materials imported from outside the agricultural sector, increased exponentially, from 0.1% of DC of materials, an insignificant figure, to 9.4% in 2008. Inputs from outside the agricultural sector increased 138 times in dry weight only; the weight was much bigger when taking into account embodied energy. Biggest increases in DE were precisely associated with phases of greater use and importance of external inputs. They currently represent 6.4% of the Total Material Requirement (TMR) of Spanish agriculture's metabolism. In fact, the agricultural sector's materials DC has increased by 64%, which is higher than DE growth; in turn, the TMR has grown by 48%, clearly showing today's comparatively higher cost of metabolic activity compared to that of the early twentieth century. However, the SVB, i.e., the plant biomass transferred to society, increased by only 26%, from 21 to 26 million tons of dry matter. This data clearly shows that the Spanish agricultural sector has specialized in livestock, responsible for the increase of both DC and TMR of agricultural metabolism, driven mainly by Reused Biomass (RB) and by feed imports.

The metabolic profile of the agricultural sector has declined sharply since 1900. In that year, the size of Spain's agricultural metabolism was 1.5 t of dry matter per capita. By 2008, it had fallen to 1 ton. There are two explanations for this drop: the growth of the Spanish population, which had multiplied by 2.5 by 2008, and, as we saw earlier, because of the partial outsourcing of the metabolic effort by importing energy and materials from outside the sector and even from outside the country. This behavior is common to other industrialized countries, despite the ever-increasing land costs of animal feed (González de Molina et al. 2017; InfanteAmate et al. 2018). Despite this, the metabolism of Spanish agriculture has increased its pressure on agroecosystems since the TMR/ha has risen by almost 50%, due to the intensification and specialization process.

Table 6.4 reflects the process of production intensification since 1900. Two distinct periods can be distinguished. The first period corresponds to the first half of the twentieth century, in which extraction of biomass intensified for all land uses, whether

Table 6.4 Evolution of productivity per hectare according to land uses (t/ha)

	1900	1933	1950	1970	1990	2008
DE/ha	1.1	1.3	1.3	1.2	1.4	1.5
DE crops/ha	0.5	0.7	0.6	0.9	1.2	1.1
DE primary crops/ha	0.2	0.4	0.3	0.5	0.8	0.8
DE pasture/ha of pasture	0.7	1.0	1.3	0.4	0.3	0.5
DE wood and firewood/ha of woodland	0.9	0.9	0.9	0.5	0.5	0.5
1900 = 100						
DE/ha	100	123	125	116	133	138
DE crops/ha	100	133	117	180	219	212
DE primary crops/ha	100	149	124	220	316	336
DE pasture/ha of pasture	100	144	186	57	48	69
DE wood and firewood/ha of woodland	100	110	103	61	56	60

Source author's compilation

for agricultural, forestry or livestock uses. This intensification was a logical response as the sector was still linked to the territory and depended on it for animal and human food as well as for providing raw materials to industry. It also depended on it to supply the bulk of domestic energy, i.e., to satisfy the basic needs of a population that had grown from 18 to 28 million inhabitants over the period. Except for some fertilizers that were already produced by the chemical industry based on non-renewable sources, these needs were largely met using biomass extracted from the territory; the economy was still of a basic nature and essentially organic, within a metabolic arrangement under industrial transition. The second period corresponds to developments from the 1950s to 2008: forest and pasture DE decreased substantially, while biomass extraction became more intense in croplands, especially in the main parts of the crops. This is the logical consequence of the energy transition and the use of disproportionate imported livestock feed in relation to the land's capacity to sustain it.

Consequently, Spanish agroecosystems have undergone a significant process of production intensification and specialization. The specialization has been twofold. On the one hand, some of the country's autonomous communities, which traditionally concentrated crops and occupied the territory in a relatively balanced way, currently display a higher degree of concentration of some crops and in a less balanced way. On the other hand, crops are grown now less dispersed and more concentrated in the autonomous communities than in the past, when they were more evenly balanced over the territory. The Gini index on Final Agricultural Production of seven types of uses (vegetables, fruits, wine, oil, eggs, meat, milk) shows this. It went from 0.285 in 1959 to 0.383 in 2000. That is, it grew by 34.4%, reflecting a significant increase of production specialization. In terms of uses, we observe that all, without exception, show a strong degree of territorial concentration based on the Gini index of distribution over Spanish regions. The case of vegetables stands out. Vegetables are a star product of intensification and external commercialization and its degree of specialization has grown by almost 70%.

During the first half of the twentieth century, this intensification involved practically all areas and uses. In the second half of the century, the Spanish agrarian sector specialized in a group of intensively managed crops and intensive livestock, due to its dissociation from the land, among other factors. This fundamental change in the agrarian sector required, as we have seen, the injection of large quantities of external inputs, the scope of which is not properly reflected in dry weight ton measurements. When analyzing Spanish agriculture's metabolic activity from an energy perspective, more obvious conclusions can be drawn. The energy efficiency of agricultural production has declined considerably as we saw in Chap. 5.

6.4 The Pace of Intensification and Specialization ($I + S$)

The intensification process, however, did not unfold steadily nor was it boosted by the same drivers over the study period. We used decomposition analysis to better differentiate the phases underwent by Spanish agriculture since 1900. The decomposition analysis method is based on the proposal by Ang (2005) for additive decompositions. It allows to estimate the variation over time of a given variable (it is generally used to study changes in energy consumption) and then quantifies the weight on such variation of the variation in other types of variables generally expressed in other units (GDP, population, efficiency, etc.). The final result shows the effect of these variables expressed in the measure unit of the variable that is under analysis. In our case, we wished to analyze the change in the DE of crops at the state level, measured in tons of dry matter. For this, we estimated the variation at two different moments in time. We assumed that changes in DE change can be explained, first, by changes in the agricultural area: the larger the surface area, the larger the extraction, and vice versa (the cultivated area is expressed as A in Eq. 1). However, it is possible that Extraction per area unit changes over time. We capture this effect by incorporating intensification (I), which is estimated as the inputs (measured in embodied energy) per hectare. Finally, we incorporated efficiency (E) in the use of these inputs. It is possible that more inputs be added but that the response in the form of biomass production is ever smaller. We synthesize this equality in Eq. 1.

$$\mathrm{DE} = A * I * E \tag{1}$$

Thus, the change in DE between year T and year 0 is equivalent to the sum of the changes in the variables considered:

$$\Delta\mathrm{DE}_{\mathrm{tot}} = \mathrm{DE}^{T} - \mathrm{DE}^{0} = \Delta A + \Delta I + \Delta E$$

Graph 6.4 shows the result of the analysis throughout the period in the right column. It confirms that the use of inputs was the main factor of increase in DE of arable

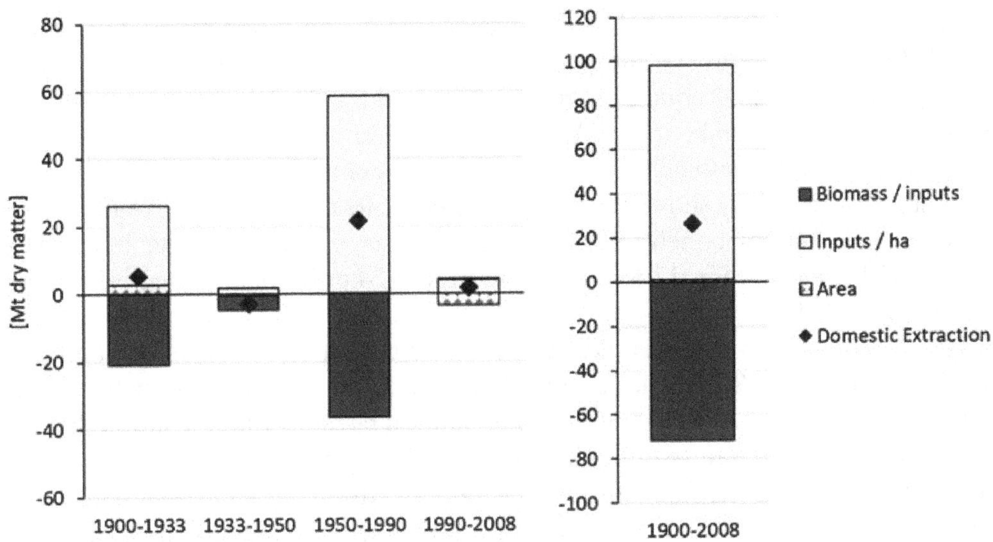

Graph 6.4 Analysis of the decomposition of DE. *Source* author's compilation

lands. It also shows that biomass input production efficiency considerably reduced DE, reflecting its progressive decrease, as observed in EROI behavior. Variations in croplands ultimately scarcely explain DE behavior. The analysis also allows to distinguish four different periods. The first period corresponds to the first third of the twentieth century, when DE increased moderately, due not so much to the cultivation of new lands, despite the incorporation of almost four million hectares, but to the use of industrial inputs, whose volume in terms of energy multiplied by 6.6. Animal traction increased due to the growth of cultivated areas and the increase in work associated with more intensive management; but it was the use of fertilizers that grew the most and, to a lesser extent, that of new irrigation systems. In previous chapters, we highlighted, in fact, the key role of chemical fertilizers in agricultural growth during the first third of the twentieth century, especially phosphate fertilizers, the expansion of irrigated land and improvements in their water provision. However, efficiency per input decreased and this had a negative effect on DE.

The second period covers the years of Early Francoism, characterized by the fall in crop DE. The analysis mainly attributes efficiency loss to extracted biomass reductions. Although the use of inputs declined significantly in the forties, at the beginning of the fifties, the expansion of irrigation and the increase of energy invested in traction, still mostly animal-based, boosted inputs per hectare. However, inefficiencies caused by the need to allocate more biomass to animal feed, i.e., raising the amount of reused biomass, brought down efficiency levels. The third period corresponds to the forty-year period between the fifties and nineties. The results of the analysis clearly show the effects of industrialization on Spanish agriculture: a very sharp growth of crop DE essentially due to the use of external inputs that multiplied by almost 11 between both dates. Among these inputs, industrial inputs grew the most, to a similar extent, while non-industrial inputs grew little, because labor reduction partially offset the increase in the use of feed, which multiplied by 30 during that period. Among the industrial inputs, mechanization, irrigation and crop protection inputs grew the most, although the role of chemical fertilizers was decisive during the first decades. As can be inferred from the land-use figures described in Chap. 2, croplands remained relatively stable and therefore, can hardly explain DE behavior. The analysis also clearly shows that DE increased while the efficiency of agricultural production (biomass obtained per TJ) was significantly reduced. As seen in Chap. 5, the EFEROI (External Final EROI) shrunk, from 9.24 units of extracted biomass in 1950 to 1.56 in 1990.

The decomposition analysis indicates a fourth period between the beginning of the nineties until 2008, of scant crops DE growth. The croplands went from 20.1 million hectares to 17.2, losing almost three million hectares for cultivation that did not, as in the past, go into swelling larger farms. This partly explains that the use of inputs continued to grow, especially animal feed, and livestock activity with it, whose impact is not fully reflected in crops DE. Decline in labor is particularly striking. In contrast, industrial inputs grew very little and sometimes even fell, as in the case of chemical fertilizers. Consequently, the stabilization and even the relative decrease of crop DE are related to livestock specialization and the transfer of the biomass necessary to sustain it to other territories.

In short, during the twentieth century, the functionality of agrarian metabolism changed substantially in the Spanish economy as a whole: from supplying an essential part of energy and materials, it became a biomass supplier for human or animal food and raw materials for industry. This brought about an increase in the demand for biomass in absolute terms leading to sustained production intensification and specialization over time. At first, the process took place more or less over the whole territory and for all uses. Later, it concentrated in croplands, especially those with better access to water and fertile soils. Since the end of the nineties, the twofold nature of the agrarian sector has become more pronounced. Currently, large underutilized or neglected territories, especially in the interior pasture lands and drylands, coexist with croplands or highly specialized landless livestock activities in which production intensification continues. It is because some of the production pressure is transferred to third country territories through international trade that increasing domestic biomass consumption is compatible with the abandonment or underutilization of a portion of Spanish agroecosystems, thus deepening its double-sided nature. This explains DE stagnation as DC continues to grow. How did this situation come about? We have already seen that the process was driven by intensification and production specialization (hereon $I + S$). But what were the drivers or underlying forces of intensification and specialization themselves?

6.5 The Drivers of $I + S$

Based on the previous sections, the reasons must be sought both within the agricultural sector, that is, on the supply side, as well as outside the sector, taking into account industry and agri-food sector biomass demand. This is all the more justified if we consider that, as we have just seen, for a long time, DE and DC did not co-evolve. Next, we examine the factors that explain $I + S$ processes, depending on whether they originated inside or outside the agricultural sector.

6.5.1 Supply Side Drivers of $I + S$

In Chap. 1, we presented the main hypothesis of this book, i.e., that the adoption of $I + S$ strategies was due to the difficulties of many agrarian households to maintain and reproduce themselves socially. Chapter 4 was devoted to the study of monetary flows, and we saw that $I + S$ strategies were common among small farmers but for a number of reasons eventually spread to all types of farms. Given the limited data at our disposal, we can only indirectly confirm these hypotheses for the first third of the twentieth century. Data from official land registry records (*Avance Catastral*), collected and studied by Carrión, confirm the overwhelming weight of small properties that were unable to reach the minimum levels of GVA required to cover the country's average consumption basket. This explains the need to maximize agricultural income, either by specializing in the production of crops with better market outlets, or by intensifying the production of subsistence crops. We also know that salaried workers, threatened by seasonal unemployment and the lack of alternative employment outside the sector, developed strategies to strengthen their position within the labor market through unions and social protest. The strategy resulted in wage increases that eventually affected the rest of the farmers who were relatively dependent on the external workforce. Given the difficulties in replacing human labor with machines, the most feasible strategy was to apply chemical fertilizers, increasing yields per unit area and compensating for the rise in labor costs. In this way, the $I + S$ strategy was adopted by practically all farmers. In fact, this explanation is supported by our decomposition analysis performed on the drivers of DE increase, in the absence of more precise data on agricultural macromagnitudes.

On the other hand, our hypothesis can be better verified for the second half of the twentieth century, since we dispose of the sector's accounts and other useful statistical information. The agrarian sector's intensification was measured in different ways, either through indicators such as DE/ha or DE/ha of cropland. Nevertheless, we advanced a hypothesis on the drivers of $I + S$ attributing a decisive weight to agrarian income and its capacity to cover average agricultural household expenditure. Therefore, it would be appropriate to approach the weight of the drivers of intensification in monetary terms rather than in biophysical terms. This can be achieved by using a proxy variable as reliable as possible. Intermediate consumption (IC), in monetary terms and reflected in the sector's accounts, effectively expresses the costs of intensification since its beginnings: as we have seen, ever since external inputs were used, the greater the use of inputs, the more intensely the agroecosystems were managed.

Let us remember that, as we saw in Chap. 4, the amount of agricultural income depended on the magnitude of intermediate consumption. Furthermore, in Chap. 5, we related the increasing use of inputs, fossil fuels in particular, to agroecosystem environmental damage over the last decade that endangered the reproduction of the most important fund after the agrarian population: the land and its capacity to produce biomass sustainably. As we saw, the use of inputs increased dramatically since the 1950s multiplying by 11 in terms of energy; this increase persisted throughout the study period, albeit at different rates. Except in the years prior to the current economic-financial crisis, farmers had to devote an increasing share of the production value to face the costs, significantly affecting income. These expenses reflect how $I + S$ efforts were conducted to compensate for falling paid prices and achieve sufficient income to cover average Spanish household expenditure. Agrarian household expenditure, apart from in the expected case of "Entrepreneurs with workers", has always found itself below the national average since 1958. That is, most of the agrarian population's access to goods and services has fallen below the rest of the country's levels of access, especially that of agricultural laborers and small landowners with no employees.

IC has therefore contributed to making agrarian activity less viable. So much so, that it has had to resort to other mechanisms to compensate income decline: on the one hand, public subsidies, at first from the State and then from the EU; on the other hand, the mobilization of professional agricultural organizations and trying to raise the prices paid or increase the amount of aid. On an individual basis, farmers have tried to bring down labor costs, by reducing employment, and when this was not possible, by leaving the sector. In many cases, it was possible to maintain levels of income per employee and even increase them, at least until early this century. Labor costs dropped practically by half, representing 60.4% of costs in 1964–5 and 31.9% in 2008. To discover the weight of the main drivers of intermediate cost behaviors, we performed a decomposition exercise presented below.

We again followed the decomposition proposal of Ang (2005) as described above. In this case, we analyzed intermediate consumption variation, measured in euros of the year 2000, between 1962 and 2008, distinguishing other intermediate periods. The variables considered in the model are described as follows: first, the number of farms or agricultural holdings (expressed as F in Eq. 2) reflects the evolution of the sector itself and provides information on the abandonment of activity, given that the number of holdings logically influences the total amount of inputs used; second, hectares per holding (ha/F) measures their size and captures the increase in the size of agricultural holdings that has taken place as a result of the drop in their number and the aim of reaching a threshold of minimum profitability by increasing the size. This has had consequences on input use since it has usually led to the replacement of labor by machines and chemical means, raising productivity; third, income per hectare (I/ha) shows the profitability of each surface area unit and captures the behavior of farmers who have tried to increase income by producing more and, therefore, using more inputs. Finally, the fourth variable refers to the intermediate consumption share of total agricultural income (IC/I). It reflects the vicious circle produced by farmers' intentions of offsetting IC increases by producing more thus paradoxically being forced to use more IC (Graph 6.5).

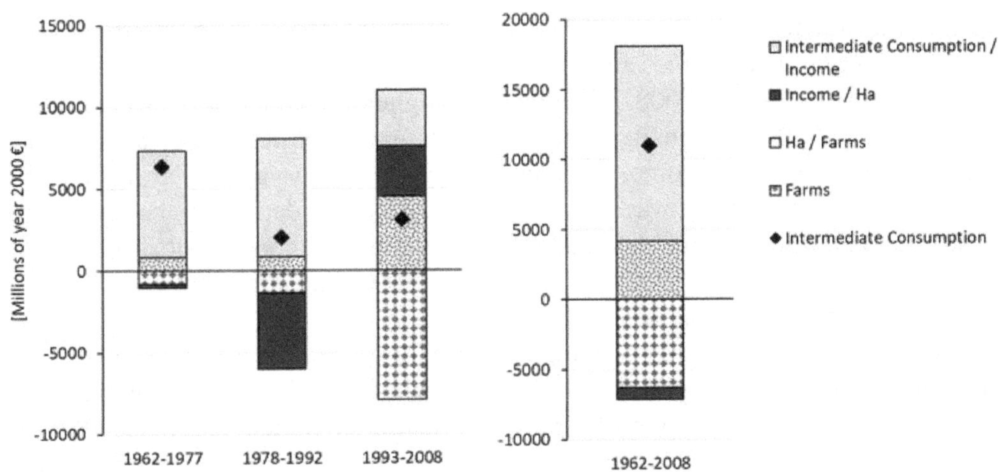

Graph 6.5 Decomposition analysis of Intermediate Consumptions (IC) explained by the number of agricultural holdings—farms—(F), hectares per farm (ha/F), income per hectare (I/ha) and the intermediate consumption share of income (IC/I) 1962–2008

$$IC = F * \frac{ha}{F} * \frac{I}{ha} * \frac{IC}{I} \tag{2}$$

The results for the whole period (1962–2008) show an IC increase of 11 billion euros. The increase was the result of the opposing strengths of the selected variables. The fall in income per hectare and farm income halted the rise of intermediate consumption, especially farm income. Farm abandonment led to a 6.3 billion euro drop in intermediate consumption. However, increases in farm size, measured as hectares per farm, as well as the growing weight of intermediate consumption over total income, were responsible for a much larger increase. The increases in farm size account for an increase of 4.2 billion. The weight of intermediate consumption in total income was, however, the main driver of total intermediate consumption increase. Its impact is estimated at 13.9 billion euros. Overall, agrarian intensification, measured as an intermediate consumption increase, is explained by forces that pushed in opposite directions, notably the abandonment of farms, which caused a drop in their numbers but, in turn, the loss of the sector's profitability forced farms to increase their consumption that, in turn, grew in size due to property concentration. This latter factor also pushed up intermediate consumption.

It is possible to find differentiate historical phases when analyzing the effects of decomposition The first period runs between 1962 and 1977, during which IC was mostly driven by its increasing share of agricultural income and to a much lesser extent by farm size increases. Farm numbers hardly dropped and farms hardly contributed to IC variations. A second period can be distinguished between 1977 and 1992 when IC barely rose. It continued to grow in relation to agricultural income, but the growth was mostly offset by the decline of income per hectare. This was during the oil crisis and transition to democracy. The first democratic governments adopted agrarian policies that were sensitive to the pressures of the professional agricultural organizations and workers' unions: they slowed down the process of labor substitution by machines and farm number reductions (Herrera 2007). A third period spans from 1993 to 2008. IC growth was higher than in the previous period. In this case, the growth was equally driven by the rise in the percentage of intermediate consumption over total income, farm size increase, and the growth in income per hectare. The overall growth was partially offset by a sharp decline in the number of farms. In other words, in this last period, we observe an unprecedented process of farming abandonment and relative increase in intermediate costs that was offset by intensifying farming and increasing farms size.

In view of the results of our decomposition exercise and the behavior of employment and farm numbers, agricultural macromagnitudes, it is worth breaking down the last period on the evolution of Spanish agriculture's metabolism following the last two periods—the first third of the twentieth century and Early Francoism—into two subperiods. The first sub-period starts in the sixties and runs until the early nineties; we can refer to it as the period of the industrialization of Spanish agriculture. The second runs from the early nineties until today and can be understood as the globalization of Spanish agriculture.

At the beginning of the 1960s, agrarian activity provided sufficient income to cover average Spanish household expenditure; soon the continued fall in income and the increase in average household expenditure significantly deteriorated farmers' living standards. Most were able to confront the situation by increasing production and reducing costs. The technologies associated with agricultural industrialization, i.e., fertilizers, phytosanitary products, improved and hybrid seeds, irrigation and mechanization made it possible to increase productivity, even in the least productive farms. In parallel, they tried to compensate for the increase in intermediate costs by reducing labor costs, that is, replacing work with machines and chemical means. Despite these efforts, the strategy did not yield the desired result and agrarian income remained insufficient to cover average Spanish household expenditure (Graph 6.6).

The second period, starting at the beginning of the nineties, coincided with the full implementation of the CAP. Agricultural income grew above household expenditure, though this was due to job destructions and numerous farm closures. Agricultural income did not improve, it continued to dwindle in constant terms, but was distributed among fewer farmers and fewer salaried workers. In fact, there was little possibility of increasing productivity by greater use of inputs or by substituting labor with machines. The marginal utility of technologies that had played a leading role in the industrialization of Spanish agriculture was reduced, especially for farms with low yields that could barely earn more income by incorporating inputs to increase production. Improving labor productivity and increasing farm sizes continued to be an effective strategy to offset this trend; but this was not possible for many farmers who had to abandon their activity or who did not dispose of any generational handover. The smaller farms were the most affected, and the farms under 20 ha represented the bracket with the highest number of closures. Terminations of activity were especially intense in the country's interior, in areas of low productivity and limited capacity for intensification.

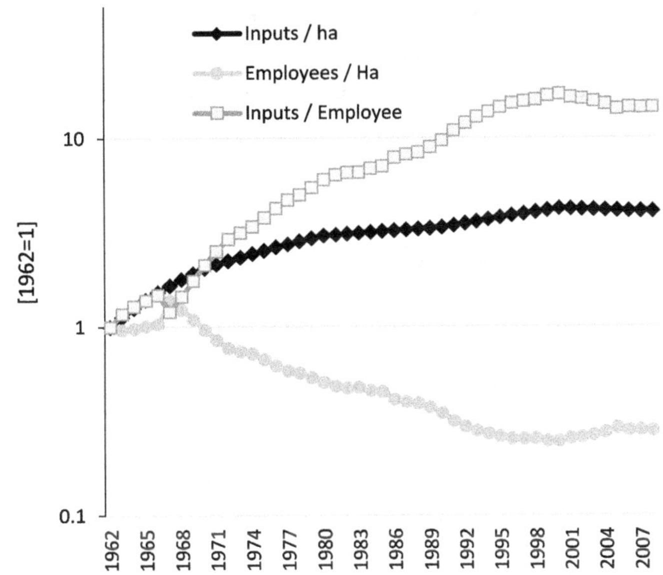

Graph 6.6 Relationship between input use and cultivated land employees

In a context of declining paid prices, farms that were able to intensify their production attempted to raise their income by increasing their surface size or by orienting production towards utilizations with higher gross margins. The data presented in Chap. 4 show that gains in European size farm units (ESU) were clearly achieved thanks to intensive livestock and forced cultivation under plastic covers. Both tendencies led to enhancing the use of external inputs in terms of energy, that is, they have broadened the scope of $I + S$ in Spanish agriculture. In the case of pigs and poultry production, high gross margins were made possible thanks to massive imports of very cheap feed based on corn and soybeans. Overcoming the profitability crisis has therefore persistently relied on $I + S$, but only a tiny share of farms was concerned. Spanish agriculture has thus branched off in two directions: on the one hand, the sector continues to intensify and specialize, associated with intensive livestock farms, that are highly industrialized and integrated into the agri-food industry, together with farms based on forced cultivation under plastic; and on the other hand, a more extensive sector unable to reach these $I + S$ levels has abandoned its activity or subsists thanks to size gains, CAP subsidies, or organic farming. Converting to certified organic production has been a way out as agri-environmental measures generate income supplements that has allowed them to increase the number of ESU.

As illustrated, employment in the sector has declined continually since the middle of the century, accelerating in recent decades. Job destruction appears to be non-ending. Despite $I + S$ efforts, monetary flows have visibly been unable to ensure the reproduction of this fund element for the functioning of the agroecosystem. In 1950, more than 5.2 million people were employed in the agricultural sector, that is, almost half of total employees and 18.6% of the Spanish population. This figure has come down to 774,500 in 2016, accounting for 4% of employees and barely over 2% of the population. As we saw in Chap. 4, around two-thirds of farms have disappeared since the 1960s. The aging of farmers has taken on worrying proportions, calling into question both the agrarian nature of households and the survival of farms. This explains the phenomenon of depopulation of Spain's interior, reflecting the non-viability of the industrialized agricultural model and the institutional arrangement in which it operates.

6.5.2 Demand Side Drivers of I + S

Changes in consumption patterns and population increase were two key drivers of agrarian intensification. Once again, we conducted a decomposition analysis to understand the drivers of cropland demand. The variable under analysis was actual demand for cropland and was calculated by adding the land embodied in imports to the country's croplands minus the land embodied in exports. To estimate the land embodied in imports, we made an estimation of the imported biomass produced in croplands in third countries. Then we applied a land demand factor to each case (for each product and in each exporting country), which varied according to whether it was a primary or processed product, if there was joint production in each crop, etc. A detailed description of the calculations can be found in Infante-Amate et al. (2018). We proceeded in the same way for exports. As mentioned throughout our study, the results show the Spanish economy's increasing land demand which is mainly due to the transfer of land use to other countries. At present, approximately 11 million hectares (Mha) of total surface area is used outside the country, the majority of which, about 10 Mha, is cultivated. Current cultivated areas accounted for approximately 17.1 Mha in 1900; however, actual land demand was 22.8 Mha, i.e., 1.4 times more. The greatest acceleration has been taking place since 1960. Imported land then amounted to 1.9 Mha and exported land barely exceeded 1.0 Mha. Net imports represented only 4% of the country's cultivated areas and total imports represented 11.6%. Five decades later, in 2008, net imports multiplied by 7, accounting for 37.7% of real demand. Total land imports represent 64.0% of the country's croplands. Given that a substantial part of that area is destined for export, we can fairly say that Spain requires almost as much surface area within its borders than outside due to biomass consumption activities.

But what drivers pushed up demand? We propose a new decomposition analysis to explain the land demands of croplands (*L*) taking into account: the population increase (*P*); changes in consumption patterns, especially diet, since most of the cropland's biomass consumption is destined to food (*D*, estimated as the kcal consumed per inhabitant) and land yields (*Y*, estimated as land required to produce each kcal).

$$L = P * D * Y \tag{3}$$

The twentieth century has been characterized by profound changes to the three factors under study. Land productivity has significantly increased, generating considerable savings in demanded area. According to Soto et al. (2016) production per hectare (dry matter) of the cultivated areas has multiplied by 3.2, going from 1.8 to 5.7 t/ha between 1900 and 2008. However, other variables have weighed on demand for land in the opposite direction. The population has multiplied by 2.5, increasing from 18.6 million inhabitants to 46.5 million. Consumption patterns have also changed towards more land-demanding models. Not only has direct consumption increased (biomass has grown by 26% per inhabitant) but the diet has changed, with a greater presence of animal products, that are more land-use intensive. How has each of these changes affected the increase in acreage demands documented above?

Graph 6.7 shows the results of the decomposition analysis. Throughout the study period, cultivated area demands grew by 6.0 Mha. Production intensification allowed to save 27.1 Mha, however, the population increase required 17.6 Mha and the consumption patterns change required 15.6 Mha. Production and technological change could have been enough to continue feeding a growing population, however, changes in food consumption patterns made that impossible.

These drivers have behaved unevenly throughout the study period. In this sense, three major periods can be distinguished: between 1900 and 1933 these drivers increased demand for land by 3.8 Mha, made it drop by 0.7 Mha between 1933 and 1960 and pushed it up again by 2.9 Mha between 1960 and 2008. In the first period, the increase was mainly motivated by population growth. Although production intensification succeeded in saving enough land to face consumption changes, it was insufficient to sustain the rapid population growth. The second period was marked by lower population growth and the atypical behavior both of land intensification (productivity decreased) and consumption patterns (per capita consumption decreased). This was due to autarkic politics during Early Francoism. During the third period, the increase in total demand was somewhat lower than during the first, however, the drivers had stronger impacts: the population grew much faster (demanding 10.9 Mha), soil intensification accelerated (saving up to 30.9 Mha), and consumption

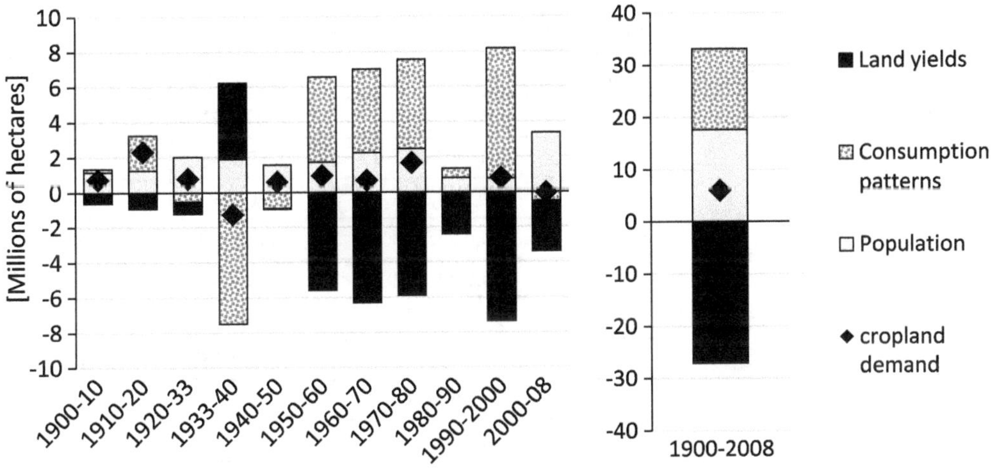

Graph 6.7 Decomposition of the variables that affect cropland demand. The variables considered are population changes, consumption pattern variations and land yields. All variables are expressed in millions of hectares

patterns became much more land-demanding (requiring 23.6 Mha). The second half of the twentieth century was thus characterized by an unprecedented acceleration of land demand drivers. In the last decade, however, demand has stabilized due to changes in consumption patterns.

The observed evolution can be interpreted as a "rebound effect" case: improvements in efficiency (i.e., land yields) were absorbed by changes in consumption patterns (i.e., an increase in food intake and waste by inhabitant and change in food consumed) so that aggregate consumption continued to grow. Globally, population increase between 1960 and 2005 was a more determining driver than diet. In fact, according to Kastner et al. (2012), southern Europe along with East Asia were the only territories where a diet change was a more important driver of land demand than population change.

What has this diet change consisted of and how did it come about? In previous analyses (González de Molina et al. 2013, 2014, 2017) we estimated apparent food consumption from 1900 to 2008. The results show a differentiated eating behavior between the first and the second half of the twentieth century. Table 6.5 shows the amount of both vegetal and animal biomass aimed at endosomatic metabolism in tons. Total biomass multiplied by 3.3 during the entire study period, apparent consumption grew significantly between 1900 and 1933 and grew again, to an even greater degree, from 1960 until today, almost doubling the amount of consumed biomass. The major driver of this growth was animal biomass that multiplied by 7, while vegetal biomass increased by a factor of 2.6. While animal biomass contributed just over 16% of total consumed biomass at the beginning of the century, that percentage had risen to 35% in 2008. The trends are easier to identify when analyzed in per capita terms.

As shown in Table 6.6, per capita consumption increased by 39.1%, that is, demand for food biomass grew not only because of population growth but also because of diet changes. The change was led by animal biomass: while the consumption of

Table 6.5 Apparent net biomass consumption (deducting losses) in t of fresh edible food (1900–2008)

Year	Vegetal biomass	Animal biomass	Total biomass
1900	8,809,163	1,722,193	10,531,356
1910	9,216,040	2,076,013	11,292,053
1922	10,872,487	2,378,680	13,251,168
1933	12,584,553	2,701,503	15,286,055
1940	10,655,118	2,913,993	13,569,112
1950	10,803,102	3,298,027	14,101,128
1960	13,930,829	4,546,685	18,477,614
1970	17,015,648	7,120,757	24,136,405
1980	20,064,341	9,513,501	29,577,841
1990	22,310,436	10,792,822	33,103,259
2000	21,676,647	12,119,428	33,796,074
2008	22,931,836	12,250,486	35,182,323

Source Author's compilation based on agrarian statistics

Table 6.6 Apparent consumption per food group (g/per capita/day in fresh edible food)

	1900	1933	1950	1970	1990	2008
Cereals	320.5	326.0	224.7	216.4	161.6	180.8
Legumes	46.6	49.3	35.6	35.6	16.4	16.4
Roots and tubers	241.1	383.6	249.3	263.0	235.6	131.5
Vegetables	263.0	276.7	238.4	293.2	405.5	345.2
Fruits	101.4	93.2	98.6	172.6	265.8	219.2
Nuts	13.7	8.2	8.2	5.5	8.2	8.2
Oilseeds	2.7	2.7	5.5	8.2	13.7	21.9
Alcoholic drinks	265.8	216.4	145.2	254.8	315.1	328.8
Oil	30.1	43.8	30.1	49.3	79.5	87.7
Sugar	16.4	30.1	24.7	82.2	71.2	71.2
Meat + fat	38.4	54.8	32.9	106.8	224.7	243.8
Eggs	8.2	11.0	11.0	27.4	32.9	24.7
Dairy products	197.3	216.4	252.1	391.8	449.3	419.2
Fish	11.0	24.7	27.4	52.1	54.8	65.8
Honey	1.1	0.8	0.5	0.8	1.6	1.9
Vegetal biomass	1301.4	1430.1	1060.3	1380.8	1572.6	1411.0
Animal biomass	255.9	307.7	323.8	578.9	763.3	755.3
Total	1557.3	1737.8	1384.1	1959.7	2335.9	2166.3

Source Author's compilation based on agrarian statistics

vegetal biomass per capita grew by only 8%, animal biomass tripled. This increase was constant over time, including during Franco's Autarky, but has been much more intense since the 1960s. The apparent consumption of animal biomass per capita grew modestly during the first half of the century (26% since 1900); but between 1960 and 2000, consumption more than doubled reaching 827.4 g/person/day having slowed down in the last decade. While animal biomass barely represented 16% of total consumed biomass in 1900, it currently reaches almost 35%: a transition from a plant-based diet to a diet where livestock products play a major part has undoubtedly taken place. The same table disaggregates previous data per food groups, revealing a substantial decrease in the consumption of cereals, legumes, roots and tubers and, conversely, a significant increase in the consumption of meat, dairy products, fish, oil, sugar and alcoholic beverages.

Table 6.7 shows the energy value expressed in calories per person per day. Consumed calories increased in line with biomass consumed, i.e., by 30% between 1900 and the year 2000, the year of maximum intake. Except in the forties and fifties, the amount of biomass loosely satisfied basic energy requirements. These requirements were calculated by Cussó (2005; Cussó et al. 2017) and determined at around 2260 for 1900; 2314 for 1960 and 2434 for 2011. The most significant fact, however, is that this increase is mainly due to food intake of animal origin. The cereals group,

Table 6.7 Apparent consumption of biomass in calories, deducting losses (1900–2008)

Year	Vegetal biomass		Animal biomass		Total biomass	
	Calories	%	Calories	%	Calories	%
1900	2328	91.2	224	8.8	2552	100.0
1910	2370	90.4	251	9.6	2621	100.0
1922	2588	90.2	281	9.8	2869	100.0
1933	2646	90.6	276	9.4	2922	100.0
1940	1959	88.6	251	11.4	2209	100.0
1950	1888	87.4	272	12.6	2160	100.0
1960	2400	86.6	374	13.4	2774	100.0
1970	2406	81.7	538	18.3	2944	100.0
1980	2409	78.5	659	21.5	3069	100.0
1990	2398	74.6	816	25.4	3214	100.0
2000	2434	72.8	908	27.2	3342	100.0
2008	2401	74.1	841	25.9	3242	100.0

Source Author's compilation based on agrarian statistics

including legumes and potatoes, used to form the basis of the diet and shifted from accounting for 40% of ingested energy in 1970 to just over 27% at present. In contrast, meat, eggs, and dairy products used to provide 17% of energy in 1970 increasing to 23% today. In the year 2000, both food groups provided a similar percentage of energy: 24 and 25%, respectively. Oil consumption has also increased and now provides almost a quarter of the calories in 2008. If we add oil, mainly olive oil, both groups of foods, accounting for 47% of calories, today form the basis of the Spanish diet (González de Molina et al. 2014).

Growth of DE and SB thus allowed feeding the Spanish population until the civil war undoubtedly on an essentially vegetarian diet. Caloric intake in the thirties was similar to that provided by the German or Austrian diet and higher than the average diet in Holland, France, Italy or Greece (Cussó 2005, 353). Table 6.7 shows the depth of the food crisis that Spain experienced as a result of the agrarian policy of successive Francoist governments until the beginning of the 1960s. In contrast with the idea that hunger and malnutrition were a thing of the past, overcome during Franco's dictatorship and thanks to the economic progress favored by the regime, the data persistently shows that "the hunger years" were an exclusively Francoist phenomenon, caused by the dictatorship, its economic policy and fierce repression after the end of the war. It would take two decades to overcome the crisis. Indeed, shortly before the Civil War began, the amount of calories per capita ingested by a Spanish citizen per day was 29% higher than needs, an amount that would not be reached until the beginning of the seventies.

During the last four decades of the twentieth century, there has been a major increase in calorie intake (20%), higher than that between 1900 and 1933 (14.5%), excessively beyond needs. Perhaps the most striking fact is that this increase has been achieved through the rising intake of products of animal origin (see Table 6.7). In fact, the amount of calories provided by vegetables has fallen continually since the thirties, when they contributed the most (2646 calories).

Graph 6.8 shows the protein composition of food consumed throughout the twentieth century. According to calculations by Cussó (2005, 345) the Spanish population's protein requirements, between 35.6 and 35.8 gr per person per day, were amply covered. A steady and prolonged tendency to substitute vegetable proteins with animal proteins can be observed despite the fact that during most of the twentieth century it was vegetables that provided the bulk of the proteins. Currently, two-thirds come from animal biomass. The graph shows the major role of meat and dairy products in protein intake in recent years to the detriment of cereals, legumes, and potatoes.

Graph 6.9 shows the composition of consumed foods in lipids or fats. A distinctive sign of Mediterranean consumption patterns has been the intake of vegetable fats, among which olive oil stands out. However, there has been a growth in animal origin fat consumption and it now reaches over a third. The contribution of fats today comes basically from olive oil, meat, and dairy products. The percentages provided by the nutritional assessment of the Spanish diet, based on official data provided by the Panel of Food Consumption (Varela Moreiras et al. 2008, 48) are quite similar. Meat consumption has more than quadrupled, from 56 g/capita/day in the 1960s to 243 g/capita/day at present, pork and chicken meat having grown the most. Milk consumption increased from 291 g/capita/day to 488 g/capita/day and that of eggs from 15 g to 25 g/per capita/day.

The data analysis allows distinguishing three different periods in the evolution of Spanish diets. A first period runs from the beginning of our study, in 1900, until the Civil War, a period in which the transition towards a typical Mediterranean diet initiated long before reached its peak. A second phase, between the forties and the seventies, runs during the Franco dictatorship: after having overcome a long and deep food crisis, levels and patterns of consumption proper to the 1930s were gradually

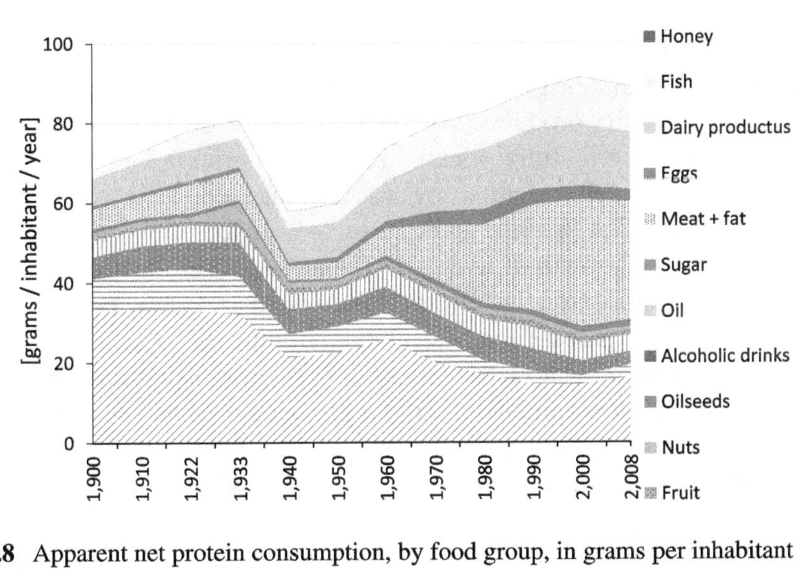

Graph 6.8 Apparent net protein consumption, by food group, in grams per inhabitant per day

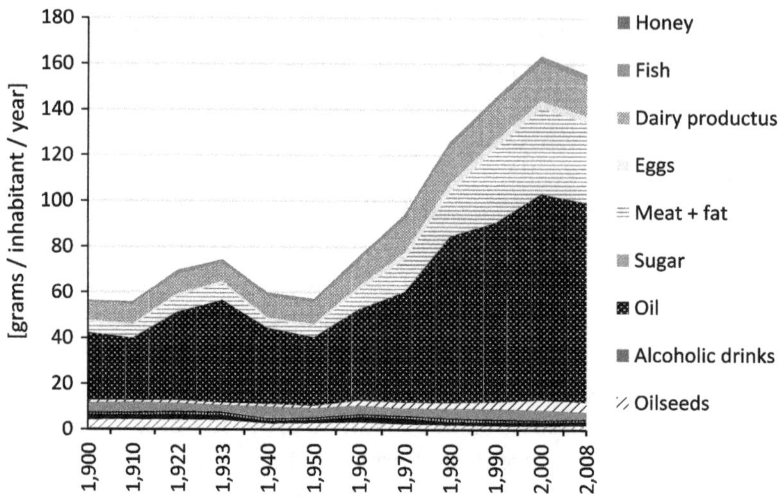

Graph 6.9 Lipid contribution of apparent consumption, grams per inhabitant per day

recovered. Thus, until the 1970s, typical Mediterranean diet consumption patterns would predominate in Spain, implying adaptations to the conditions and dynamics of Spanish agroecosystems (González de Molina et al. 2014). However, since that decade, typical developed country food consumption patterns have been adopted (European Commission 2015), moving increasingly away from the WHO recommendations (Rodríguez Artalejo et al. 1996; Nicolau and Pujol 2011), a phenomenon that has been called the 'westernization' of the diet (Kearney 2010). Del Pozo de la Calle et al. (2012) calculated the so-called Mediterranean Diet Score (MDS) and found that Spain obtained a score of 4 in 2008, on a scale of 0–9, where 9 is the maximum adaptation to the Mediterranean diet. These habits explain why 60.9% of the Spanish population is overweight (39.3%) or obese (21.6%) (Aranceta-Bartrina et al. 2016). They are also associated with degenerative diseases (Tilman and Clark 2014) and colorectal cancer (De Marco et al. 2014, 69). The diet is based, as we have seen, on high consumption levels of livestock products, on the excessive intake of proteins and fats of animal origin and on the increasing deficit of carbohydrates.

Demand for meat, dairy products, and eggs, has thus especially increased, fundamentally changing the agricultural sector's production orientation: since the 1960s, production has been largely oriented towards animal feed. This trend has intensified in recent decades. The Spanish agrarian sector reacted between 1960 and 2008 through the spectacular growth of livestock, the massive introduction of inputs, the concentration of biomass extractive efforts in cultivated areas and, paradoxically, the relative abandonment of pasture and forest lands. But these changes in food demand have been met only in part by domestic production. Livestock and changes to its composition, with monogastric animals playing a greater role, has been made possible thanks to growing imports of biomass for animal feed from other European Union and Latin American countries (Infante and González de Molina 2013). Foreign trade is therefore key in the Spanish agri-food system: on the one hand, foreign trade makes the specialization of Spanish agriculture (in oil, horticultural products and pig meat) possible, providing outlets in international markets, especially in Europe; and on the other hand, foreign trade supports Spain's growing consumption of meats and dairy products, providing a very important percentage of animal feed. This phenomenon is consistent with the data obtained by research on the evolution of the nitrogen cycle in Spain between 1961 and 2010, which showed Spanish livestock's growing dependence on imported protein, especially from Latin America (Bouwman et al. 2013; Lassaletta et al. 2014).

In short, until the 1960s, the Spanish diet was "coupled" with DE and SB satisfied the bulk of food demand. We could say that the domestic market and food demand were the main drivers of agricultural production and its production orientation. In other words, the $I + S$ of Spanish agriculture was stimulated by population growth, but also, as we will see next, by diet improvements compared to 1900. This linkage had dramatic consequences, causing an unprecedented food crisis when the Francoist regime's international isolation and its autarchic policy reduced the flow of SB. Spanish food production enjoyed, until then, a high degree of autonomy and what we would call today food security, in direct contrast with the situation today, where foreign trade is decisive and food autonomy has declined considerably: the livestock sector (and its supply of meat and dairy products) is currently dependent on feed imports. The sector is also dependent on Central European market, a preferential outlet for its fruit and vegetable production, the main specialization of Spanish agriculture. The issue can also be approached the other way around: domestic agricultural production accounted for Spaniards' food consumption to a very high degree, and therefore, changes in production can largely explain changes in food consumption habits.

What caused these important diets changes and, consequently, the demand for vegetable products and above all animal products? The relationship between per capita income increase and the increase in energy content and animal proteins in diets is well known (European Commission 2015, 8, for a review, see Tilman and Clark 2014). This certainly occurred in Spain, facilitated by cheaper food (Kearney 2010) and the loss of relative weight of food expenses in household budgets, which went from 48.7% in 1960 to 16.8% in 2015 (Martín Cedeño 2016, 222). But income growth only explains increases in meat and dairy product consumption as well as the gradual loss of the Mediterranean diet. It does not explain, however, why this increase in meat has been based on monogastric livestock, dependent on imported quality grains and not on pastures or harvest residues. Graph 6.10 compares the evolution of the prices paid by consumers for pork and chicken meat and other foods of animal origin, with the evolution of selected groups of basic vegetable foods. We can observe that foods of animal origin have become progressively cheaper, while vegetables have become more expensive. This explains why pork and chicken meats, eggs, milk, and yogurt have eventually become as affordable as bread, cereals, legumes, fruits and vegetables. The cheapening of pork is especially striking, due to the economies of scales of increasingly concentrated intensive farms and the import of cheap grains (corn and soybeans), which has cheapened end prices of these meats. In 2015, Spain even turned into the biggest pork exporter in the EU (Rousseau 2016).

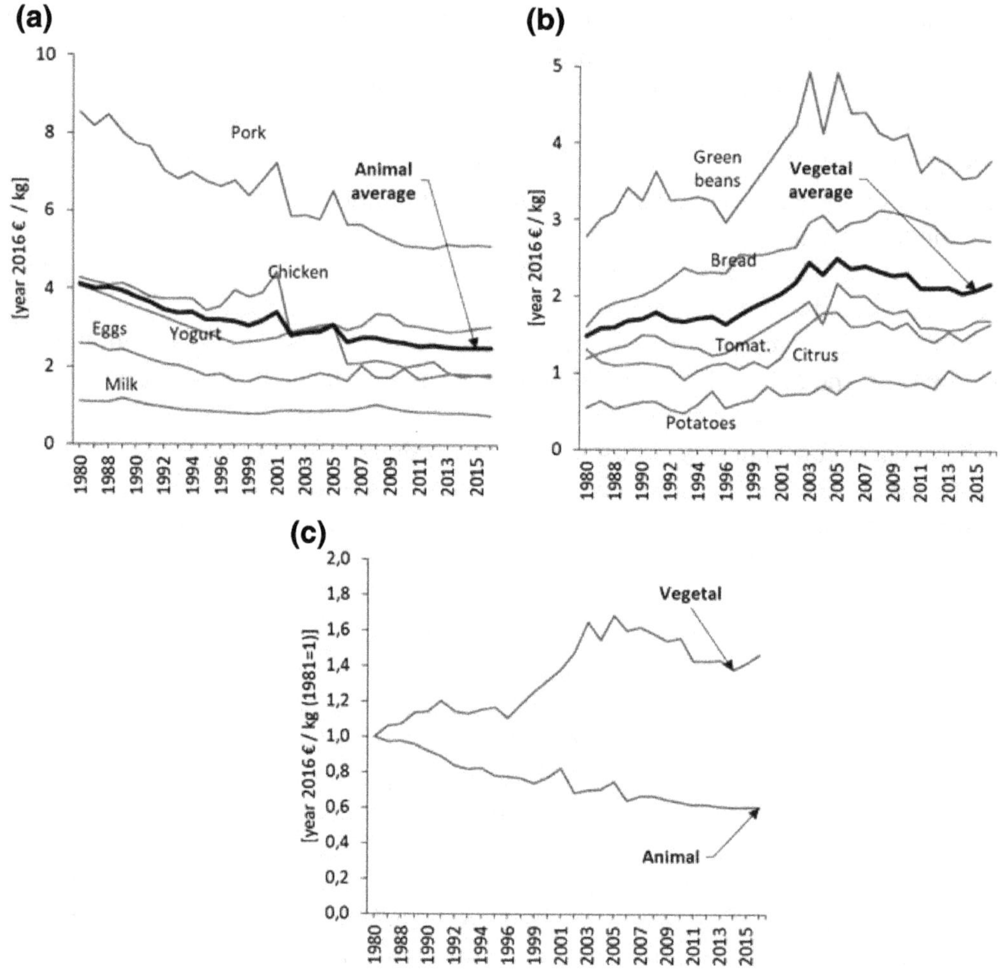

Graph 6.10 Comparison of the prices paid by consumers for some animal (**a**) and vegetable products in 2016 constant euros (**b**), in 2016 constant euros and in percentage of 1981 prices (**c**). *Source* Household Budgets Survey and Continuous Household Budgets Survey (INE 1980–2105)

Moreover, the acute processes of livestock farm concentration, production process industrialization (Domínguez Martín 2001; Clar, 2005, 2008; Clar et al. 2015) and vertical integration in the agri-food industry, explains increases in offer and reduced production and prices. Both pig and poultry farming are a good illustration of this major transformation (Segrelles Serrano 1993; Clar 2010; Fundación Cajamar 2011; MAPAMA 2013 and 2016a, b).

As we have seen, the globalization of food markets has turned Spain into a net importer of biomass, favored by the comparatively lower prices of agricultural commodities (soybean, corn, etc.) in international markets, the basis of intensive livestock feed (Mayer et al. 2015; Falconí et al. 2016). What has actually happened, as we have shown above, has been a shift towards third countries with lower production costs for parts of the land consumed by the Spanish agri-food system. Our results suggest that foreign trade has saved c. 18 million hectares (data for 2010) that would have been cultivated in the country's interior territory in case it had not been possible to access those markets and maintain current levels of consumption.

6.6 Conclusions

The traditional narrative judges the agrarian transformations that took place during the twentieth century positively, especially since the 1960s with the sector's industrialization. It emphasizes that Spanish agriculture underwent a significant intensification process leading to a threefold production increase and multiplying its value by five based on constant prices. The secret to this success was production intensification and specialization, achieved thanks to yield increases per unit area. Growth of cereals, fruits and vegetables, forage plants and olive groves were especially intense and reflected Spanish agriculture's progressive specialization in these crops. Greater still was livestock production growth, which multiplied by a factor of 8.2 over the same period. Its overall weight in the sector increased, reaching almost 17% of agricultural production value. Meat and egg production was the major player in this unusual livestock growth. Milk production multiplied by five, though it represented the major livestock production in 1900. The picture emerging from Spanish agriculture's biophysical analysis is less bright and downplays the scale of this growth, weighing it down with the effects on the environment and farmers, that is, on the sector itself. Conventional discourse barely contemplates the centrality of the sector as a whole: it attributes a subordinate role to agriculture and focuses on assessing agriculture's contribution to the country's economic growth.

From a biophysical perspective, which necessarily considers NPP as a whole, biomass grew by only 28.5%, a figure far from that of evaluations in monetary values and fresh weight. How can we account for the difference between both narratives? Between 1900 and 2008, Spanish agriculture transitioned from an organic metabolic regime to an industrial regime, and this process has accelerated in recent decades. The transformation consisted of a greater appropriation of biomass produced by agroecosystems for human use to the detriment of other species. This has been possible by transferring extraction efforts from pastures and forests to croplands; translocating the photosynthetic capacity of plants grown from straw to grain (arable crops) or from the trunks and leaves to the harvestable fruit (woody crops) simplifying the multifunctionality of the crops; reducing rotations and breaking the integration between different land uses; and shifting the production orientation from human consumption to animal consumption (livestock farming), driven by changes in the diet. These changes have prevented the completion of physical–biological cycles and have required the use of large amounts of external inputs manufactured and driven by fossil fuels. Production limitations proper to organic-based societies have been apparently overcome (Wrigley 2016), allowing for substantial Spanish population growth and, above all, for increasing levels of consumption.

These transformations, however, have been possible at the cost of a deterioration of the fund elements not only of Spanish agroecosystems but also of third countries. Biophysical funds have mostly undergone negative changes: not only have the funds that allow to maintain ecosystem services in good conditions such as soil fertility, biodiversity, carbon sequestration, water quality, etc. deteriorated, but they have become an active source of pollution and lead to the depletion of scarce resources such as fossil fuels or sources of mineral nutrients. In contrast, the weight or dissipative effects of livestock have risen disproportionately, consuming almost 60% of both DE and foreign trade biomass and requiring a similar amount of inputs. Livestock has lost its ties to the land and its maintenance is almost decoupled from it. Livestock represents the biggest source of emissions in the sector and contamination by slurry. The relationship between both funds has been almost completely lost, weakly maintained by extensive and semi-intensive livestock, whose production and consumption weight is hardly relevant.

Perhaps the most significant deterioration is the ongoing shrinking of the agrarian population. The existing population is aging and threatened by a lack of generational replacement. It is, however, a key fund for the future viability of sustainable agrarian activity. On the other hand, the technical means of production, i.e., the other social fund has increased in size. As we saw in Chap. 3. It has acquired excessive weight and has turned the metabolism of Spanish agriculture into a structure of high entropy dissipation, which requires a constant and growing supply of energy, mostly from fossil sources. The congruence between the two social funds, population and technical means of production, has also been broken: not only from the perspective of their respective sizes but also regarding the information flows making it possible to manage the technical means. These flows barely come from the farmers themselves, but from the companies that supply the inputs. Agrarian activity has become a lucrative market for input industries that promote technologies that are remote for farmers whose main purpose is to ensure the continuity of the business. The congruence between physical funds and social funds has thus been broken, but also the congruence between them: the activity does not ensure the reproduction of the agrarian population and hinders the reproduction or replacement of the technical means of production. Its use helps to produce more, but only in a limited way, and yet it has a negative impact on the territory and generates monetary costs that further depress the income that the other funds must support. The industrial metabolic regime, which was imposed from the 1960s, has implemented a form of operation that compromises not only the environmental health of agroecosystems but also the viability of the agricultural activity itself as we know it today.

The significant contribution of the agrarian sector to a country's economy should certainly be relevant, if only because it produces the food required to support the population, providing employment and maintaining an ecological infrastructure that is essential for the functioning of society. But a legitimate question is whether increasing levels of land or labor productivity should be achieved at the expense of the deterioration of the fund elements that make agrarian activity itself possible and whether successive increases in labor and land productivity can be maintained indefinitely. What limits should we set to the constant transfer of capital, income, labor, to the activity's profitability losses, job destruction, lack of generational change, etc.? The hypothesis defended in this book proposes that the prevailing industrial model, due to its intrinsic characteristics, leads to either collapse of agriculture or to the spending of considerable public resources (subsidies) to delay its collapse. It is urgent to redefine the role of the agrarian sector in the economy and, consequently, to determine which criteria should be used to value the place of agriculture in the economy. This redefinition requires thorough discussions that have only just begun and that are likely to continue over the next few years. We hope that the conclusions of the analysis presented in this book will contribute to the debate.

References

Ang BW (2005) The LMDI approach to decomposition analysis: a practical guide. Energy policy 33(7):867–871

Aranceta-Bartrina J, Pérez-Rodrigo C, Alberdi-Aresti G, Ramos-Carrera N, Lázaro-Masedo S (2016) Prevalencia de obesidad general y obesidad abdominal en la población adulta española (25–64 años) 2014-2015: estudio ENPE. Revista Española de Cardiología 69(6):579–587

Bouwman L, Klein Goldewijk K, van der Hoekc KW, Beusen A, van Vuuren DP, Willems J, Rufino MC, Stehfest E (2013) Exploring global changes in nitrogen and phosphorus cycles in agriculture induced by livestock production over the 1900–2050 period. PNAS 110(52):20882–20887

Clar E (2005) Del cereal alimento al cereal pienso. Historia y balance de un intento de autosuficiencia ganadera, 1967–1972. Hist Agrar 37:513–544

Clar E (2008) La soberanía industrial: Industrias del complejo pienso-ganadero e implantación del modelo de consumo fordista en España: 1960–1975. Rev Hist Ind 36:133–165

Clar E (2010) A world of entrepreneurs: the establishment of agribusiness during the Spanish pork and poultry boom, 1950-2000. Agric Hist 84(2):176–194

Clar E, Pinilla V, Serrano A (2015) El comercio agroalimentario español en la segunda globalización, 1951–2011. Hist Agrar 65:195–228

Costanza R, Graumclih LJ, Steffen W (eds) (2007) Sustainability or Collapse? an integrated history and future of people on earth. The MIT Press, Massachusetts

Cussó X (2005) El estado nutritivo de la población española 1900–1970. Análisis de las necesidades y las disponibilidades de nutrientes. Historia Agraria 36:329–358

Cussó X, Gamboa G, Pujol-Andreu P (2017) El estado nutritivo de la población española, 1860–2010: diferencias de género y generacionales. In: Documento presentado al XII Congreso Internacional de la Asociación Española de Historia Económica, (Salamanca, 5–9 de septiembre de 2017)

de Marco A, Velardi M, Camporeale C, Screpanti A, Vitale M (2014) The Adherence of the diet to mediterranean principle and its impacts on human and environmental health. Int J Environ Prot Policy 2(2):64–75

del Pozo de la Calle S, Cuadrado Vives C, Ruiz Moreno E, Valero Gaspar T, Ávila Torres JM, Varela Moreiras G (2012) Valoración Nutricional de la Dieta Española de acuerdo al Panel de Consumo Alimentario. Madrid. Ministerio de Agricultura y Pesca, Alimentación y Medio Ambiente

Domínguez Martín R (2001) La ganadería española: del franquismo a la CEE. Balance de un sector olvidado, Historia Agraria 23:29–52

European Commission (2015) World food consumption patterns, trend and drivers. In: EU Agricultural Markets Briefs, 6

Falconí F, Ramos-Martín R, Cango P (2016) Caloric unequal exchange in Latin America and the Caribbean. Ecol Econ 134:140–149

Fundación Cajamar (2011) El Sector del Porcino en España. Almería, Fundación Cajamar

Gierlinger S, Krausmann F (2012) The physical economy of the United States of America: extraction, trade, and consumption of materials from 1870 to 2005. J Ind Ecol 16(3):365–377

González de Molina M, Soto D, Infante-Amate J, Aguilera E (2013) ¿Una o varias transiciones?

Nuevos datos sobre el consumo alimentario en España (1900–2008). In XIV Congreso Internacional de Historia Agraria (Badajoz, 7–9 de noviembre 2013). https://doi.org/10.13140/2.1.1823.5684

González de Molina M, Soto D, Aguilera E, Infante-Amate J (2014) Crecimiento agrario en España y cambios en la oferta alimentaria, 1900–1933. Hist Soc 80:157–183

González de Molina M, Soto D, Infante-Amate J, Aguilera E, Vila Traver J, Guzmán GI (2017) Decoupling food from land: the evolution of Spanish agriculture from 1960 to 2010. Sustainability 9(12):23–48

Infante-Amate J, González de Molina M (2013) 'Sustainable de-growth' in agriculture and food: an agro-ecological perspective on Spain's agri-food system (year 2000). J Clean Prod 38:27–35

Infante-Amate J, Soto D, Aguilera E, García Ruiz R, Guzmán G, Cid A, González de Molina M (2015) The spanish transition to industrial metabolism long-term material flow analysis (1860–2010). J Ind Ecol, 19(5):866–876. Available in: https://doi.org/10.1111/jiec.12261

Infante-Amate J, Aguilera E, Palmeri F, Guzmán G, Soto D, García-Ruiz R, González de Molina M (2018) Land embodied in Spain's biomass trade and consumption (1900–2008): Historical changes, drivers and impacts. Land Use Policy 78:493–502

Kastner T, Rivas MJI, Koch W, Nonhebel S (2012) Global changes in diets and the consequences for land requirements for food. Proc Nat Acad Sci 109(18):6868–6872

Kearney J (2010) Food consumption trends and drivers. Philos Trans R Soc B 365:2793–2807

Krausmann F, Gingrich S, Eisenmenger N, Erb KH, Haberl H, Fischer-Kowalski M (2009) Growth in global materials use, GDP and population during the 20th century. Ecol Econ 68(10):2696–2705

Lassaletta L, Billen G, Romero E, Garnier J, Aguilera E (2014) How changes in diet and trade patterns have shaped the N cycle at the national scale: Spain (1961–2009). Reg Environ Chang 14(2):785–797

MAPAMA (Ministerio de Agricultura, Pesca, Alimentación y Medio Ambiente) (2013) Informe 2012 sobre el estado del Patrimonio Natural y de la Biodiversidad en España. Madrid

MAPAMA (Ministerio de Agricultura, Pesca, Alimentación y Medio Ambiente) (2016) Caracterización del Sector Porcino Español, año 2015. Madrid

MAPAMA (Ministerio de Agricultura, Pesca, Alimentación y Medio Ambiente) (2016) Informe del consumo de alimentación en España, año 2015. Madrid

Martín Cerdeño V (2016) Cincuenta años de alimentación en España. Mercasa, Madrid

Mayer A, Schaffartzik A, Haas W, Rojas-Sepúlveda A (2015) Patterns of global biomass trade. Implications for food sovereignty and socio-environmental conflicts. Environ Justice Organ, Liabilities Trade, 20

Nicolau R, Pujol J (2011) Aspectos políticos y científicos del modelo de la transición nutricional nutricional: evaluación crítica y nuevas perspectivas. In: Bernabeu-Mestre J, Barona JL (eds) Nutrición, salud y sociedad: España y Europa en los siglos XIX y XX. Valencia. Universitat de València, Seminari d'Estudis sobre la Ciència, pp 19–57

Rodríguez Artalejo FJR, Banegas MA, Graciani R, Hernández Vecino Y, Rey Calero J (1996) El consumo de alimentos y nutrientes en España en el período 1940–1988: análisis de su consistencia con la dieta mediterránea. Medicina Clínica 106(5):161–168

Rousseau O (2016) Denmark loses place as top EU pork exporter. Global Meat News [web page], 24 March 2016. Available in: https://www.globalmeatnews.com/Article/2016/03/24/Denmark-loses-place-as-top-EU-pork-exporter

Schandl H, Schulz N (2002) Changes in the United Kingdom's natural relations in terms of society's metabolism and land-use from 1850 to the present day. Ecol Econ 41(2):203–221

Segrelles Serrano JA (1993) La ganadería industrial en España: cabaña porcina y avicultura de carne. Univ Alicant, Alicante

Soto Fernández D, González de Molina M, Infante-Amate J, Guzmán Casado G (2016), La evolución de la ganadería española (1752 y 2012): del uso múltiple al uso alimentario. Una evaluación de la fiabilidad de los censos y de las estadísticas de producción. Seminario Anual de la Sociedad Española de Historia Agraria, (Madrid, noviembre de 2016)

Tilman D, Clark M (2014) Global diets link environmental sustainability and human health. Nature 515:518–522

Varela Moreiras G, Ávila Torres JM, Cuadrado Vives C, Del Pozo de la Calle S, Ruiz Moreno E, Moreiras Tuny O (2008) Valoración de la dieta española de acuerdo al panel de consumo alimentario. Madrd. Ministerio de Agricultura, Alimentación y Medio Ambiente

Epilogue

To this day, the trends identified until 2008 have not only been persisted, but have intensified. The data available for 2017 show that there have been no great variations in land uses, but the trends already mentioned in Chap. 2 seem to persist: Pastures and woodlands continue to grow (by 0.8 and 0.3%, respectively), while areas under cultivation keep declining (−0.9%). Within the latter, arable crops have continued to lose surface area (−1.1%). This trend shows that land abandonment is ongoing. The number of farms has fallen from 1,069,700 in 2007 to 933,000 in 2016, according to the farm survey. UAA size per farm has remained stable, which means that farm numbers continue to drop due to the abandonment of activity, especially inland, where deagrarization and depopulation jeopardize the very management of agroecosystems.

We can also observe a decline in monetary terms, both in the volume of agricultural production and in the final production value, confirming the latest statistical data. Like its neighboring countries, Spain's agricultural production seems to be somewhat stagnating. In terms of fresh matter, the volume of agricultural production decreased between 1990 and 2015 by 5.8%. Industrial crops, forage production, and cereals declined the most. This does not mean that the domestic consumption of plant products has decreased; on the contrary, it has continued to grow. In fact, livestock production, dependent on cereals and fodder, increased by 37.6% over the same period. And, unlike agricultural production, livestock production has grown steadily until today, precisely as the domestic consumption of meat and milk has begun to slow down (del Pozo de la Calle et al. 2012). There has also been a growing tendency to replace these products with production purchased on international markets (Witzke and Noleppa 2010; Infante and González de Molina 2013; Infante-Amate et al. 2018). The balance between imports and exports shows a deficit of more than 3,794.2 million euros in the "other food" category, which includes the bulk of animal feed. Spain has become a net meat exporter, with a favorable balance in 2015 of more than 3600 million euros, i.e., 38.6% of the total net balance of foreign trade in the food, beverages, and tobacco section (MINECO 2017). The production orientations that now provide a higher gross margin of exploitation (2016 data) continue to be intensive landless livestock farms, thanks to cheap feed and vertical integration in the chain. A new specialization has been added to Spain's traditional production of fruits, vegetables, and olive oil: meat for export, especially pork. This explains that according to Spain's National Statistics (INE 2017, 384), the value of agricultural production increased by 8.3% and agricultural income by 6% since 2009 (in 2009 prices), despite stagnating agricultural production.

Spain's agricultural industry's progressive integration into global agricultural markets has been consolidated. The Spanish agri-food system requires ever more raw materials to function. Imports of biomass have therefore become decisive. Resorting to international markets explains that domestic consumption continues to increase and, at the same time, that it is compatible with the abandonment of low-productivity farmland, the underutilization of pastures, and farmers' abandonment of activity. Unless these trends are reversed, the system could collapse. Paradoxically, the quest for a decent income has led many farms that cannot follow the path of intensification for lack of soil, climate, and water capacity to turn to organic farming (OF); an agricultural orientation that a priori should reverse these trends. In fact, organic farming constitutes more than a seal of differentiated quality, as the sector's agents and institutions themselves often regard it.

Organic farming (OF) has grown significantly in recent years, in terms of both surface area and producers, associated agribusiness and market volume. In 2015, OF reached 50.9 million hectares (1% of the UAA) and accounted for 75,000 million euros worldwide, of which 11.2 Mha (6.2% of the UAA) and 27.100 M€ corresponded to the European Union (EU) (Willer and Lernaud 2017). Spain is the fourth country in the world in size of certified ecological managed surface area, after Australia, Argentina, and the USA (Willer and Lernaud 2017). It is the first country in Europe with 2.02 Mha in 2016 (8.7% of the UAA) (MAPAMA 2017a, b) (Graph 7.1) and a market of 1.5 million euros (tenth in the 2015 world ranking) (Willer and Lernaud 2017).

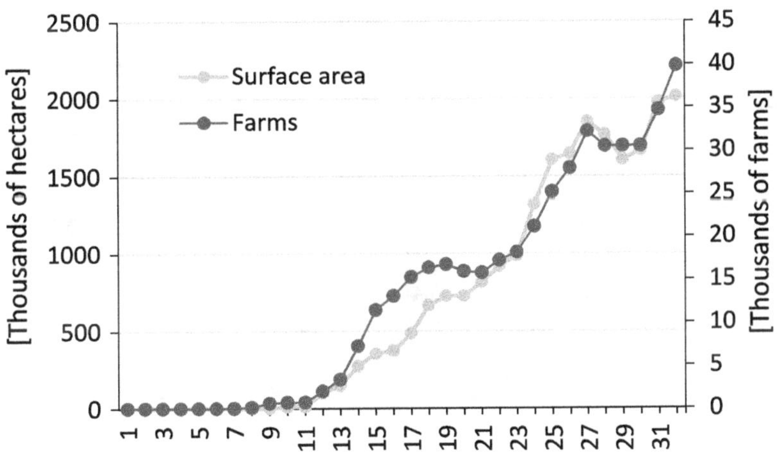

Graph 7.1 Evolution of organic farming surface area and number of farms in Spain, in thousands of hectares. *Source* MAPAMA (several years)

Graph 7.2 Distribution by production orientation of certified organically managed surface areas in Spain, percentage. *Source* MAPAMA (2017a)

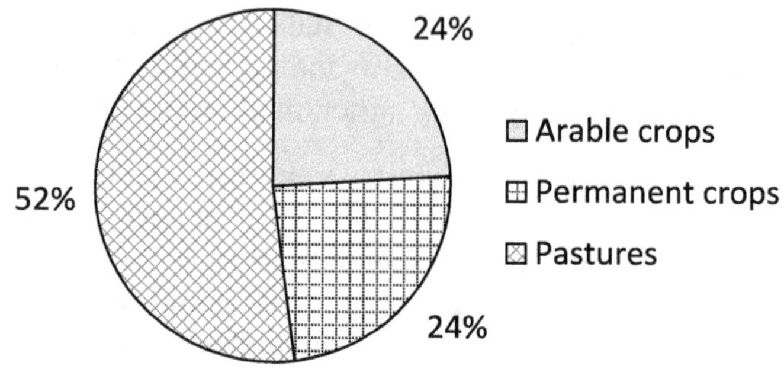

Spain's OF surface area is generally oriented toward livestock production, since 52% of the certified area corresponds to pastures and grasslands (MAPAMA 2017a). The remaining 48% is cultivated land and is distributed equally between arable and permanent crops (Graph 7.2). As a result, 5.7% of the cultivated land is classified as ecologically managed, compared to 11.2% of the pasture area (MAPAMA 2017a, b). Andalusia, with 48.4% of the certified area, and Castilla-La Mancha, with 19.4%, concentrate most of Spain's surface areas under ecological management.

OF has been a refuge for extensive livestock, especially in Andalusia, where significant agri-environmental aid has favoured its conversion. In 2015, 3.4% of ruminants were ecologically managed in Spain and 64.3% of them in Andalusia. However, monogastric animals have barely undergone conversion (<0.1%), given the almost absolute decoupling of cattle with the land. Ecological livestock strongly tied to the land through grazing contributes to reducing the risk of fires, which is essential in Mediterranean ecosystems to prevent large areas from deteriorating. Land degradation lessens its capacity to store CO_2, whether in the soil or in the biomass (Graph 7.3).

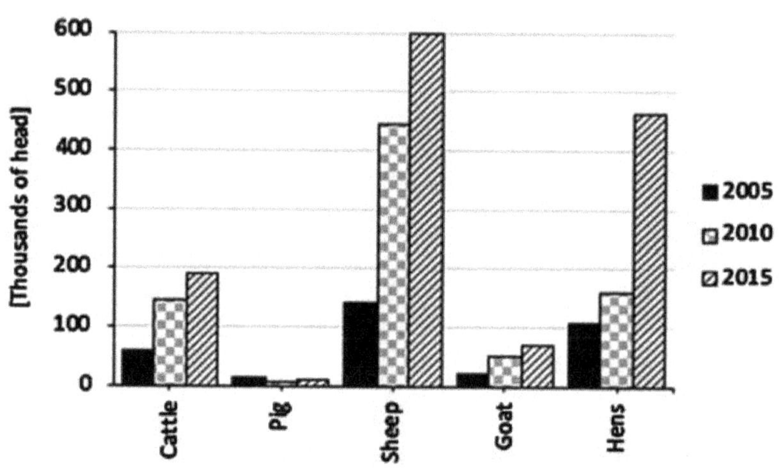

Graph 7.3 Evolution of the number of livestock heads under certified organic management in Spain, thousands of head. *Source* MAPAMA (several years)

OF does not use chemical fertilizers, so farmers must rely on organic nutrient sources (manure, crop and agri-industry residues, legumes, etc.). Using these types of fertilizers implies that the soil receives a greater amount of organic matter than in conventional systems. This drives the accumulation of organic carbon, generating a synergistic effect of adaptation and climate change mitigation by increasing climate resilience, while carbon from the atmosphere is trapped in the soil. Organic farming systems store significantly more carbon than conventional systems, but the magnitude of this sequestration depends to a large extent on the quality and quantity of organic inputs (Aguilera et al. 2013a). For example, several studies on Andalusian olive groves have shown that the chopping of pruning residues, the use of olive solid waste compost as an organic amendment, and the maintenance of vegetation cover in olive groves significantly increase the soil's organic matter and remove CO_2 from the atmosphere, these practices being common among organic olive farmers (Nieto Cobo 2011; Gómez-Muñoz 2011). On the other hand, organic fertilizers in Mediterranean conditions contribute to the reduction of direct and indirect emissions of nitrous oxide (N_2O), a potent greenhouse gas (Aguilera et al. 2013b). As a result, Spanish OF is significantly contributing to reducing GHG emissions (Aguilera et al. 2015a, b), especially in woody crops, such as olive groves, and subtropical or citrus crops (Aguilera et al. 2015a).

Organic farmers often work with a wider range of crops, including legumes. Diversification contributes to resilience against variations in climate, mainly through better control of pests and diseases, a more efficient use of natural resources and a reduction in economic risks. In addition, the use of traditional crop varieties and breeds is spreading in ecological agriculture and livestock, this latter sector being recognized as a refuge for the conservation of agrodiversity (CAP 2012). For example, 75% of outdoor organic horticulturists in Andalusia and Castilla-La Mancha use some traditional crop variety, which is related to the development of local markets as a destination for organic production (Martín Sánchez et al., in preparation). Together, Mediterranean varieties and breeds represent a huge genetic reservoir that is ideal for selecting specific adaptations in a context of changing climatic conditions (Di Falco and Chavas 2008). In addition, traditional varieties often generate a greater amount of residue available for the soil, as is the case, for example, of cereal varieties, without differences in grain yields under rainfed conditions (Carranza-Gallego et al., under review).

Spanish OF not only leads to crop diversity, but also favors wild varieties. Several comparative studies conducted in Spain show that the biodiversity of wild flora and arthropods is greater in ecological farms (Jerez-Valle et al. 2015; Ponce et al. 2011; Chamorro et al. 2016). In addition, ecological farms usually consume less non-renewable energy, mainly because they avoid the use of synthetic nitrogen fertilizers that require intensive use of fossil energy. At the same time, unlike in temperate areas, the use of machinery in Mediterranean farming systems is often no greater than in conventional systems (e.g., Kavargiris et al. 2009; Alonso and Guzmán 2010). In a study on 78 pairs of ecological and conventional farms, Alonso and Guzmán (2010) found an increase in the efficiency of non-renewable energy and a reduction in its consumption compared to their conventional counterparts. On average, organic farming uses 24% less non-renewable energy. Therefore, the use of non-renewable energy in Spanish agriculture could be reduced considerably if the area dedicated to organic farming were increased. Meanwhile, Spanish agriculture would become more autonomous. There is, however, still room to increase non-renewable energy savings in organic farming through the application of other cleaner technologies (photovoltaic, biofuels, etc.), reduction of tillage and so on.

Overall, OF can make a notable contribution to mitigating climate change, but OF also represents an adaptation to a context of increasing global energy shortages. Spain is very energy dependent, making the country especially vulnerable to possible supply cuts or price increases caused by oil peaks and the projected impact of climate change on global trade (Isbell 2006). In this regard, dependence on fossil energy in Spanish agriculture needs to be urgently reduced, and OF can help achieve this goal.

Lastly, OF is currently a viable economic alternative for many farmers and ranchers. Demand for organic food is growing even over the latest years of crisis (Willer and Lernoud 2017) driven by the population's health and environmental concerns. Alonso et al. (2008) have shown that, on average, OF obtains lower yields in Spain, but it also benefits from higher prices, higher revenues, more stable costs—upward or downward cost trends are avoided, as costs vary according to each crop—and a more favorable economic balance. The higher differential price obtained by organic farmers is related both to the product's higher end price and to receiving a greater share of those end prices, via shorter distribution channels (home deliveries, consumer associations and cooperatives, online sales, bio-trade fairs, on-farm sales and direct supply to specialized stores and social consumption centers—schools and hospitals—among others). Their income is directly influenced by yields and prices, but also by received subsidies. European organic producers have access to specific agri-environmental aids for their management. OF farmers' average age is considerably lower than that of conventional farmers, and the percentage of women heading organic farms is higher than in the rest of the sector. In Spain at least, OF is a positive factor of generational change and maintenance of farms in the country's interior, where depopulation risks are greatest.

OF is, therefore, a more sustainable alternative to industrial agriculture. It seems to reverse, in part at least, the degradation of agroecosystems' fund elements. Nevertheless, OF has to overcome some major challenges. As pointed out in another study (Ramos-García et al. 2018), OF is undergoing an important process of *conventionalization*; i.e., alternative features that distinguish it from the dominant model are being eroded, and therefore OF is losing its ability to reproduce an organic version of the conventional model. The reasons for this should be sought in its institutional framework: The practices of OF directed toward sustainability are penalized economically. OF is thus pushed toward a behavior similar to that of conventional producers, using a large amount of industrial inputs or selling production via channels controlled by large supermarket chains. If institutional obstacles are not removed and no strongly supportive public policies are implemented, the conventionalization process mentioned above may not only limit OF growth, but also seriously affect its capacity to prevent a collapse and reverse the tendencies that threaten the future viability of Spanish agriculture.

References

Aguilera E, Lassaletta L, Gattinger A, Gimeno BS (2013a) Managing soil carbon for climate change mitigation and adaptation in Mediterranean cropping systems: a meta-analysis. Agricu Ecosyst Environ 168:25–36

Aguilera E, Lassaletta L, Sanz-Cobena A, Garnier J, Vallejo A (2013b) The potential of organic fertilizers and water management to reduce N_2O emissions in Mediterranean climate cropping systems. Agric Ecosyst Environ 164:13–52

Aguilera E, Guzmán GI, Alonso Mielgo AM (2015a) Greenhouse gas emissions from conventional and organic cropping systems in Spain, I. Herbaceous crops. Agron Sustain Dev 35:713–724

Aguilera E, Guzmán GI, Alonso Mielgo AM (2015b) Greenhouse gas emissions from conventional and organic cropping systems in Spain, II. Fruit tree orchards. Agron Sustain Dev 35:725–737

Alonso Mielgo AM, Guzmán Casado GI (2010) Comparison of the efficiency and use of energy in organic and conventional farming in Spanish agricultural systems. J Sustain Agric 34:312–338

Alonso Mielgo AM, Guzmán Casado GI, Foraster Pulido L, González Lera R (2008) Impacto socioeconómico y ambiental de la agricultura ecológica en el desarrollo rural. In: Guzmán Casado GI, García Martínez AR, Alonso Mielgo AM, Perea Muñoz JM (coords.) Producción ecológica. Influencia en el desarrollo rural. Ministerio de Medio Ambiente y Medio Rural y Marino, Madrid, pp 72–267

Chamorro L, Masalles RM, Sans FX (2016) Arable weed decline in Northeast Spain: does organic farming recover functional biodiversity? Agric Ecosyst Environ 223:1–9

Di Falco S, Chavas JP (2008) Rainfall shocks, resilience and the effect of crop biodiversity on agroecosystem productivity. Land Econ 84:83–96

Del Pozo de la Calle S, Cuadrado Vives C, Ruiz Moreno E, Valero Gaspar T, Ávila Torres JM, Varela Moreiras G (2012) Valoración Nutricional de la Dieta Española de acuerdo al Panel de Consumo Alimentario. Ministerio de Agricultura y Pesca, Alimentación y Medio Ambiente, Madrid

Gómez Muñoz B (2011) Desarrollo y optimización de un protocolo de fertilización para el olivar ecológico. Universidad de Jaén, Tesis doctoral inédita, Jaén

Infante-Amate J, Aguilera E, Palmeri F, Guzmán G, Soto D, García-Ruiz R, González de Molina M (2018) Land embodied in Spain's biomass trade and consumption (1900–2008): historical changes, drivers and impacts. Land Use Policy 78:493–502

INE (Instituto Nacional de Estadística) (2017) Anuario de estadística agraria. Madrid

Isbell P (2006) La dependencia energética y los intereses de España. Real Instituto Elcano, Área: Economía y Comercio Internacional 32:1–7

Jerez-Valle C, García-López PA, Campos M, Pascual F (2015) Methodological considerations in discriminating olive-orchard management type using olive-canopy arthropod fauna at the level of order. Spanish J Agric Res 4(13):1–14

Kavargiris SE, Mamolos AP, Tsatsarelis CA, Nikolaidou AE, Kalburtji KL (2009) Energy resources' utilization in organic and conventional vineyards: energy flow, greenhouse gas emissions and biofuel production. Biomass Bioenergy 33:1239–1250

MAPAMA (Ministerio de Agricultura y Pesca, Alimentación y Medio Ambiente) (2017a) Agricultura Ecológica. Estadísticas 2016. Madrid. Available in: http://publicacionesoficiales.boe.es/

MAPAMA (Ministerio de Agricultura y Pesca, Alimentación y Medio Ambiente) (2017b) Anuario de Estadística, 2016. Madrid. Available in: http://publicacionesoficiales.boe.es/

MINISTERIO DE INDUSTRIA, COMERCIO Y TURISMO, SECRETARÍA DE ESTADO DE COMERCIO (MINECO) (2017) Estadísticas del Comercio Exterior. Available in: http://www.comercio.mineco.gob.es/es-ES/comercio-exterior/estadisticas-informes/Paginas/estadisticas-comercio.aspx. Accessed 20 July 2018

Nieto Cobo OM (2011) Propiedades de los suelos de olivar con diferentes manejos. Simulación del carbono orgánico fijado aplicando el modelo Roth. Universidad de Granada, Tesis doctoral inédita, Granada

Ramos-García M, Guzmán Casado GI, González de Molina M (2018) Dynamics of organic agriculture in Andalusia: moving towards conventionalisation? Agroecol Sustain Food Syst 42(3):328–359

Willer H, Lernaud J (2017) The world of Organic Agriculture. Statistics and emerging trends 2017. Rheinbreitbach, Germany. Research Institute of Organic Agricultura (FiLB) and IFOAM - Organic International, pp 143–148

Witzke H, Noleppa S (2010) EU agricultural production and trade: Can more efficiency prevent increasing "land grabbing" outside of Europe? Humboldt University of Berlin, Faculty of Agriculture and Horticulture, Berlin

Annex I
Calculation of the Physical Production Series of Spanish Agriculture

In this annex, we describe how we built the physical production series that supports the results described in this book. In other works, we developed the methodological aspects and theoretical foundations based on the adaptation of social metabolism methodologies to the characteristics of agriculture and livestock (Guzmán et al. 2014; Aguilera et al. 2015; Soto et al. 2016a, b; Guzmán et al. 2017). In this section, we will therefore limit ourselves to analyzing the sources used and the problems we faced when estimating series for which there is no direct information in Spanish primary sources.

A.1.1 Sources and Methodological Decisions to Calculate the Domestic Extraction of Vegetal Biomass

The examination of the biophysical evolution of Spanish agriculture requires collecting and processing a huge amount of quantitative information. This requirement limits the chronological reach of this type of research. The pivotal moment for agricultural statistics was the constitution in 1879 of the agronomic service that used provincial agronomists and was directed by the Provincial Advisory Board (later the Agronomic Advisory Board, the *Junta Consultiva Agronómica* or *JCA*, by its Spanish acronym used hereon),[1] part of the Ministry of Public Works. Spain's modern agricultural statistics began with this reorganization of the agronomic services.

The regulation approved in 1891 established, as recommended by the international organizations of the time, an indirect method to calculate production; it was

[1]The detailed history of these processes and an evaluation of their reliability can be found in GEHR (1991). A more recent review of historiographic opinions on the construction of statistics and their reliability can be consulted in Soto Fernández (2002)

obtained by multiplying average yields by the harvested surface. This indirect procedure, which remained in force until 1928, was designed to avoid concealment of production. Therefore, the reliability of the data during the first third of the twentieth century is determined by the accuracy of the calculation of the yields and by the quality of the information collected on the harvested area. Yields were obtained through field observation by provincial engineers. The harvested area was obtained thanks to all available information (assessments and cadastral information) so its reliability would also vary according to the provinces' different abilities in performing advanced cadastral work.

During the 1880s, the JCA gathered information (mainly from the second half of the decade) that culminated in the publication between 1890 and 1892 of specific reports on olives, vineyards, and cereals in addition to another on livestock.[2] The JCA began to publish annual statistical data on the main productions of the Mediterranean trilogy (cereals and legumes, vineyards, and olives) from 1891. Complete data on surface areas, production, and yields were only published as from 1898. From this moment until 1928, we dispose of two sources of information on crop production that differentiate themselves by the quality and quantity of the data provided. The first source is an annual series of cereals and legumes, vineyards, and olives. These groups represented between 63% and 68% of primary crop production in dry matter until 1936. They undoubtedly constitute the most reliable data, both in terms of temporal continuity of the series, because they represent the productive basis of Mediterranean Spain, where the state's administration had more control over the territory. We do not dispose of a continuous series for the rest of the production until 1928. We do dispose of partial information that allows us to reconstruct practically all the crop production in 1910 and some of the major production in 1900. In addition to annual statistics, the provincial engineers of the JCA had to advance annual reports on specific topics (irrigation and fertilization). Some of these advance reports offer sufficient information on several crops. In 1902, a first attempt was made to collect data on other produced crops, but they did not achieve a wide coverage.[3] In 1904 and

[2]Dirección General de Agricultura, Industria y Comercio, *Avance estadístico sobre el cultivo cereal y de leguminosas asociadas en España formado por la Junta Consultiva Agronómica, 1890, quinquenio de 1886 a 1890 ambos inclusive*, Madrid, Tipografía de L. Peant e hijos. Dirección General de Agricultura, Industria y Comercio, *Avance estadístico sobre el cultivo y producción del olivo en España formado por la Junta Consultiva Agronómica, 1888,* Madrid, Tipografía de L. Peant e hijos, 1891. Dirección General de Agricultura, Industria y Comercio, *Avance estadístico sobre el cultivo y producción de la vid en España formado por la Junta Consultiva Agronómica, 1889,* Madrid, Tipografía de L. Peant e hijos, 1891. Dirección General de Agricultura, Industria y Comercio, La ganadería en España. *Avance sobre la riqueza pecuaria en 1891 Formado por la Junta Consultiva Agronómica conforme a las memorias reglamentarias que en el citado año han redactado los ingenieros del servicio agronómico,* Madrid, Tipografía de L. Peant e hijos, 1892.

[3]In particular, potatoes, sugar beet, fodder beet, turnips, saffron, flax, hemp, orange, lemon, carob, pomegranate, almond, fig trees, and apple trees. Ministerio de Agricultura, Industria, Comercio y Obras Públicas, Dirección General de Agricultura, *Noticias estadísticas sobre la producción agrícola española por la Junta Consultiva Agronómica, 1902*, Imprenta Alemana, Madrid, un dated.

1912, records were dedicated to meadows and pastures,[4] those of 1910 to fruit trees and bushes as well as tubers,[5] and those of 1911 to horticultural and industrial plants.[6] This first stage in the history of Spanish agricultural statistics rounded up in the 1922 compendium, which for the first time succeeded in producing a global evaluation of agricultural production in Spain as well as the value of pasture and woodland production.[7]This latter publication is especially important, not only because it is the most detailed in the first third of the twentieth century, but because it offers differentiated data according to harvested and cultivated areas.

The 1922 compilation continued to be produced following new legislation in 1927 that established a semi-direct method requiring farmers to declare cultivated areas. Over this period, the Local Agricultural Information Boards, a new administrative layer, was established at the municipal level to check the veracity of information provided by farmers. The provincial agronomic section had the task of verifying the accuracy of the surface data, calculating average yields per hectare and, finally, obtaining the production figures. As a result, the *Statistical Yearbooks of Agricultural Production* started to be published as from 1928. They collected all agricultural productions annually for the first time. It would be a fundamental part of Spain's agricultural statistics until the creation, in 1972, of the Yearbooks of Agricultural Statistics (that refound agricultural, forestry, and livestock statistics).[8] Despite the fact that the 1927 rule change led to more complete information, not all reliability problems were solved. Some production data, such as horticultural production, are

---

[4]However, only that of 1912 has reasonably reconstructed the uses of land and production. The livestock report of 1891 is more complete in this sense than that of 1904. Ministry of Agriculture, Industry, Commerce and Public Works, meadows and pastures. Summary made by the Agricultural Advisory Board of the reports on this subject submitted by the chief engineers of the National Agronomic Service section, Madrid, Imprenta de los hijos de M.G. Hernández, 1905. Ministerio de Fomento, Dirección General de Agricultura, Minas y Montes, *Avance estadístico de la riqueza que en España representa la producción media anual de Pastos, prados y algunos aprovechamientos y pequeñas industrias zoógenas anexas. Resumen hecho por la Junta Consultiva Agronómica de las memorias de 1912, remitidas por los ingenieros del servicio agronómico provincial*, Madrid, Imprenta de los hijos de M.G. Hernández, 1914.

[5]Ministerio de Fomento, Dirección General de Agricultura, Minas y Montes, *Avance estadístico de la riqueza que en España representa la producción media anual de árboles y arbustos frutales. Tubérculos, raíces y bulbos. Resumen hecho por la Junta Consultiva Agronómica de las memorias de 1910, remitidas por los ingenieros del servicio agronómico provincial*, Madrid, Imprenta de los hijos de M.G. Hernández, 1913.

[6]Ministerio de Fomento, Dirección General de Agricultura, Minas y Montes, *Avance estadístico de la riqueza que en España representa la producción media anual de las plantas hortícolas y plantas industriales. Resumen hecho por la Junta Consultiva Agronómica de las memorias de 1911, remitidas por los ingenieros del servicio agronómico provincial*, Madrid, Imprenta de los hijos de M.G. Hernández, 1914.

[7]Ministerio de Fomento, Dirección General de Agricultura, Minas y Montes, *Avance estadístico de la producción agrícola en España. Resumen hecho por la Junta Consultiva Agronómica de las memorias de 1922, remitidas por los ingenieros del servicio agronómico provincial*, Madrid, Imprenta de los hijos de M.G. Hernández, 1923.

[8]All the yearbooks, as well as the previous annual statistics, are digitized in http://www.mapama.gob.es/es/estadistica/temas/publicaciones/anuario-de-estadistica/ retrieved on June 26, 2018.

expressed in non-metric units and non-convertible measures (units and bundles). In the case of forage plants, the 1922 advance describes major maize-related cultivation in the northwest that only started to be reflected in the Yearbooks in the 1940s (Soto Fernández 2006). Problems could still be found in recent decades. In 1983, José Manuel Naredo pointed out that statistics had been progressively simplified (in physical terms) and that yearbooks did not explicitly describe the methodology used to collect information making it difficult to review them critically (Naredo 1983). Spanish agricultural statistics are generally variably assessable, especially those provided in the first decades of statistical service. The consensus (GEHR 1991) is that the biggest problem is that of measuring surface area for lack of a cadastre covering the whole territory. Nevertheless, information quality has visibly improved over time, and data on Mediterranean agrarian system's hard core (cereals, vineyard, and olives) have been relatively accurate.

Either way, we have to rely on these data sets as starting information to reconstruct Spanish agriculture biomass flows. The first step (followed to calculate the NPP described above) was to calculate domestic extraction of biomass. To begin with, in order to solve the problem of incomplete annual crop statistics over the whole period, we chose to rebuild the evolution of biomass extraction based on 12 points in time (1900, 1910, 1922, 1933, 1940, 1950, 1960, 1970, 1980, 1990, 2000, 2008)[9] organized into crop yield 5-year averages whenever possible. We had to perform a data estimation for the 1900–1933 period. Horticultural crop production that was not expressed in kilograms was obtained by multiplying the surface area by average yields from the early 1950s.[10] For 1933, data from 1922 were replicated for forages not included in the yearbooks over those years. To calculate 1900 forages, we took the 1891 advance surface data information and used the same yields as in 1910. For fruit trees and industrial crops not accounted for in 1900, we replicated the 1910–1922 trend for the 1900–1910 period.

This first set of data was based on statistics as well as estimations for products not included at the start. The set enabled us to reconstruct the entire Spanish crop primary production or grain production between 1900 and 2008. It was also necessary to know the broad categories of the production's destination (human food, animal feed, seeds, industrial materials, fuel) to rebuild this information and that of biophysical flows. This helped to differentiate between broad net primary production categories (socialized plant biomass and reused biomass) and to estimate domestic extraction, as well as total biomass consumption, food consumption, and the possibilities of livestock feeding. There was not enough accurate information for the first decades of the twentieth century to estimate the destination of crop production. But since 1961,

[9] 1922 is justified as the year of publication of the 1922 advance. 1933 is chosen for two reasons. The first reason is methodological, to avoid using the means of the first years of operation of the statistical system of 1927. The second is historiographic: 1933 allows us to have a chronological point of reference portraying agriculture during the republican stage. The year 2008 can be justified by the fact that at the time we conducted calculations, the latest data published were from 2010, so we had to use the mean between 2006 and 2010.

[10] We opted for the yields of the 1950s to solve the problem of the generalized fall in yields in the postwar period observed for most crops.

the FAO[11] database offers an annual series of biomass production destinations. We used this series for the 1960–2008 period. For the previous period, we considered the same percentages as for 1960, since Spain's agricultural industrialization was just beginning.

To rebuild domestic extraction, i.e., society-appropriated biomass, we also needed to know society's use of residue production, either to generate energy (firewood) or to feed livestock, and evaluate the amount of burned biomass that was not recirculated in agroecosystems. The EW-MFA literature usually calculates residue based on fixed utilization rates based on harvest rates that are usually fixed as well (Wirsenius 2003). This method, however, cannot be applied in the case of a historical analysis for several reasons. Firstly, as pointed out in the description of the NPP calculation methodology, varietal changes led to a reduction in straw production compared to that of grains as the Green Revolution's technologies were progressively introduced. Second, changes in livestock structure have also led to a drop in the weight of straw with respect to animal feed. Finally, changes in the energy system meant that firewood lost significance as a fuel, although this evolution took place later and less steadily than usually described (Infante-Amate and IriarteGoñi 2017).

Fortunately, from the outset, Spanish agricultural statistics included information on the amount of straw harvested from cereals and legumes, which we can relate to the amount of straw actually used (mainly as animal feed). This enabled us to elaborating a series of straw harvest indexes used. In all cases, the amount of straw used gradually declined throughout the twentieth century, as shown in Table A.1.1. The statistics do not provide information on residues for the remainder of the crops. Based on the harvest indexes of Guzmán et al. (2014), we estimated a 10% use of total residues in the case of tubers and horticultural plants. In the case of woody crop firewood, the statistics offer partial and discontinuous information on some crops (vineyards and olives but not fruit). We made our own estimate on the amount of firewood from woody crops based on a systematic search for yields.[12] In the case of biomass domestic extraction, burned residue quantities are usually included in the appropriation of biomass (Eurostat 2015). There is no evidence of residue burning in Spain until after 1940. We assumed that the burning of residues began in 1950 and gradually increased until reaching a peak in 1970. Residue burning has dropped sharply since it was restricted by 1990 legislation. We used data from the *National Emissions Inventory* to calculate the percentage of residues burned before and after 1990 (MARM 2008a, b, c).

Though we could rebuild crop production reasonably well by simply using estimates for some products at the beginning of the series, we could not do the same for pasture and forestry production. There are no physical estimates of pasture production and utilization (except for a limited period between 1973 and 2003, though data are incomplete). As we explained previously, to solve this problem we calculated pastures' NPP based on representative studies of three major Spanish agroclimatic regions. These productivity data per hectare were related to the evolution of land uses

[11] http://www.fao.org/faostat/es/#home. Retrieved on June 29, 2018.

[12] This estimate is detailed in Infante-Amate et al. 2014a, b).

Table A.1.1 Indices to calculate the straw used (units of straw used per unit of grain) of major cereals and legumes

	1900	1910	1922	1933	1940	1950	1960	1970	1980	1990	2000	2008
Rice	0.8	0.8	0.7	0.7	0.8	0.9	0.7	0.6	0.0	0.0	0.0	0.0
Barley	1.2	1.2	1.1	1.2	1.2	1.3	1.2	1.1	0.9	0.6	0.5	0.5
Rye	1.7	1.7	1.7	1.7	1.7	1.6	1.4	1.4	0.9	0.7	0.7	0.5
Corn	1.2	1.2	1.2	1.2	1.2	1.2	1.1	1.0	0.0	0.0	0.0	0.0
Wheat	1.5	1.5	1.4	1.4	1.5	1.5	1.4	1.1	0.8	0.7	0.5	0.4
Chickpeas	1.0	1.0	0.9	1.2	1.2	1.5	1.1	1.1	0.9	0.6	0.4	0.3
Green peas	1.4	1.4	1.3	1.3	1.3	1.3	1.1	1.2	1.0	0.5	0.4	0.3
Broad beans	1.1	1.1	1.0	1.2	1.2	1.3	1.0	1.2	1.0	0.8	0.5	0.4
Beans	1.3	1.3	1.5	1.5	1.4	1.7	1.0	1.1	1.0	1.0	0.5	0.4
Lentils	1.2	1.2	1.4	1.4	1.4	1.4	1.1	1.2	1.1	0.6	0.4	0.3

Source Authors' compilation based on the Yearbooks of Agricultural Production and the Agricultural Statistics Yearbooks broken down per province over the entire period. To calculate the biomass actually grazed, we applied a livestock feed balance model similar to that used in other studies (Krausmann et al. 2008). The estimates are detailed below since they are directly related to livestock data. We calculated, on the one hand, the livestock feeding needs from the livestock censuses; on the other hand, we calculated the availability of food for livestock from crops and residues, as well as food from foreign trade. The grazed biomass is the difference between the availability of food from crops and trade and stockfeed needs.

Firewood and wood statistics were published later than agricultural data. They began in 1946 with the publication of Spain's Forestry Statistics (subsumed as from 1972 in the Yearbooks of Agrarian Statistics). But the data from the first years are problematic, especially regarding firewood. Serial data are only provided from the late 1950s. We had to make several estimates for the previous period (Iriarte, 2013; Infante et al. 2014c). Wood extraction was considered to be equal to consumption (for which we dispose of serial data since the nineteenth century) deducting exports plus imports. We calculated firewood extraction by systematizing and homogenizing the evolution of land uses at a provincial level. We then applied yield data per hectare taking into account regional variability and historical changes in forest production. To this end, a systematic review of various Spanish sources was conducted.

A.1.2 The Reliability of Livestock Censuses

The criticism of Spanish livestock statistics that recurs the most is that found at the time by GEHR (1991) in the Spanish agricultural statistics compilation between 1865 and 1935. The group gathered all livestock counts carried out between 1865 and 1933, only 6 of which provided some complementary information beyond the number of heads.[13] Along with the quantitative information, the reports of the Agronomic Advisory Board of 1891 and 1917 provide a huge amount of highly valuable complementary information on the changes in races and average weights, livestock management, and productions (*Ministerio de Fomento*, 1892, 1920). In most cases, such precarious information makes it very difficult to know when to trust the censuses. Furthermore, in pre-Civil War censuses, we cannot be sure whether offspring were included except for the counts in 1865, 1917, 1920, 1924, 1929, 1931, 1933, and 1935. These problems among others have caused many authors to doubt the validity of the 1891 census and, even that of the early twentieth century censuses (Simpson 1989; Jiménez Blanco 1986; Gallego 1986), and to seriously question the evolution of Spanish livestock based on this data.

[13]They are the censuses of 1865, 1917, 1920, 1924, 1929, and 1933, to which we added the corrections made to the census of 1929 in 1931, published in the Statistical Yearbook of Agricultural Production of that year, plus the census of 1935 that the GEHR does not include (Sierra & Sierra, 1938).

This does not much improve in the years immediately after, neither in terms of quantity of data nor reliability of the figures. Quite the reverse. Between 1939 and 1959, we only dispose of counts for 6 years: 1939, 1940, 1942, 1948, 1950, and 1955. The last three censuses do not include offspring aged under one year. The first three, in turn, offer numbers that are relatively similar to those of the first half of the 1930s and fairly higher than those of the 1960s. These figures do not seem very trustworthy, given the negative impact that the war and postwar period should have had on labor and transport cattle raising as well as income livestock. In fact, the statistics show a significant drop in livestock production (meat and milk), consistent with food shortages confirmed by several studies, including one of our own analyses on Spaniards' apparent consumption (González de Molina et al. 2013). However, total herd heads were as numerous as in the 1930s and there is no relationship between the total number of heads and their productivity in terms of meat and milk. If pre-Civil War censuses have been criticized for underestimating livestock, later ones can be criticized for having overestimated them. We can corroborate this understanding, however, by comparing availabilities and food needs, enabling us to know whether the livestock reflected in postwar census could be fed with the feed available.

The situation would change substantially after 1960. As of that date, we dispose of annual livestock censuses, with the sole exception of 1961. The methodology and production dates (the vast majority are situated between September and November) are consistent throughout providing a wide range of information beyond the number of heads. There is a reasonable degree of certainty about the validity and homogeneity of the series.

Another difficulty was that of converting census figures, expressed in number of heads, and applying a single indicator expressed in kg or t, in live weight, which allows comparing the different censuses and establishing a livestock series over time. Virtually, no census offers live weight data, except for 1917 in the pre-war period and several years during the 1950s and 1960s. For the period after 1950, it is relatively easy to obtain live weight data or data on the average weight of manure-producing species in the censuses. For the period before that, however, the issue is much more problematic. The majority of historiographic authors (GEHR 1978, 1979; García Sanz 1991; Muñoz Rubio 2015) have opted to use the average weight data provided by Flores de Lemus (1951).

However, this choice does not seem adequate: It does not take into account livestock breed changes brought about during the first third of the twentieth century (Fernández Prieto 1992), nor the stockfeed transformations, i.e., the greater presence of forage and grain compared to the pasture that prevailed in the nineteenth century. We tried to solve this problem by analyzing the information available in the 1891 and 1917 advances or records for all Spanish provinces, constructing two series of average weights for each of these years (Table A.1.2). As shown in Table A.1.2, our figures are similar to those of Flores for 1917, but not for 1891, suggesting that significant changes took place between 1891 and 1917, especially regarding cattle. We performed our own calculations based on the 1891 and 1917 compendium, the post-Civil War censuses, the 1865 census, and estimates by García Sanz for 1752. Using this estimate, we drew a linear evolution of the live weight of Spanish livestock

Table A.1.2 Average livestock weights according to different sources, in kilograms

	1891	1917	Flores de Lemus
Cattle	270	375	371
Sheep	33	33	30
Goats	33	33	34
Pigs	62	69	77
Horses	325	325	326
Mules	325	359	326
Donkey	182	182	172

Source 1891: Author's compilation based on the *Ministerio de Fomento*, 1892, 1917: Author's compilation based on the *Ministerio de Fomento*, 1920, Flores de Lemus, 1951

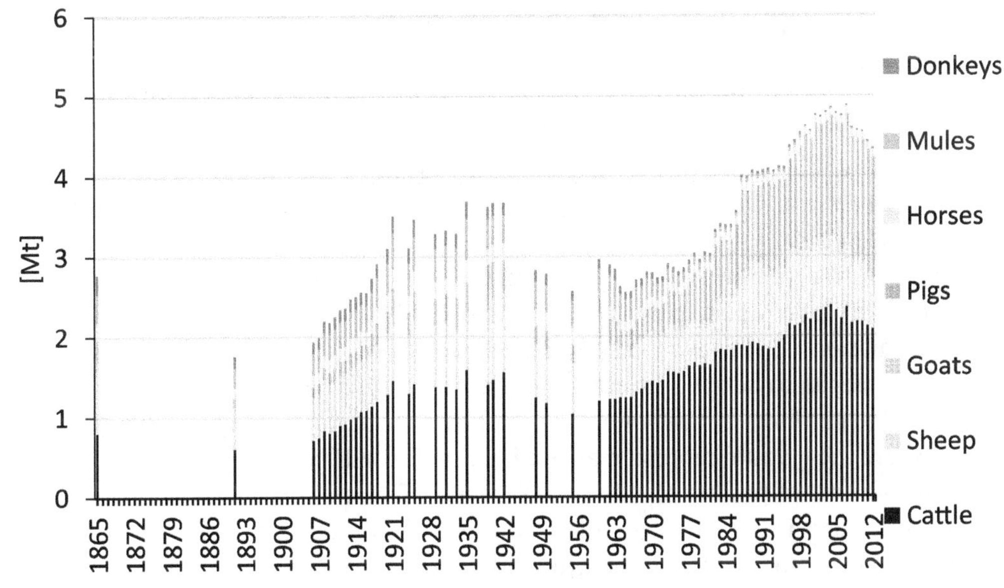

Graph A.1.1 Live weight of livestock in Spain, censuses data, in tons. *Sources* Author compilation based on GEHR (1991), Sierra & Sierra, 1938, Spanish Statistics Yearbooks, Agricultural Statistics Yearbooks

from the mid-eighteenth century to 2012, excluding poultry or rabbits. The results are shown in Graph A.1.1.

The evolution shown in the graph reveals the problems related to livestock data sources. The drop between 1865 and 1891 seems excessive. Figures for the last decades fail to include poultry or changes to pig livestock that produced offspring practically all year round when the sector became industrialized, leading us to believe total livestock live weight is underestimated for those years. However, it is one thing to point out the problems of censuses and quite another to establish corrective mechanisms. Despite this, we tried different alternatives to correct the figures. In the following section, we explain the comparison mechanisms relating to livestock in the first half of the twentieth century. The results show that the only years of the first half of the twentieth century that seem to include a significant census underestimation are those of the first two decades of the twentieth century (the 1891 census would also be logically underestimated). Based on this same method, the livestock census figures of the 1920s and the first half of the 1930s are unlikely to be underestimated, because apparent consumption is fairly balanced given livestock needs. As a result, we increased the number of heads showed in the censuses between 1891 and 1917 to balance needs with apparent consumption. It is not possible to carry out the same exercise for the whole of Spain for 1865, since we lack detailed information on agricultural production. Nevertheless, according to most historiographic sources, this census is not overestimated.

As we have seen, a major postwar census problem is the exclusion of offspring in 1948, 1950, and 1955 censuses. To solve this problem, we increased the census of 1948, 1950, and 1955, applying the factor of correction proposed by Galindo (1969), that is, adding a number of offspring equivalent to a percentage of the total heads of each species: 23% for sheep, 22% for cattle, 19% for goats, 42% for pigs, 11% for horses, 3.5% for mules, and 5.7% for horses. When taking this adjustment into account, the census figures are still somewhat higher than those of the 1960s, although lower than those of 1939, 1940, and 1942.

The last problem we addressed is the exclusion of poultry and rabbits. Very few censuses offered data throughout the twentieth century. We have some information on poultry for the years 1908–12, 1929, and 1933. The censuses after 1955 provide the number of layers and, between 1955 and 1970, the number of adult animals. This information is, however, insufficient since it does not reflect the importance that poultry began to adopt as meat providers. To correct these deficiencies, we first calculated the number of adult animals, increasing the percentage of layers according to what share they would represent all adult animals for 1955–1970. We added to this figure the number of poultry slaughtered for meat. This estimate obviously has many weaknesses, but it is the only possible estimate we have today that allows evaluating the role of the poultry sector in the growth of Spanish livestock. For the series prior to 1908, we assumed that chickens accounted for the same percentage of livestock as in 1908–12.

Rabbits were less important quantitatively. We also counted the number of slaughtered animals and increased this figure by a percentage based on an estimate of breeding rabbits, taken from the years for which we did have this information. There were no data for this category before 1940, so we assumed that rabbits' share of livestock was the same as the average in 1940–43. The results of both estimates are shown in Graph A.1.2 for the 1891–2012 period. There is no doubt the census adjustments proposed in this work should be approached with great caution and used exclusively as an initial estimate to be compared with a more disaggregated territorial analysis. However, we do believe that the data provide a more accurate picture of the evolution of livestock than the raw data found in the censuses.

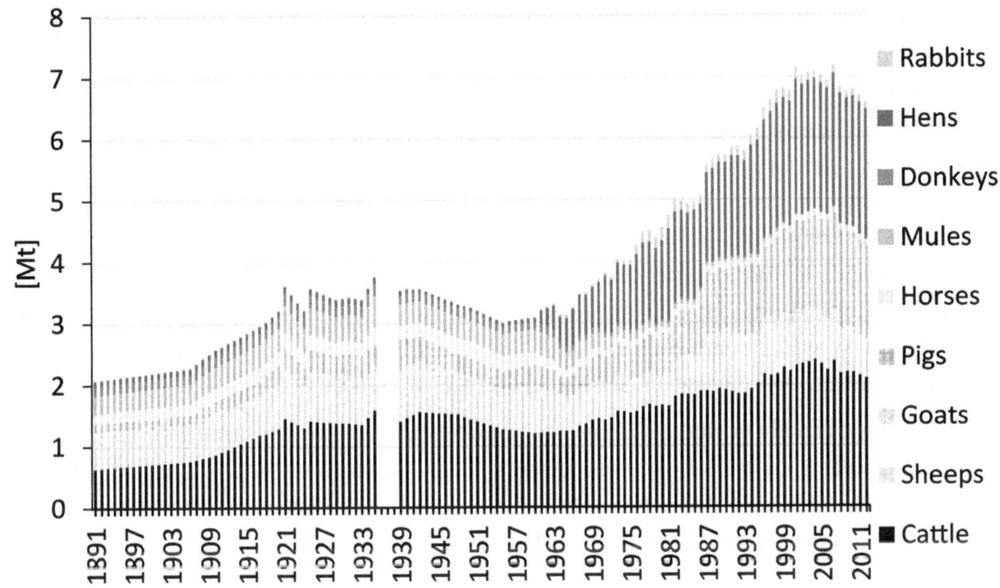

Graph A.1.2 Live weight of Spanish livestock. Estimate based on data presented in this annex, in tons. *Source* See text

A.1.3 Adjusting Spanish Livestock in the First Third of the Twentieth Century

As explained above, we have reasons to doubt the reliability of livestock censuses of the first half of the twentieth century. Studying the mechanisms used to draw up the censuses, which is only possible in some cases, does not allow making any reasonable hypothesis about how to correct the data. That is why we chose to compare the needs of livestock with livestock feed availability. Censuses in 1865 were only established for some provinces, and among these, Cordoba is the province offering the greatest quantity and quality of information. We are thus going to focus on Cordoba and perform an in-depth study of the livestock censuses, thus drawing conclusions that will be of utility for the whole of Mediterranean Spain. Also worthy of note, we consider that the statistical sources for the following agricultural crops be reliable: cereals, legumes, forage foods, and meadows, as explained in Annex I.

The objective is to adjust the livestock area density and use these figures to calculate the amount of plant biomass from the pasture lands and natural meadows during that period consumed by livestock, in kg of dry matter. The working hypothesis is that the 1906 to 1917 livestock censuses were underestimated. Our adjustments to livestock density for Cordoba Province were based on the following four assumptions:

1. Farm livestock should be reasonably adjusted to traction needs.
2. Livestock feeding needs should be consistent with available feed at all times.

3. Given its economic function and its management, the stabled cattle (adult equines and among adult cattle that dedicated to labor, milk production, and stallions) would consume mostly quality foods (grains, straw, and fodder); therefore, these stockfeed needs should also correspond to food supply.
4. Livestock density on pasture land, reserved mainly for income animals (sheep, goats, pigs, and beef cattle)[14] and young stabled cattle animals (horses and beef cattle) should correspond to their maintenance capacity. Specifically, we considered that a realistic percentage of use of pastures would be a 40–60% use of its potential, taking into account that below 40% of use, a low-cost resource available would be squandered, and above 40%, overgrazing could occur, given that wild herbivores also consume the same resource.

The heads of cattle collected in livestock censuses have been transformed into food needs (gigajoules of metabolizable energy). Likewise, feedstock supplies (grains, straw, fodder, grass) were transformed into metabolizable energy. The feed supplies include what is produced in the area under study added to the corresponding share of the balance of imports minus exports of livestock feed. In no case did we include the residues from agroindustries (beer, flour, sugar, olive, wine, etc.), nor any vegetal or potato residues, nor any grazing of rainfed or irrigated fruit trees. We did not include either any other by-products such as those deriving from fruit tree pruning (including olive groves.).

Livestock energy needs were calculated according to species, sex, age, livestock dedication (labor, meat, milk, etc.), and the type of management (stabled versus grazing). Given the significant energy cost of dairy production, the amount of milk production was also taken into account for milk cows

For 1929–30, the supply of livestock cattle labor is adjusted to the needs. The size of the available food supply is also adjusted to demand, implying a pasture density of 0.20 LU/ha, accounting for 50% of Cordoba Province's potential. Therefore, the livestock census seems to have provided relatively reliable figures.

We did not calculate supply and demand of work animals for 1865, since there are no data on land uses and production to work on. We believe the census is reliable, based on historiographic conclusions cited at the beginning of this annex. Furthermore, the livestock figures really do not seem to be underestimated, since the amount of available quality food represents 62% of the needs of the stabled cattle (in other considered years, this share would be above 100%) and the pasture grazing density would be the largest in the series (0.26 LU/ha) (Graphs A.1.3 and A.1.4). This year is an important year of reference.

For 1902, 1910, and 1917, we can confirm that the supply of work animals is well below the needs, implying that censuses for those years underestimated the figures. Given the suitable quality of the basic data available, supply and demand are adjusted (adult work animals multiplied by 2.04 in 1902, by 1.508 in 1910, and by 1.158 in

[14] These decisions are based on the cattle census records of 1917. *Ministerio de Fomento. Dirección General de Agricultura, Minas y Montes, Estudio de la ganadería en España*. Summary by the *Junta Consultiva Agronómica de las memorias de* 1917, *remitidas por los ingenieros del Servicio Agronómico Provincial, Madrid, Imprenta de los hijos de M.G. Hernández, 1920.*

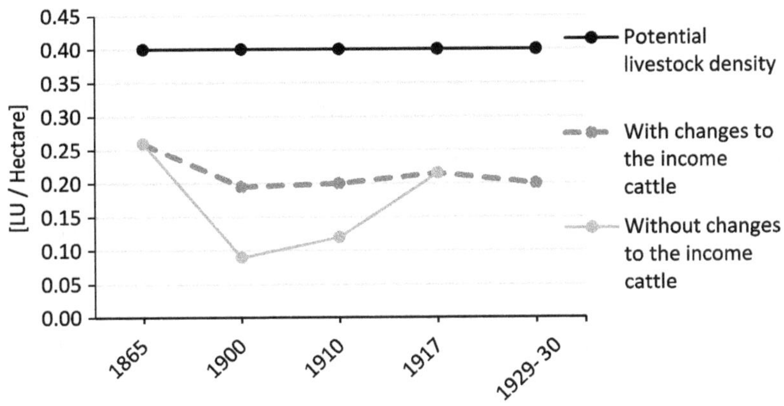

Graph A.1.3 Potential livestock density of Cordoba Province pastures and real cattle density adjusted to work animals in 1900, 1910, and 1917. Income livestock is modified in 1900 and 1910 in one scenario and not in the other. *Source* Own estimation

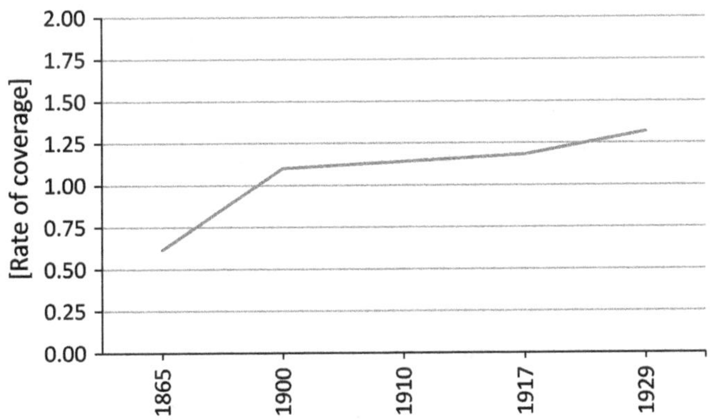

Graph A.1.4 Rate of coverage of stabled cattle needs for quality feed (work animals are corrected). *Source* Own estimation

1917. Work animals offspring multiplied by 1.289 in 1902, by 1.131 in 1910, and by 1.039 in 1917). We calculated the food needs of the modified cattle herd based on this data (Graph A.1.5).

With this modification, the 1917 livestock feed needs are consistent with the amount of available food, giving rise to a livestock density of 0.21 LU/ha. However, the 1902 and 1910 livestock figures would give us very low livestock pasture density (0.09 and 0.12, respectively), which would represent a sharp drop from the 0.26 LU/ha figure provided for 1865. This situation seems implausible, since it would also imply that agricultural land was used to feed livestock, while pastures remained underutilized. Therefore, we arbitrarily increased the income livestock for those two years, multiplying it by 2.5 in 1902 and by 2 in 1910. In this way, the pasture livestock density would remain at 0.20 LU/ha, similar to 1910 and 1917 figures (Graph A.1.3). It would lower the 1865 livestock density, but no longer dramatically (Graph

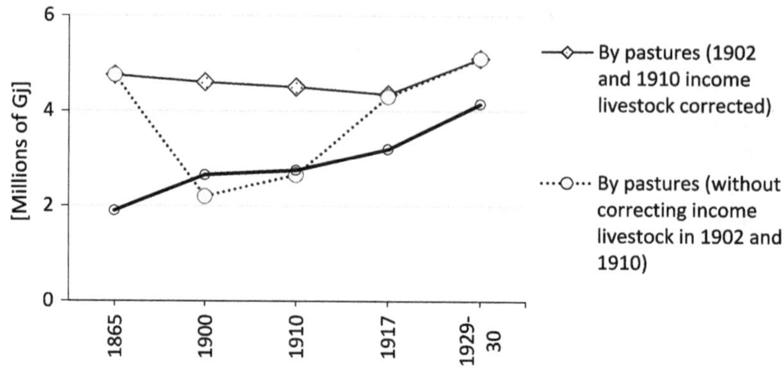

Graph A.1.5 Metabolizable energy provided by pastures and quality food to meet the needs of livestock, after modifying the livestock in 1900, 1910, and 1917. *Source* Own estimation

A.1.3). This less dramatic fall would be consistent with pasture area reductions to the benefit of cropland. On the other hand, in energy terms, pastures' contribution to metabolizable energy would remain almost constant over the period, which is a reasonable assumption. If we accept this data, we can conclude that income cattle concealment was proportionately greater than that of work animals in 1900 and 1910.

Based on Cordoba's case, we can draw further conclusions after modifying livestock figures applicable to the rest of the Mediterranean area (both dry and semiarid regions).

The share of quality livestock feed trended upward (29% in 1865, and 36, 38, 42, 45% in 1929) compared to that of pasture feed, which in absolute values remained at around 4.5–5 million GJ, while its contribution in relative terms dropped (Graph A.1.5). It went from 71% in 1985 to 55% in 1929. This makes the adjustment between quality food and stabled cattle unstable. Thus, if the coverage ratio of the metabolizable energy of quality food was 0.62 of stabled cattle needs in 1865, it was 1.08% in 1902, 1.12% in 1910, 1.17% in 1917, and finally, 1.31 in 1929 (Graph A.1.4).

The stocking rate remained at 50% of pasture density capacity; therefore, overgrazing does not seem to have caused the overexploitation of woodlands. It would reach 65% only in 1865. Deforestation was undoubtedly due to the collection of firewood, although livestock could have had a negative effect, even at very low density, on reforestation due to feeding on young shoots of sprouts and seedlings.

Based on the Cordoba livestock study, we performed livestock adjustments for dry and semiarid regions of Spain. We contemplated similar criteria for the Spanish Mediterranean coast, though we were limited by missing data for 1865, and we could not adjust employment supply and demand either. Therefore, we mainly relied on feedstock balance and followed the logic established in the case of Cordoba.

Considering the 1929 census is more dependable, we obtain results similar to those of Cordoba for the same year. Pasture stocking density would be 0.13, approximately 50% of the potential density (approximately 0.3) (Graph A.1.6). Likewise, the likely percentage of quality food metabolizable energy would be 136% of stabled cattle needs (Graph A.1.7), thus following the same pattern as that of Cordoba in that year.

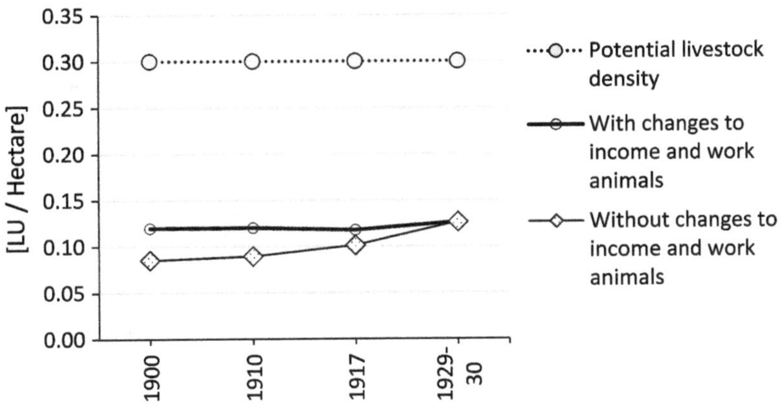

Graph A.1.6 Potential livestock density of pastures in dry and semiarid Spanish regions, and real livestock density adjusted for work animals in 1900, 1910, and 1917 and income livestock in 1910 and 1917. *Source* Own estimation

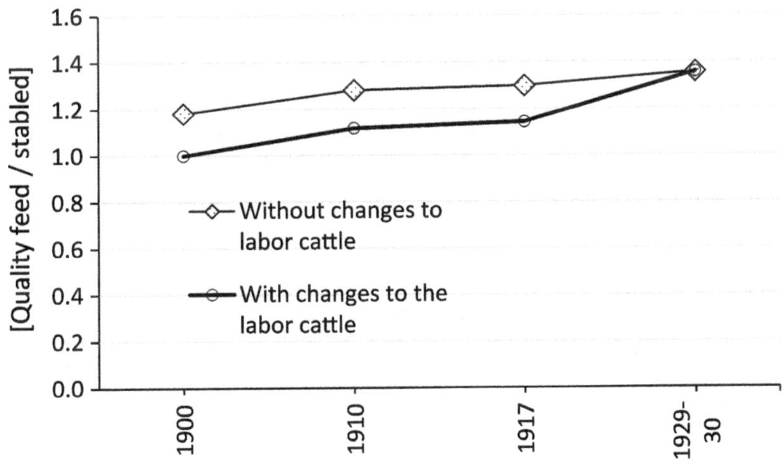

Graph A.1.7 Ratio of stabled livestock needs covered by quality feed. *Source* Own estimation

By 1900, 1910, and 1917, pasture livestock density would be low, that is: 0.09 LU/ha for 1900 and 1910, and 0.1 LU/ha for 1917. On the other hand, the ratio of quality feed with respect to stabled cattle needs would be 1.18% in 1902, 1.28% in 1910, and 1.30% in 1917. These values are higher than those of Cordoba Province. Both indicators, cattle grazing density and adjustment to quality food needs, would suggest that income livestock and stabled cattle in these years were underestimated, but as we will see, to a much lesser extent than in the province of Cordoba for the years 1902 and 1910.

Graph A.1.8 shows the contribution of metabolizable energy from pastures and quality food to the maintenance of labor livestock. The model presents a major difference with that of Cordoba. In the case of Cordoba, pastures would contribute more than quality food, although this difference would progressively narrow down. Based

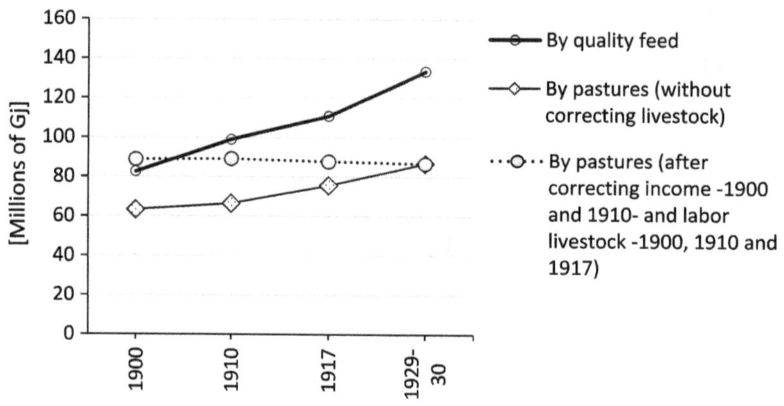

Graph A.1.8 Metabolizable energy provided by pastures and quality feed to satisfy livestock needs. *Source* Own estimation

solely on (unmodified) livestock census data, this situation of pasture preeminence would never have existed throughout the Mediterranean coast. It would only come about at the start of the century if we accept the modifications to the censuses of 1900, 1910, and 1917. In any case, the gap between both contributions grew, showing that the increase in livestock had to be based on the consumption of food of agricultural origin, a trend that has continued until today. Cordoba's lesser dependence (in percentage) on agricultural food is logical considering the higher quality of its pastures and grasslands compared to the Mediterranean coast average.

Considering the same percentage of labor livestock concealment in 1917 (let us recall that it was not necessary to adjust the income livestock in Cordoba that year), the pasture livestock density would be 0.12 LU/ha, i.e., similar to that of 1929, and this would be plausible as it would represent 40% of its capacity. In addition, the adjustment of quality food to stabled livestock needs would be 1.14% that would be similar to the case of Cordoba for the same year.

In 1902, the percentage increase applied in the case of Cordoba would be excessive if used for the whole of the Spanish Mediterranean coast, generating values of 0.3 LU/ha of pastures (100% of the potential). In 1910, if we applied the Cordoba percentage of increase of labor and income livestock, we would obtain a pasture density of 0.23 that also seems high and would mean that in the following years it would undergo a sharp fall reaching 0.13. This drop would be inexplicable. Therefore, it is conceivable that concealment in Mediterranean provinces as a whole was less of a significant phenomenon than in Cordoba. Taking into account the two adjustment criteria of food demand to needs (livestock density in pastures and adjustment of the stabled livestock to quality feed supply), the correction of the livestock censuses of 1902, 1910, and 1917 would be as described in the following paragraphs.

In 1902, the rise was 20% of that of Cordoba's for labor livestock and would be multiplied by 1.2 for income livestock. By 1910, the increase would be one-third of Cordoba's for livestock while the income livestock in the census would multiply by 1.15. In 1917, labor livestock would multiply by the same percentage as in the case of Cordoba for the same year. That year's income livestock would not change. Specifically, adult work animals was multiplied by 1.208 in 1902, by 1.169 in 1910, and by 1.158 in 1917. The number of working cattle was multiplied by 1.058 in 1902, by 1.044 in 1910, and by 1.039 in 1917. Income cattle increased by 1.2% in 1902 and by 1.15% in 1910.

With these increases, the pattern of livestock density (that was stable over the period) and the coverage ratio of the quality food needs of stabled cattle would be similar to those found for Cordoba. The results shown in Graph A.1.8 would also be more adjusted, since it would seem nonsensical that the agricultural area aimed at animal feed would increase significantly, while pastures remained underutilized.

Finally, we must remember that we have been very conservative with our estimates of livestock modifications. On the one hand, as mentioned, we did not include some of the feed (by-products from agri-industry, rainfed fruit pastures, etc.) on the supply side. On the other hand, we applied a very cautious level of livestock density (50% of the potential) (Table A.1.3).

For the Atlantic provinces, we do not suggest any modifications. It would be necessary to add the livestock of Galicia, Santander, Asturias, Vizcaya, and Guipúzcoa.

We applied the same methodology to evaluate the reliability of the 1940 livestock census, based on Cordoba data. As indicated, a historiographical consensus exists on the decline of livestock census figures during the Civil War. It is therefore surprising that this fall is not reflected in the 1940 census. However, our data checks led us to discover that there was no major problem in feeding the livestock in 1940. It is true that the livestock density is higher than that figuring in the 1940 censuses, but this density does not imply there was an excessive pressure on pastures, accounting for 33.9% in the case of Cordoba and 35% if we extrapolate the data to the whole of Spain. Therefore, based on the information obtained from matching feed availability on the one hand and labor livestock needs on the other, the 1940 census change is not justified (Table A.1.4).

How can we explain this behavior? Our hypothesis is that this large amount of livestock is the result of the strategy of allowing a high share of offspring to reach adulthood without being slaughtered in order to replace the livestock losses during the war. Indeed, the number of young (nonproductive) animals is considerably higher in 1940 than in subsequent censuses (Graph A.1.9). This reflects a change in the structure of livestock and does not invalidate the thesis of a livestock decline during the Civil War. Most of these animals are therefore unproductive, and this is compatible with the fall in livestock production (meat and milk) reflected in statistics from the forties onward.

Table A.1.3 Resulting livestock on the Spanish Mediterranean coast in 1900, 1910, 1917 and 1929, in number of head

Cattle

Year	Total animals	0–3 years	>3 years			
			Stallions + Bulls	Labor	Milk cows	Other cows
1906	1,291,847	359,120	24,842	533,864	187,586	186,435
1910	1,375,109	383,739	26,900	559,448	203,134	201,888
1917	1,641,887	458,359	32,286	665,133	243,802	242,306
1929-30	1,737,986	499,840	36,581	650,790	276,235	274,540

Sheep

Year	Total animals	<2 years	>2 years (castrated)	>2 years	
				Male	Female
1906	15,670,529	4,310,752	1,017,962	414,580	9,927,235
1910	17,288,932	4,755,953	1,123,094	457,396	10,952,489
1917	15,767,847	4,337,523	1,024,284	417,154	9,988,886
1929-30	18,206,834	5,008,456	1,182,721	481,680	11,533,977

Goat

Year	Total animals	<2 years	>2 years castrated males	>2 years	
				Males	Females
1906	2,883,596	760,867	123,329	89,113	1,910,288
1910	3,351,654	884,369	143,347	103,578	2,220,361
1917	3,254,639	858,770	139,198	100,580	2,156,091
1929–30	4,132,057	1,090,286	176,724	127,695	2,737,352

Pigs

Year	Total animals	<12 months	>1 year		
			Males	Reproductive females	Feed
1906	1,696,652	613,352	21,171	193,442	868,688
1910	2,032,105	734,621	25,357	231,688	1,040,440
1917	2,724,080	984,775	33,991	310,582	1,394,732
1929–30	3,732,008	1,349,148	46,568	425,500	1,910,792

Horses

Year	Total animales	0–3 years	Labor	Mounting
1906	427,464	75,530	326,121	25,813
1910	486,094	87,102	368,825	30,167
1917	501,318	90,121	379,835	31,362
1929–30	495,819	96,403	364,559	34,857

(continued)

Table A.1.3 (continued)

Mules

Year	Total animals	0–3 years	Labor	Mounting
1906	919,709	121,718	739,462	58,529
1910	983,871	132,148	787,326	64,396
1917	1,058,022	142,593	845,609	69,821
1929–30	1,135,595	166,276	884,727	84,592

Donkeys

Year	Total animals	0–3 years	Labor	Mounting
1906	839,543	85,576	698,667	55,300
1910	905,242	93,695	750,188	61,359
1917	926,749	96,260	767,147	63,342
1929-30	918,793	104,004	743,682	71,107

Source Own estimation

Table A.1.4 Food balance of Cordoba livestock

	Total labor livestock	Income livestock	Use of pastures %	Spain's average %
Needs	3,544,263	3,019,709	33.9%	35.0%
Available	3,386,296	8,898,420		
Deficit/surplus	−157,967	5,878,711		
% represented by the deficit/surplus compared to produced feed	−5%	66%		

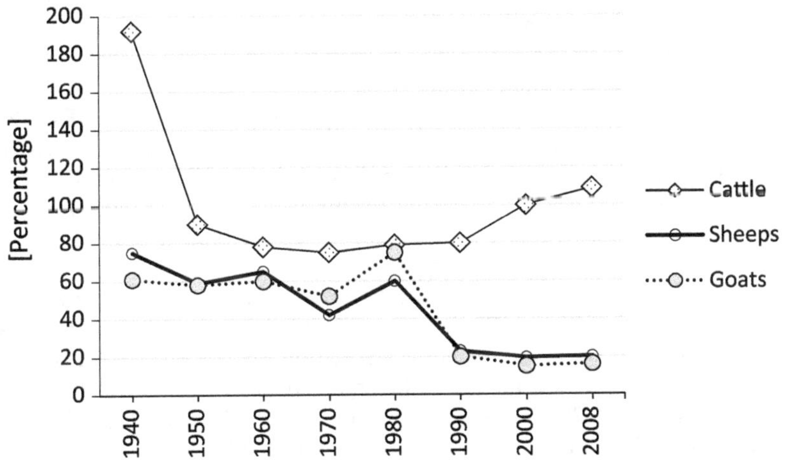

Graph A.1.9 Percentage of offspring per mother according to type of livestock

References

Aguilera E, Guzmán GI, Infante-Amate J, Soto D, García-Ruiz R, Herrera A, Villa I, Torremocha E, Carranza G, González de Molina M (2015) Embodied energy in agricultural inputs. Incorporating a historical perspective. In: Documentos de Trabajo de la Sociedad Española de Historia Agraria, nº 15-07

EUROSTAT (2015) Economy-wide Material Flow Accounts (EW-MFA). Compilation Guide 2013. European Statistical Office, Luxembourg

Fernández Prieto L (1992) Labregos con ciencia. Estado e sociedade e innovación tecnolóxica na agricultura galega, 1850-1936. Xerais, Vigo

Gallego D (1986) La producción agraria de Álava, Navarra y La Rioja desde mediados del siglo XIX a 1935. Universidad Complutense de Madrid, Madrid

García Sanz A (1991) La ganadería española entre 1750-1865: los efectos de la reforma agraria liberal. In: Agricultura y Sociedad, nº 72, pp 81–120

GEHR (Grupo de Estudios de Historia Rural) (1978) Contribución al análisis histórico de la ganadería española, 1865-1929. In: Agricultura y Sociedad, nº 8, pp 129–182

GEHR (Grupo de Estudios de Historia Rural) (1979) Contribución al análisis histórico de la ganadería española, 1865–1929. In: Agricultura y Sociedad, nº 10, pp 105–169

GEHR (Grupo de Estudios de Historia Rural) (1991) Estadísticas Históricas de la producción agraria española, 1859-1935. Ministerio de Agricultura Pesca y Alimentación, Madrid

González de Molina M, Soto D, Infante-Amate J, Aguilera E (2013) "¿Una o varias transiciones? Nuevos datos sobre el consumo alimentario en España (1900–2008)". In: XIV Congreso Internacional de Historia Agraria (Badajoz, 7-9 de noviembre 2013). https://doi.org/10.13140/2.1.1823.5684

Guzmán GI, Aguilera E, Soto D, Cid A, Infante-Amate J, García-Ruiz R, Herrera A, Villa I, González de Molina M (2014) Methodology and conversión factors to estimate the net primary productivity of 112 historical and contemporary agro-ecosystems (I). In: Documento de Trabajo de la Sociedad Española de Historia Agraria, nº 14-06. Disponible en: www.seha.info

Guzmán GI, González de Molina M, Soto D, Infante-Amate J, Aguilera E (2017) Spanish agriculture from 1900 to 2008: a long-term perspective on agroecosystem energy from an agroecological approach. Reg Environ Change 149:335–348

Iriarte Goñi I (2013) Forests, fuel wood, pulpwood, and lumber in Spain, 1860–2000: a non-declensionist story. Environ Hist 18(2),333–359

Infante-Amate J, Aguilera E, González de Molina M (2014a) La gran transformación del sector agroalimentario español. Un análisis desde la perspectiva energética (1960–2010). In: Documento de Trabajo de la Sociedad de Estudios de Historia Agraria, nº 14-03. Available in: https://ideas.repec.org/p/seh/wpaper/1403.html

Infante-Amate J, Soto D, Iriarte Goñi I, Aguilera E, Cid A, Guzmán G, García Ruiz R, González de Molina M (2014b) La producción de leña en España y sus implicaciones en la transición energética. Una serie a escala provincial (1900-2000). In: Sociedad Española de Historia Agraria, Documento de Trabajo nº 14-16. Available in: http://econpapers.repec.org/paper/ahedtaehe/1416.htm

Infante-Amate J, González de Molina M, Vanwalleghem T, Soto Fernández D, Gómez Calero JA (2014c) Reconciling Boserup with Malthus. Agrarian change and soil degradation in Olive Orchards in Spain (1750-2000). In: Fischer-Kowalski M, Reenberg A, Schaffartzik A, Mayer A (eds) Ester Boserup's legacy on sustainability: orientations for contemporary research. Springer, New York, pp 99–116. Available in: https://doi.org/10.1007/978-94-017-8678-2_7

Infante-Amate J, Iriarte Goñi I (2017) Las bioenergías en España. Una serie de producción, consumo y stocks entre 1860 y 2010. In: Documento de Trabado de la Sociedad de Estudios de Historia Agraria, nº 17-02

Krausmann F, Erb KE, Gringrich S, Lauk C, Haberl H (2008) Global patterns of socioeconomic biomass flows in the year 2000: a comprehensive assessment of supply, consumption and constraints. Ecol Econ 65:471–487

Krausmann F, Fischer-Kowalski M, Schandl H, Eisenmenger N (2008) The global sociometabolic transition: past and present metabolic profiles and their future trajectories. J Ind Ecol 12(5–6):637–656

MARM (Ministerio de Medio Ambiente, Medio Rural y Marino) (2008) Inventario de emisiones de gases de efecto invernadero de España 1990–2006. Ministerio de Medio Ambiente, Medio Rural y Marino, Madrid

MARM (Ministerio de Medio Ambiente, Medio Rural yMarino) (2008) Anuario de Estadística Agraria 2008. Madrid

MARM (Ministerio de Medio Ambiente, Medio Rural y Marino), (2008): *Encuesta sobre superficies y rendimientos de cultivos*. Madrid

MF (Ministerio de Fomento) (1892) La ganadería en España. Avance sobre la riqueza pecuaria en 1891, formada por la Junta Consultiva Agronómica, conforme a las memorias reglamentarias que en el citado año han redactado los ingenieros del Servicio Agronómico. Ministerio de Fomento, Madrid

MF (Ministerio de Fomento) (1920) Estudio de la ganadería en España. Madrid

Muñoz Rubio M (2015) El transporte ferroviario de ganado y el mercado entre 1848 y 1913. Historia Agraria 67:79–109

Naredo JM (1983) Reflexiones con vista a una mejora de las estadísticas agrarias. Agricultura y Sociedad 29:239–254

Soto Fernández D (2002) As estatísticas para o estudio da agricultura galega no primeiro tercio do século XX: análise crítica". In: Documentos de Traballo do IDEGA. Historia, n° 14, pp 1–41

Soto Fernández D (2006) Historia dunha agricultura sustentábel. Transformación sprodutivas na agricultura galega contemporánea. Santiago de Compostela. Xunta de Galicia, Consellería do Medio Rural

Soto Fernández D, González de Molina M, Infante-Amate J, Guzmán Casado G (2016) La evolución de la ganadería española (1752 y 2012): del uso múltiple al uso alimentario. Una evaluación de la fiabilidad de los censos y de las estadísticas de producción. In *Seminario Anual de la Sociedad Española de Historia Agraria,* (Madrid, noviembre de 2016)

Soto Fernández D, Infante-Amate J, Guzmán GI, Cid A, Aguilera E, García-Ruiz R, González de Molina M (2016) The social metabolism of biomass in Spain, 1900–2008: From food to feed-oriented changes in the agro-ecosystems. Ecol Econ 128:130–138

Wirsenius S (2003) The biomass metabolism of the food system. A model-based survey of the global and regional turnover of food biomass. J Ind Ecol 7(1):47–80

Annex II
Historical Evolution of the Spanish Agrarian Metabolism and the Spanish Economy Metabolism

A.2.1 Historical Evolution of the Spanish Agrarian Metabolism

The tables of the first part of this annex are referred to as the historical evolution of the Spanish Agrarian Metabolism (1900–2008) in the units specified in each case. The different series come from the sources described in Annex I (Tables A.2.1.1, A.2.1.2, A.2.1.3, A.2.1.4, A.2.1.5, A.2.1.6, A.2.1.7, A.2.1.8, A.2.1.9, A.2.1.10, A.2.1.11, A.2.1.12, A.2.1.13, A.2.1.14, A.2.1.15, A.2.1.16, A.2.1.17, A.2.1.18, A.2.1.19, A.2.1.20, A.2.1.21, A.2.1.22, A.2.1.23, A.2.1.24, A.2.1.25, A.2.1.26, A.2.1.27, and A.2.1.28).

A.2.2 Historical Evolution of Spanish Economy Metabolism

The tables of the second part of this annex are referred to as the evolution of the metabolism of the Spanish economy in terms of biomass (1900–2008) in the units specified in each case. The different series come from the sources described above in Annex I (Tables A.2.2.1, A.2.2.2, A.2.2.3, A.2.2.4, and A.2.2.5).

Table A.2.1.1 Land uses, in millions of hectares

	1900	1910	1922	1933	1940	1950	1960	1970	1980	1990	2000	2008
Cultivated land	16,479	17,228	18,899	20,368	18,782	19,856	20,413	20,885	20,499	20,172	18,304	17,271
Woodland	2836	2980	3336	3558	3759	3697	4929	6240	6741	7189	7460	8307
Coppice	7126	6872	6549	5843	6115	5619	5076	4640	4824	4979	5055	5146
Open woodland	2989	2814	2702	2586	2626	2776	3320	3835	4033	3636	3893	4342
Pastures	16,901	16,437	14,845	13,976	15,049	14,383	12,594	11,190	10,691	10,746	11,645	11,294
Improductive	4168	4169	4169	4169	4169	4169	4169	3710	3712	3777	4143	4139
Total	50,500	50,500	50,500	50,500	50,500	50,500	50,500	50,500	50,500	50,500	50,500	50,500

Table A.2.1.2 Net primary productivity, in megatons of dry matter

	1900	1910	1922	1933	1940	1950	1960	1970	1980	1990	2000	2008
Accumulated biomass	11.7	11.3	11.5	10.8	12.2	12.0	17.0	21.7	25.4	23.4	24.4	23.7
Crop aerial biomass	1.1	1.1	1.2	1.4	1.5	1.7	1.9	2.0	2.0	2.0	2.0	2.1
Crop root biomass	0.5	0.5	0.5	0.5	0.6	0.7	0.7	0.7	0.7	0.7	0.6	0.6
Woodland aerial biomass	0.5	0.3	0.2	−0.1	0.6	0.7	4.6	9.3	12.5	10.4	10.9	9.8
Woodland root biomass	9.6	9.4	9.5	9.1	9.5	9.0	9.8	9.7	10.2	10.4	10.8	11.1
Unharvested biomass	183.4	186.3	191.9	184.3	188.5	184.2	187.5	203.5	196.7	202.9	217.4	222.1
Crop aerial biomass	19.0	20.6	23.4	26.9	23.3	19.9	18.7	15.5	15.0	20.6	25.7	27.0
Crop root biomass	21.1	23.4	26.2	29.6	24.4	22.5	23.4	21.2	19.8	22.1	23.3	22.1
Pastures aerial biomass	53.6	53.3	52.8	44.9	48.6	49.9	50.4	64.3	61.3	60.3	62.5	64.6
Pasture root biomass	54.2	54.2	54.3	49.4	57.4	58.0	55.4	56.3	52.7	52.2	56.5	57.8
Woodland aerial biomass	16.3	15.9	16.1	15.4	16.0	15.4	17.5	19.3	20.0	20.0	21.0	21.6
Woodland root biomass	19.2	18.8	19.0	18.1	18.8	18.5	22.1	26.9	27.8	27.6	28.5	29.0
Reused biomass	28.4	31.0	33.5	37.6	37.9	39.9	41.9	35.8	36.6	41.3	43.6	41.9
Crops	14.3	16.6	18.4	20.6	14.6	17.3	23.1	29.7	32.0	36.5	35.5	34.3
Pastures	14.1	14.4	15.1	16.9	23.2	22.6	18.8	6.1	4.6	4.9	8.1	7.6
Socialized vegetal biomass	21.2	21.7	22.4	23.1	21.9	21.8	21.6	22.2	21.5	24.9	25.3	26.6
Food	4.4	5.0	5.7	6.6	4.9	5.3	7.1	8.0	9.3	10.0	9.9	9.4
Raw materials	0.7	0.5	0.8	0.8	0.6	0.6	0.8	0.8	1.1	1.5	2.6	2.6
Wood	1.0	1.3	1.5	1.6	1.9	1.9	2.4	4.7	4.1	6.5	6.3	8.2
Crop firewood	10.1	9.8	9.9	9.6	9.2	8.8	6.9	3.0	1.1	1.0	1.1	1.0
Forest firewood	5.0	5.1	4.5	4.5	5.3	5.2	4.4	5.7	5.9	5.9	5.4	5.4
Total net primary productivity	244.6	250.2	259.2	255.8	260.5	257.9	268.0	283.3	280.2	292.6	310.7	314.2

Table A.2.1.3 Domestic extraction, in megatons of dry matter

	1900	1910	1922	1933	1940	1950	1960	1970	1980	1990	2000	2008
Primary crops	11.18	12.63	14.94	16.69	12.71	13.84	19.95	24.81	30.80	35.58	38.66	37.55
Cereals	5.47	6.30	6.79	7.87	5.03	5.87	7.56	10.67	13.11	16.54	18.63	20.19
Leguminous	0.47	0.57	0.66	0.67	0.48	0.55	0.67	0.61	0.34	0.23	0.30	0.21
Roots and tubers	0.54	0.71	0.77	1.09	0.79	0.81	1.04	1.16	1.29	1.17	0.70	0.50
Vegetables	0.24	0.26	0.31	0.43	0.64	0.34	0.43	0.54	0.66	0.90	0.95	0.92
Fruits	0.54	0.71	0.88	1.10	0.88	0.91	1.11	1.16	1.39	1.59	1.71	1.79
Vineyards	1.10	0.79	1.12	0.96	0.86	0.74	0.77	1.19	1.77	1.55	1.79	1.57
Olive groves	0.59	0.58	0.87	0.98	0.84	0.86	1.05	1.12	1.40	1.58	2.82	3.38
Textile and Oilseeds	0.19	0.10	0.12	0.10	0.09	0.11	0.23	0.28	0.63	1.58	1.21	0.92
Sugar crops	0.25	0.32	0.40	0.56	0.39	0.53	1.02	1.59	1.73	1.90	2.00	1.40
Condiments	0.00	0.00	0.00	0.00	0.00	0.00	0.00	0.00	0.00	0.00	0.00	0.00
Other industrial	0.00	0.00	0.00	0.00	0.00	0.00	0.00	0.01	0.01	0.02	0.01	0.01
Pastures and cultivated grasslands	1.80	2.30	3.02	2.93	2.71	3.12	6.08	6.50	8.46	8.52	8.53	6.66
Used crop residues	13.17	14.56	14.33	15.80	12.81	14.62	15.42	19.45	17.47	18.21	14.68	14.18
Cereal straw	7.63	8.77	9.01	10.37	6.80	8.30	9.53	11.35	8.76	8.33	7.50	7.24
Leguminous straw	0.51	0.64	0.76	0.84	0.61	0.79	0.77	0.69	0.33	0.16	0.13	0.07
Potatoes and vegetable residues	0.07	0.08	0.09	0.12	0.10	0.10	0.13	0.17	0.19	0.23	0.21	0.19
Vineyard fuelwood	2.72	2.68	1.98	1.78	1.93	1.95	1.62	2.09	2.24	2.05	1.52	1.78
Olive grove fuelwood	1.03	1.17	1.32	1.42	1.79	1.69	1.48	2.02	1.87	1.72	1.65	1.65
Fruit fuelwood	1.21	1.22	1.18	1.28	1.58	1.59	1.33	1.59	1.80	2.09	2.22	2.00
Burned residues	0.00	0.00	0.00	0.00	0.00	0.20	0.56	1.54	2.28	3.62	1.44	1.26

(continued)

Table A.2.1.3 (continued)

	1900	1910	1922	1933	1940	1950	1960	1970	1980	1990	2000	2008
Forest and pastures	25.18	25.46	26.58	28.20	34.27	33.23	28.13	13.81	12.64	14.00	17.22	16.75
Grazed biomass	14.08	14.36	15.12	16.94	23.23	22.55	18.84	6.10	4.61	4.88	8.09	7.60
Wood	0.95	1.27	1.52	1.65	1.85	1.86	2.40	4.67	6.92	8.07	8.00	8.16
Fuelwood	10.14	9.83	9.94	9.62	9.19	8.82	6.89	3.04	1.11	1.04	1.12	1.00
Total domestic extraction	49.52	52.65	55.85	60.70	59.79	61.69	63.50	58.07	60.91	67.79	70.55	68.48

Table A.2.1.4 Net primary productivity and animal socialized biomass, in terajoules

	1900	1910	1920	1930	1940	1950	1960	1970	1980	1990	2000	2008
NPPact (a+c+d+e)	4,366,701	4,459,715	4,630,949	4,569,559	4,650,367	4,609,400	4,800,059	5,072,909	4,986,858	5,215,922	5,566,456	5,625,189
Socialized vegetal biomass (SVB)(a)	400,170	408,350	436,541	458,468	411,673	417,700	425,184	412,028	396,373	478,206	491,564	505,661
Socialized vegetal biomass (crops)	164,879	174,146	195,683	222,312	181,175	195,085	234,616	264,716	300,139	341,307	356,544	341,303
Socialized vegetal biomass (woodland)	235,290	234,204	240,858	236,156	230,498	222,615	190,569	147,312	96,234	136,899	135,020	164,358
Socialized animal biomass (SAB)(b)	9,333	11,184	13,575	15,048	12,686	14,380	20,157	35,758	54,876	72,569	97,691	105,869
Socialized biomass (SB)(a+b)	409,503	419,534	450,116	473,516	424,359	432,080	445,341	447,786	451,249	550,775	589,256	611,530
Reused biomass (RuB)(c)	501,739	554,967	602,111	682,618	676,989	705,379	746,671	563,447	609,599	714,879	832,006	854,664
Unharvested biomass (UhB)(d)	3,235,392	3,275,863	3,367,919	3,215,947	3,324,727	3,254,799	3,300,618	3,671,279	3,480,496	3,560,737	3,762,342	3,798,384
Aerial unharvested biomass (AUhB)	1,540,562	1,548,910	1,584,224	1,475,528	1,522,243	1,479,476	1,482,719	1,778,637	1,659,234	1,713,080	1,800,514	1,825,377
Root unharvested biomass (RUhB)	1,694,829	1,726,953	1,783,694	1,740,419	1,802,484	1,775,323	1,817,899	1,892,642	1,821,262	1,847,658	1,961,828	1,973,007
Reused biomass (RcB)(c+d)	3,737,130	3,830,831	3,970,030	3,898,565	4,001,716	3,960,178	4,047,289	4,234,726	4,090,096	4,275,616	4,594,348	4,653,048
Accumulated biomass (AB)(e)	229,401	220,534	224,378	212,526	236,978	231,522	327,586	426,155	500,389	462,101	480,543	466,480

Table A.2.1.5 Net primary productivity of cropland, in terajoules

	1900	1910	1920	1930	1940	1950	1960	1970	1980	1990	2000	2008
NPPact	1,109,788	1,226,893	1,376,411	1,560,143	1,281,140	1,253,538	1,371,287	1,426,425	1,471,174	1,715,584	1,821,935	1,793,445
Socialized vegetal biomass (SVB)	164,879	174,146	195,683	222,312	181,175	195,085	234,616	264,716	300,139	341,307	356,544	341,303
Reused biomass (RuB)	254,257	302,660	336,467	384,952	268,778	309,027	415,610	456,273	528,539	628,004	686,282	719,606
Aerial unharvested biomass (AUhB)	292,327	312,256	354,408	399,001	368,009	312,861	267,260	283,029	244,192	307,763	319,000	292,004
Root unharvested biomass (RUhB)	369,500	409,639	458,990	518,602	426,961	396,200	411,991	373,179	348,298	388,219	409,708	388,481
Accumulated biomass (AB)	28,825	28,191	30,863	35,277	36,216	40,364	41,811	49,227	50,006	50,291	50,401	52,050

Table A.2.1.6 Net primary productivity of pastures, in terajoules

	1900	1910	1920	1930	1940	1950	1960	1970	1980	1990	2000	2008
NPPact	2,142,041	2,142,041	2,146,586	1,955,278	2,271,047	2,293,947	2,188,885	2,227,802	2,084,407	2,061,839	2,234,369	2,284,455
Socialized vegetal biomass (SVB)	0	0	0	0	0	0	0	0	0	0	0	0
Reused biomass (RuB)	247,482	252,307	265,644	297,666	408,211	396,352	331,060	107,174	81,060	86,875	145,724	135,058
Aerial unharvested biomass (AUhB)	942,541	937,716	926,904	788,599	853,482	878,063	884,988	1,130,494	1,076,944	1,058,591	1,095,592	1,134,084
Root unharvested biomass (RUhB)	952,018	952,018	954,038	869,012	1,009,354	1,019,532	972,838	990,134	926,403	916,373	993,053	1,015,313
Accumulated biomass(AB)	0	0	0	0	0	0	0	0	0	0	0	0

Table A.2.1.7 Net primary productivity of woodland, in terajoules

	1900	1910	1920	1930	1940	1950	1960	1970	1980	1990	2000	2008
NPPact	1,114,871	1,090,781	1,107,952	1,054,137	1,098,181	1,061,916	1,239,886	1,418,681	1,431,276	1,438,500	1,510,151	1,547,289
Socialized vegetal biomass (SVB)	235,290	234,204	240,858	236,156	230,498	222,615	190,569	147,312	96,234	136,899	135,020	164,358
Reused biomass (RuB)	0	0	0	0	0	0	0	0	0	0	0	0
Aerial unharvested biomass (AUhB)	305,694	298,938	302,912	287,927	300,753	288,552	330,472	365,113	338,098	346,726	385,922	399,289
Root unharvested biomass (RUhB)	373,311	365,295	370,667	352,805	366,168	359,591	433,070	529,328	546,562	543,066	559,067	569,212
Accumulated biomass(AB)	200,576	192,343	193,515	177,249	200,762	191,158	285,775	376,927	450,383	411,809	430,142	414,430

Table A.2.1.8 Farm machinery census, in thousands of units

	1900	1910	1922	1933	1940	1950	1960	1970	1980	1990	2000	2008
Locomobiles	0.31	0.38	0.48	0.52	0.19	0.00	0.00	0.00	0.00	0.00	0.00	0.00
Tractors	0.00	0.01	0.32	5.04	6.95	14.53	61.69	260.24	517.19	737.49	903.11	1027.08
Rototillers	0.00	0.00	0.00	0.00	0.00	0.11	2.85	73.35	218.82	281.54	282.43	281.48
Harvesters	0.00	0.00	0.00	0.00	0.00	0.07	2.16	29.05	43.80	51.60	54.32	59.27
Other engines	0.30	0.93	3.53	12.03	15.63	33.80	76.39	64.70	69.67	69.70	62.74	59.93
Threshers	0.30	0.73	2.11	5.47	6.91	9.73	17.85	19.98	10.63	6.64	3.85	2.49

Table A.2.1.9 Life expectancy of tractors, rototillers, harvesters, and other engines, in years

	1900	1910	1922	1933	1940	1950	1960	1970	1980	1990	2000	2008
Locomobiles	20.0	20.0	20.0	20.0	20.0							
Tractors		20.0	20.0	20.0	20.0	20.0	20.0	19.8	20.0	29.1	38.0	44.5
Rototillers					11.4	11.4	11.4	11.2	16.0	24.0	32.6	39.8
Harvesters					11.4	11.4	11.4	11.2	17.0	25.4	34.0	41.9
Other engines	25.0	25.0	25.0	25.0	25.0	25.0	25.0	25.0	25.0	25.0	25.0	25.0
Threshers	20.0	20.0	20.0	20.0	20.0	20.0	20.0	20.0	20.0			

Table A.2.1.10 Average rated power per machinery unit, in HP/unit

	1900	1910	1922	1933	1940	1950	1960	1970	1980	1990	2000	2008
Locomobiles	10.0	10.0	10.0	10.0	10.0							
Tractors		25.0	25.0	25.2	25.5	27.7	35.2	44.8	52.8	56.7	61.0	63.7
Rototillers						7.2	7.0	10.0	13.2	12.9	13.0	12.7
Harvesters						28.3	39.9	63.9	83.5	91.4	100.7	107.3
Other engines	5.1	5.1	5.1	5.2	5.2	5.7	6.5	6.9	9.4	10.6	11.4	11.9

Table A.2.1.11 Number of registrations and removals of tractors, rototillers, threshers, harvesters, other engines, tillage machinery, and other farm implements, in thousands of units/year

		1900	1910	1922	1933	1940	1950	1960	1970	1980	1990	2000	2008
Tractors	Ups	0.00	0.00	0.08	0.84	0.17	1.86	11.71	26.00	31.29	19.46	22.54	15.04
	Downs	0.00	0.00	0.00	−0.03	−0.03	−0.10	−0.42	−2.99	−2.34	−2.21	−3.36	−1.74
Rototillers	Ups	0.00	0.00	0.00	0.00	0.00	0.03	1.16	11.66	13.50	3.66	1.17	0.59
	Downs	0.00	0.00	0.00	0.00	0.00	0.00	−0.02	−0.49	−1.39	−2.99	−1.43	−0.69
Threshers	Ups	0.04	0.10	0.28	0.69	0.35	1.15	1.42	1.00	0.76	0.00	0.00	0.00
	Downs	−0.02	−0.04	−0.11	−0.27	−0.30	−0.48	−0.54	−1.29	−2.18	−0.37	−0.22	−0.14
Harvesters	Ups	0.00	0.00	0.00	0.00	0.00	0.02	0.68	3.18	1.36	1.06	0.86	0.68
	Downs	0.00	0.00	0.00	0.00	0.00	0.00	−0.03	−0.41	−0.52	−0.83	−0.35	−0.08
Other engines	Ups	0.05	0.14	0.52	1.60	0.81	4.01	8.29	0.00	3.30	1.50	0.00	0.00
	Downs	−0.01	−0.04	−0.14	−0.44	−0.27	−1.35	−3.04	−2.68	−1.89	−1.88	−0.36	−0.34
Tillage machinery	Ups	159.34	169.16	181.48	189.84	176.78	158.63	148.82	93.03	49.84	59.81	45.56	45.56
	Downs	−132.46	−143.05	−154.57	−209.07	−224.93	−201.83	−161.85	−249.85	−110.83	−68.63	−45.56	−45.56
Other farm implements	Ups	21.80	23.16	24.86	27.94	37.38	57.38	70.34	61.45	118.87	129.57	106.93	106.93
	Downs	−18.86	−20.21	−21.90	−11.96	−6.03	−9.25	−81.61	−70.55	−67.12	−154.46	−106.93	−106.93

Table A.2.1.12 Rated power of the average machines, registered machines, and removed machines, including tractors, rototillers, harvesters, and other engines, in HP/unit

		1900	1910	1922	1933	1940	1950	1960	1970	1980	1990	2000	2008
Tractors	Census average		25.0	25.0	25.2	25.5	27.7	35.3	44.8	52.8	56.7	61.0	63.7
	Registrations		25.0	25.0	25.9	26.5	31.8	39.5	50.2	56.3	69.6	74.6	94.8
	Removals		25.0	25.0	25.0	25.0	25.0	25.0	31.1	40.7	38.9	42.0	42.7
Rototillers	Census average					7.2	7.2	7.0	10.0	13.2	12.9	13.0	12.7
	Registrations					7.2	7.2	6.7	15.4	9.8	15.2	11.7	10.7
	Removals					7.2	7.2	7.2	6.7	6.2	8.0	9.2	9.8
Harvesters	Census average					28.3	28.3	39.4	63.8	83.5	91.3	100.7	107.2
	Registrations					28.3	28.3	48.7	96.8	148.9	138.3	172.8	219.9
	Removals							28.3	48.8	49.1	53.3	57.5	57.5
Other engines	Census average	5.1	5.1	5.1	5.2	5.2	5.7	6.5	6.9	9.4	10.6	11.4	11.9
	Registrations	5.1	5.1	5.1	5.3	5.5	6.0	6.8	7.2	9.4	10.6	11.4	11.9
	Removals	5.1	5.1	5.1	5.1	5.1	5.1	5.4	5.6	6.7	6.7	8.3	8.8

Table A.2.1.13 Direct energy of the combustibles, in terajoules/ year

	1900	1910	1922	1933	1940	1950	1960	1970	1980	1990	2000	2008
Coal	24.4	41.7	118.0	124.7	73.9	0.0	0.0	0.0	0.0	124.8	0.0	0.0
Gasoline	1.5	11.7	98.7	1247.1	1480.8	2236.9	510.1	108.2	0.0	0.0	0.0	0.0
Fuel	0.6	1.3	10.5	102.6	182.7	391.5	1500.1	502.3	2102.2	3842.3	2128.8	1271.8
Diesel oil	0.0	0.0	0.0	0.0	223.1	745.6	10918.6	62439.0	87814.6	61165.3	71082.5	74556.3
LPG	0.0	0.0	0.0	0.0	0.0	0.0	0.0	0.0	0.0	1082.6	3109.0	2060.4
Natural gas	0.0	0.0	0.0	0.0	0.0	0.0	0.0	0.0	0.0	4411.7	2514.5	10339.2
Biomass	0.0	0.0	0.0	0.0	0.0	0.0	0.0	0.0	0.0	0.0	221.0	1510.5
Biogas	0.0	0.0	0.0	0.0	0.0	0.0	0.0	0.0	0.0	0.0	1.8	75.0
Other renewable	0.0	0.0	0.0	0.0	0.0	0.0	0.0	0.0	0.0	138.0	180.2	219.4
Total	26.5	54.7	227.2	1474.4	1960.5	3374.0	12928.8	63049.5	89916.8	70764.7	79237.7	90032.5
Irrigation liquid fuels	0.1	1.7	18.3	341.6	444.7	1202.1	2922.3	5191.8	8225.2	8444.2	8334.3	7987.3
Irrigation coal	0.4	3.3	35.9	39.8	51.8	0.0	0.0	0.0	0.0	0.0	0.0	0.0
Irrigation combustible	0.5	4.9	54.2	381.4	496.4	1202.1	2922.3	5191.8	8225.2	8444.2	8334.3	7987.3
Livestock combustible	0.0	0.0	0.0	0.0	0.0	0.0	0.0	0.0	0.0	5632.3	6026.4	14204.4
Traction combustible	26.0	49.7	173.0	1093.0	1464.1	2171.8	10006.6	57857.7	81691.6	56688.3	64877.0	67840.7

Table A.2.1.14 Nitrogen fertilizer consumption (N), in thousands of tons of nutrient

	1900	1910	1922	1933	1940	1950	1960	1970	1980	1990	2000	2008
Guano	0.8	0.4	0.1	0.1	0.0	0.0	0.0	0.0	0.0	0.0	0.0	0.0
Chilean nitrate	0.3	5.1	9.9	11.9	10.2	13.1	19.4	10.9	4.8	2.6	1.4	2.9
Cyanamide	0.0	0.0	0.1	0.4	0.0	0.3	0.8	0.0	0.0	0.0	0.0	0.0
Ammonium sulfate	6.5	13.3	24.1	52.6	12.5	37.0	172.9	157.8	89.6	99.0	80.1	75.7
Ammonium nitrate	0.0	0.0	0.0	0.0	0.0	0.0	0.0	0.0	103.5	139.8	115.5	41.3
Calcium ammonium nitrate	0.0	0.0	0.0	0.0	0.0	10.3	40.2	198.8	256.1	207.4	256.1	199.8
Urea	0.0	0.0	0.0	0.0	0.0	0.0	0.1	51.3	162.1	236.9	291.7	244.5
NPK	0.0	0.0	0.0	0.0	0.0	0.0	0.0	108.5	182.8	255.3	313.4	202.8
Other	0.1	0.5	2.5	9.1	9.6	8.0	40.7	74.5	75.3	102.0	95.3	116.5
Total N	7.7	19.3	36.9	74.0	32.3	68.8	274.1	601.9	874.3	1042.9	1153.5	883.5

Table A.2.1.15 Phosphate fertilizer consumption (P_2O_5), in thousands of tons of nutrient

	1900	1910	1922	1933	1940	1950	1960	1970	1980	1990	2000	2008
Guano	0.8	0.4	0.1	0.1	0.0	0.0	0.0	0.0	0.0	0.0	0.0	0.0
Rock phosphate	0.0	0.0	1.2	2.0	0.0	0.0	0.0	0.0	0.0	0.0	0.0	0.0
Thomas meal	0.5	2.1	3.0	1.4	0.2	0.0	6.3	3.6	1.0	0.6	0.4	1.2
Superphosphate	17.9	64.0	110.3	150.3	99.9	149.7	291.7	223.3	113.7	61.0	39.0	27.6
NPK	0.0	0.0	0.0	0.0	0.0	0.0	0.0	176.2	323.0	457.0	545.6	324.3
Other	0.5	2.1	7.6	9.8	9.2	0.0	1.4	12.3	0.0	3.5	27.8	22.9
Total (guano excluded)	18.9	68.1	122.1	163.6	109.3	149.7	299.4	415.5	437.7	522.1	612.7	376.1

Table A.2.1.16 Potassium fertilizer consumption (K_2O), in thousands of tons of nutrient

	1900	1910	1922	1933	1940	1950	1960	1970	1980	1990	2000	2008
Guano	0.2	0.1	0.0	0.0	0.0	0.0	0.0	0.0	0.0	0.0	0.0	0.0
Potassium chloride	1.4	8.1	14.9	15.1	25.9	42.2	78.5	63.3	19.1	92.4	116.0	96.7
Potassium sulfate	1.1	6.8	13.7	10.2	21.6	13.9	14.0	33.9	20.8	15.0	16.2	17.5
NPK	0.0	0.0	0.0	0.0	0.0	0.0	0.0	119.7	231.7	261.6	356.1	221.2
Other	0.0	0.3	1.2	2.2	1.4	0.0	0.0	0.0	0.0	2.6	0.0	0.1
Total (guano excluded)	2.5	15.3	29.8	27.6	48.9	56.1	92.5	217.0	271.6	371.6	488.3	335.6
Total	29.2	102.7	188.8	265.2	190.5	274.5	665.9	1234.3	1583.7	1936.6	2254.6	1595.1

Table A.2.1.17 Irrigated area, in millions of hectares

	1900	1910	1922	1933	1940	1950	1960	1970	1980	1990	2000	2008
Surface	0.89	1.04	1.18	1.10	0.96	0.93	1.39	2.19	2.24	1.95	1.47	1.09
Surface intermittent	0.34	0.26	0.19	0.29	0.40	0.51	0.62	0.00	0.00	0.00	0.00	0.00
Sprinkler	0.00	0.00	0.00	0.00	0.00	0.00	0.02	0.23	0.50	0.91	0.79	0.74
Trickle	0.00	0.00	0.00	0.00	0.00	0.00	0.00	0.00	0.02	0.23	1.10	1.54
Irrigated area	1.23	1.30	1.37	1.39	1.36	1.44	2.03	2.42	2.76	3.09	3.35	3.37
Effectively irrigated surface	1.00	1.12	1.24	1.19	1.10	1.10	1.62	2.42	2.76	3.09	3.35	3.37

Table A.2.1.18 Number of irrigation engines, in thousands of units

	1900	1910	1922	1933	1940	1950	1960	1970	1980	1990	2000	2008
Combustion	0.0	0.2	2.1	14.5	18.6	41.6	88.2	148.6	171.5	157.3	144.3	132.4
Electric	0.0	0.1	1.3	12.3	15.4	16.3	19.7	22.5	25.3	93.8	130.5	146.5
Total	0.0	0.3	3.4	26.8	34.0	57.9	107.9	171.1	196.7	251.1	274.8	278.9

Table A.2.1.19 Average rated power of irrigation engines, in kilowatts/unit

	1900	1910	1922	1933	1940	1950	1960	1970	1980	1990	2000	2008
Combustion	3.8	3.8	3.8	3.8	3.9	4.2	4.8	5.1	6.9	7.8	8.4	8.7
Electric	3.4	3.4	3.4	3.5	3.5	3.8	4.5	4.8	6.9	7.7	8.3	8.7
Total	3.7	3.7	3.6	3.6	3.7	4.1	4.7	5.0	6.9	7.7	8.3	8.7

Table A.2.1.20 Total installed power of irrigation engines, in megawatts

	1900	1910	1922	1933	1940	1950	1960	1970	1980	1990	2000	2008
Combustion	0.1	0.7	7.8	55.2	71.8	173.9	422.8	751.2	1190.1	1221.7	1205.8	1156
Electric	0.0	0.3	4.5	42.4	53.9	61.8	87.8	107.6	174.0	723.1	1082.1	1270
Total	0.1	1.0	12.4	97.6	125.7	235.7	510.6	858.8	1364.0	1944.9	2287.9	2425

Table A.2.1.21 Direct fuel consumption in irrigation from unconventional water sources, total electricity consumption and total direct energy use, in terajoules / year

	1900	1910	1922	1933	1940	1950	1960	1970	1980	1990	2000	2008
Liquid	0.1	1.7	18.3	341.6	444.7	1202.1	2922.3	5191.8	8225.2	8444.2	8334.3	7987.3
Coal	0.4	3.3	35.9	39.8	51.8	0.0	0.0	0.0	0.0	0.0	0.0	0.0
Total	0.5	4.9	54.2	381.4	496.4	1202.1	2922.3	5191.8	8225.2	8444.2	8334.3	7987.3
On-farm electricity	7.4	17.9	76.5	202.5	158.4	472.7	1355.2	3736.9	7594.2	12405.8	17255.7	19383.0
Upstream electricity	0.0	0.0	0.0	0.0	0.0	0.0	0.0	0.0	1091.9	1358.3	2267.9	3715.6
Total electricity	7.4	17.9	76.5	202.5	158.4	472.7	1355.2	3736.9	8686.1	13764.1	19523.6	23098.7
Total direct energy	7.9	22.8	130.7	583.9	654.8	1674.8	4277.5	8928.7	16911.3	22208.2	27858.0	31086.0

Table A.2.1.22 Direct energy consumption for irrigation, in terajoules/year

	1980	1990	2000	2008
Diversions	1091.9	1243.1	1428.8	1494.5
Desalinated	0.0	97.2	680.7	1789.1
Reused	0.0	18.0	158.4	432.0
Total	1091.9	1358.3	2267.9	3715.6

Table A.2.1.23 Input summary, water in tons/year, and the other inputs in kilotons/year

	1900	1910	1922	1933	1940	1950	1960	1970	1980	1990	2000	2008
Machinery	2.6	3.1	4.6	10.3	5.4	18.2	72.5	166.7	177.6	219.4	184.1	187.2
Fuels	0.6	1.3	4.1	15.7	25.1	75.1	274.8	1380.4	1967.3	1545.3	1739.8	2023.2
N fertilizer	7.7	19.3	36.9	74.0	32.3	68.8	274.1	601.9	874.3	1042.9	1153.5	883.5
P fertilizer	18.9	68.1	122.1	163.6	109.3	149.7	299.4	415.5	437.7	522.1	612.7	376.1
K Fertilizer	2.5	15.3	29.8	27.6	48.9	56.1	92.5	217.0	271.6	371.6	488.3	335.6
Pesticides	1.2	11.0	28.4	40.0	25.2	28.7	32.6	31.9	47.3	63.3	59.8	62.7
Greenhouses	0.0	0.0	0.0	0.0	0.0	0.0	0.0	2.0	75.0	544.9	831.5	835.0
Water use	9000.0	10050.0	11100.0	12150.0	12750.0	12375.0	14987.5	17600.0	20925.0	24000.0	23870.0	24400.0
Water consumption	5400.0	6131.3	6862.7	7594.0	8288.0	8353.0	10336.5	12320.0	14648.0	17400.0	18499.0	20163.0
% (consumption/use)	60.0	61.0	61.8	62.5	65.0	67.5	69.0	70.0	70.0	72.5	77.5	82.6

Table A.2.1.24 Pesticide consumption, in tons

	1900	1910	1922	1933	1940	1950	1960	1970	1980	1990	2000	2008
Copper and sulfur	5393	11,008	21,419	39,960	29,255	29,078	30,420	23,869	23,869	23,869	23,869	23,869
Synthetic pesticides	0	0	0	0	0	531	2142	8035	23,410	39,435	35,942	38,803
Total	5393	11,008	21,419	39,960	29,255	29,609	32,562	31,904	47,279	63,304	59,811	62,672

Table A.2.1.25 Protected crop areas, in thousands of hectares

	1900	1910	1922	1933	1940	1950	1960	1970	1980	1990	2000	2008
Greenhouses	0.0	0.0	0.0	0.0	0.0	0.0	0.0	0.0	2.8	27.4	49.4	49.5
Tunnels	0.0	0.0	0.0	0.0	0.0	0.0	0.0	0.1	1.1	9.7	12.5	12.9
Plastic mulches	0.0	0.0	0.0	0.0	0.0	0.0	0.0	0.1	7.3	65.4	111.6	49.6
Sand mulches	0.0	0.0	0.0	0.0	0.0	0.0	0.0	0.1	2.6	9.3	10.8	11.6
Total	0.0	0.0	0.0	0.0	0.0	0.0	0.0	0.3	13.8	111.8	184.3	123.5

Table A.2.1.26 Embodied Energy of the external inputs, in terajoules

Embodied Energy		1900	1910	1922	1933	1940	1950	1960	1970	1980	1990	2000	2008
Machinery	Implements	482	519	556	615	620	666	991	2042	3848	7768	8934	9290
	Motorized	22	41	103	353	440	733	2268	9170	17,419	16,718	15,737	15,220
Traction Fuels	Production	3	7	30	252	338	515	2150	12,677	17,334	11,985	14,545	16,254
	Direct	26	50	173	1093	1464	2172	10,007	57,858	81,692	56,688	64,877	67,841
Irrigation	Infrastructure	1339	1322	1284	949	868	830	1081	1697	2309	3485	4945	4977
	Fuels production	0	1	7	70	90	233	597	1097	1682	1721	1802	1884
	Direct Fuels	1	5	54	381	496	1202	2922	5192	8225	8444	8334	7987
	Electricity	70	61	162	331	247	966	2916	9760	26,382	45,841	63,894	59,047
Fertilizers	N	784	1783	3546	8256	3378	5441	24,715	55,765	74,575	83,727	88,601	63,681
	P	1047	3766	6762	9065	6052	8885	12,600	15,804	14,504	14,698	14,607	7,992
	K	54	318	604	544	946	1086	1454	3671	4763	6563	8830	5945
Crop protection	Pesticides	0	115	229	434	789	567	632	872	2258	7086	13,777	14,486
	Greenhouses	0	0	0	0	0	0	0	0	59	1571	11,860	15,708
Livestock	Fuels direct	0	0	0	0	0	0	0	0	0	5632	6026	14,204
	Fuels production	0	0	0	0	0	0	0	0	0	1191	1351	3403

Table A.2.1.27 Embodied energy in the external inputs, in terajoules

	1900	1910	1920	1930	1940	1950	1960	1970	1980	1990	2000	2008
External inputs (EI)(a+d)	23,723	29,327	38,042	43,722	38,166	46,748	92,690	248,832	401,945	352,368	482,101	510,260
Non-industrial inputs (a) = b+c	20,289	21,406	24,089	20,798	22,216	20,174	30,505	68,799	137,283	61,149	146,955	196,434
Feed imported (b)	1585	2747	5476	2230	3694	1698	12,074	53,403	124,922	51,822	140,663	193,177
Human labor (c)	18,705	18,659	18,613	18,568	18,522	18,477	18,431	15,396	12,361	9327	6292	3257
Industrial inputs (d) = e+f+g+h	3434	7922	13,953	22,924	15,950	26,573	62,185	180,033	264,662	291,219	335,146	313,826
Traction (e)	529	594	775	2131	2949	5681	12,764	81,696	124,785	103,296	116,318	129,666
Irrigation (f)	995	1231	1687	2132	2135	5096	9769	20,712	37,435	58,565	77,110	75,407
Fertilizers (g)	1884	5867	10,912	17,866	10,376	15,143	38,768	75,241	93,842	104,988	112,039	77,618
Crop protection (h)	26	229	579	795	489	653	884	2385	8599	24,369	29,679	31,135

Table A.2.1.28 Economical and agroecological EROIs

	1900	1910	1920	1930	1940	1950	1960	1970	1980	1990	2000	2008
EROIs economical												
Final EROI (FEROI)	0.78	0.72	0.70	0.65	0.59	0.57	0.53	0.55	0.45	0.52	0.45	0.45
Final External EROI (EFEROI)	17.26	14.31	11.83	10.83	11.12	9.24	4.80	1.80	1.12	1.56	1.22	1.20
Final Internal EROI (IFEROI)	0.82	0.76	0.75	0.69	0.63	0.61	0.60	0.79	0.74	0.77	0.71	0.72
EROIS agroecological												
NPPact-EROI	1.16	1.16	1.16	1.16	1.15	1.15	1.16	1.13	1.11	1.13	1.10	1.09
Final EROI agroecological (AE-FEROI)	0.11	0.11	0.11	0.12	0.11	0.11	0.11	0.10	0.10	0.12	0.12	0.12
EROI-Biodiversity	0.86	0.85	0.84	0.82	0.82	0.81	0.80	0.82	0.77	0.77	0.74	0.74
Woodenig-EROI	0.06	0.06	0.06	0.05	0.06	0.06	0.08	0.10	0.11	0.10	0.09	0.09

Table A.2.2.1 Final destination of domestic extraction of biomass, in megatons of dry matter

	1900	1910	1922	1933	1940	1950	1960	1970	1980	1990	2000	2008
Food	4.4	5.0	5.7	6.6	4.9	5.3	7.1	8.0	9.3	10.0	9.9	9.4
Feed	27.8	30.3	32.7	36.6	37.2	39.0	40.5	33.2	33.2	36.6	41.1	39.4
Seeds	0.6	0.7	0.8	0.9	0.6	0.7	0.9	1.1	1.1	1.2	1.0	1.3
Wood and firewood	16.0	16.2	15.9	15.7	16.3	15.9	13.7	13.4	11.2	13.4	12.9	14.6
Raw materials	0.7	0.5	0.8	0.8	0.6	0.6	0.8	0.8	1.1	1.5	2.6	2.6
Burned residues	0.0	0.0	0.0	0.0	0.0	0.2	0.6	1.5	2.3	3.6	1.4	1.3
Total	49.5	52.7	55.8	60.7	59.8	61.7	63.5	58.1	58.1	66.3	68.9	68.5

Table A.2.2.2 Biomass importations, in megatons of dry matter

	1900	1910	1922	1933	1940	1950	1960	1970	1980	1990	2000	2008
Food	0.1	0.1	0.2	0.1	0.5	0.2	0.7	0.6	1.1	2.8	4.4	6.3
Feed	0.1	0.2	0.3	0.1	0.2	0.1	0.6	3.0	6.9	5.6	10.4	13.3
Seeds	0.0	0.0	0.0	0.0	0.1	0.0	0.1	0.0	0.1	0.1	0.4	0.6
Wood and firewood	0.4	0.4	0.3	0.5	0.1	0.1	0.5	1.3	1.9	5.2	9.6	9.8
Raw materials	0.2	0.2	0.3	0.3	0.1	0.1	0.3	0.3	0.5	0.6	1.4	2.0
Burned residues	0.8	0.8	1.2	1.0	0.9	0.5	2.2	5.3	10.5	14.3	26.2	31.9

Table A.2.2.3 Biomass exportations, in megatons of dry matter

	1900	1910	1922	1933	1940	1950	1960	1970	1980	1990	2000	2008
Food	0.1	0.2	0.2	0.3	0.1	0.4	0.5	0.9	1.1	1.9	3.7	4.0
Feed	0.0	0.0	0.1	0.0	0.0	0.0	0.1	0.2	0.3	1.4	1.4	1.9
Seeds	0.0	0.0	0.0	0.0	0.0	0.0	0.0	0.0	0.0	0.1	0.1	0.1
Wood and firewood	0.1	0.1	0.1	0.1	0.0	0.0	0.1	0.3	1.1	1.8	4.4	6.2
Raw materials	0.1	0.1	0.1	0.0	0.0	0.0	0.0	0.1	0.3	0.3	0.4	0.5
Burned residues	0.3	0.4	0.5	0.4	0.2	0.5	0.6	1.5	2.8	5.4	9.9	12.7

Table A.2.2.4 Physical trade balance, in megatons of dry matter

	1900	1910	1922	1933	1940	1950	1960	1970	1980	1990	2000	2008
Food	0.0	−0.1	0.0	−0.2	0.3	−0.1	0.2	−0.3	0.0	0.9	0.7	2.3
Feed	0.1	0.1	0.3	0.1	0.1	0.0	0.5	2.8	6.5	4.2	9.0	11.4
Seeds	0.0	0.0	0.0	0.0	0.1	0.0	0.1	0.0	0.0	0.0	0.3	0.5
Wood and firewood	0.3	0.3	0.2	0.4	0.1	0.0	0.4	1.0	0.8	3.5	5.2	3.5
Raw materials	0.1	0.1	0.2	0.2	0.1	0.1	0.3	0.2	0.2	0.4	1.0	1.5
Burned residues	0.5	0.5	0.7	0.6	0.7	0.0	1.6	3.8	7.6	8.9	16.3	19.3

Table A.2.2.5 Biomass domestic consumption, in megatons of dry matter

	1900	1910	1922	1933	1940	1950	1960	1970	1980	1990	2000	2008
Food	4.4	4.9	5.7	6.4	5.3	5.1	7.3	7.7	9.3	10.9	10.6	11.7
Feed	27.8	30.4	32.9	36.7	37.4	39.0	41.0	36.0	39.7	40.7	50.1	50.8
Seeds	0.6	0.7	0.8	1.0	0.7	0.7	1.0	1.1	1.1	1.2	1.3	1.7
Wood and firewood	16.3	16.5	16.2	16.2	16.4	16.0	14.2	14.5	12.0	16.9	18.1	18.1
Raw materials	0.7	0.6	0.9	0.8	0.7	0.7	0.8	0.9	1.4	1.8	3.0	3.1
Burned residues	0.0	0.0	0.0	0.0	0.0	0.2	0.6	1.5	2.3	3.6	1.4	1.3
Total	50.0	53.1	56.4	61.1	60.4	61.7	64.8	61.7	65.9	75.1	84.6	86.7

Permissions

All chapters in this book were first published in TSMSATMWTI, by Springer; hereby published with permission under the Creative Commons Attribution License or equivalent. Every chapter published in this book has been scrutinized by our experts. Their significance has been extensively debated. The topics covered herein carry significant information for a comprehensive understanding. They may even be implemented as practical applications or may be referred to as a beginning point for further studies.

The contributors of this book come from diverse backgrounds, making this book a truly international effort. We would like to thank all the contributing authors for lending their expertise to make the book truly unique. They have played a crucial role in the development of this book. Without their invaluable contributions this book wouldn't have been possible. They have made vital efforts to compile up to date information on the varied aspects of this subject to make this book a valuable addition to the collection of many professionals and students.

This book was conceptualized with the vision of imparting up-to-date and integrated information in this field. To ensure the same, a matchless editorial board was set up. Every individual on the board went through rigorous rounds of assessment to prove their worth. After which they invested a large part of their time researching and compiling the most relevant data for our readers.

The editorial board has been involved in producing this book since its inception. They have spent rigorous hours researching and exploring the diverse topics which have resulted in the successful publishing of this book. They have passed on their knowledge of decades through this book. To expedite this challenging task, the publisher supported the team at every step. A small team of assistant editors was also appointed to further simplify the editing procedure and attain best results for the readers.

Apart from the editorial board, the designing team has also invested a significant amount of their time in understanding the subject and creating the most relevant covers. They scrutinized every image to scout for the most suitable representation of the subject and create an appropriate cover for the book.

The publishing team has been an ardent support to the editorial, designing and production team. Their endless efforts to recruit the best for this project, has resulted in the accomplishment of this book. They are a veteran in the field of academics and their pool of knowledge is as vast as their experience in printing. Their expertise and guidance has proved useful at every step. Their uncompromising quality standards have made this book an exceptional effort. Their encouragement from time to time has been an inspiration for everyone.

The publisher and the editorial board hope that this book will prove to be a valuable piece of knowledge for students, practitioners and scholars across the globe.

Index

www.ingramcontent.com/pod-product-compliance
Lightning Source LLC
LaVergne TN
LVHW080642210926
848485LV00003B/305